New Fragments

JOHN TYNDALL

CAMBRIDGE
UNIVERSITY PRESS

CAMBRIDGE UNIVERSITY PRESS

Cambridge, New York, Melbourne, Madrid, Cape Town,
Singapore, São Paolo, Delhi, Tokyo, Mexico City

Published in the United States of America by Cambridge University Press, New York

www.cambridge.org
Information on this title: www.cambridge.org/9781108038447

© in this compilation Cambridge University Press 2011

This edition first published 1892
This digitally printed version 2011

ISBN 978-1-108-03844-7 Paperback

CAMBRIDGE LIBRARY COLLECTION

Books of enduring scholarly value

Physical Sciences

From ancient times, humans have tried to understand the workings of
the world around them. The roots of modern physical science go back to
the very earliest mechanical devices such as levers and rollers, the mixing
of paints and dyes, and the importance of the heavenly bodies in early
religious observance and navigation. The physical sciences as we know them
today began to emerge as independent academic subjects during the early
modern period, in the work of Newton and other 'natural philosophers',
and numerous sub-disciplines developed during the centuries that followed.
This part of the Cambridge Library Collection is devoted to landmark
publications in this area which will be of interest to historians of science
concerned with individual scientists, particular discoveries, and advances in
scientific method, or with the establishment and development of scientific
institutions around the world.

New Fragments

Born in Leighlinbridge in Ireland, John Tyndall (1820–93) was a brilliant
nineteenth-century experimental physicist and gifted science educator. He
worked initially as a draughtsman, then spent a year teaching at an English
school before attending the University of Marburg to study physics and
chemistry. Tyndall carried out important research on magnetism, light and
bacteriology. Among his many significant achievements, he demonstrated
the greenhouse effect in Earth's atmospheric gases using absorption
spectroscopy. He was a skilled and entertaining educator and as Professor of
Natural Philosophy at the Royal Institution he gave many public lectures and
demonstrations of science. In this engaging potpourri of essays published in
1893, Tyndall's prose enlivens subjects as diverse as the life of Louis Pasteur,
observing the Sabbath, the prevention of phthisis (tuberculosis), personal
experiences of Alpine mountaineering, and the science of rainbows.

Cambridge University Press has long been a pioneer in the reissuing of out-of-print titles from its own backlist, producing digital reprints of books that are still sought after by scholars and students but could not be reprinted economically using traditional technology. The Cambridge Library Collection extends this activity to a wider range of books which are still of importance to researchers and professionals, either for the source material they contain, or as landmarks in the history of their academic discipline.

Drawing from the world-renowned collections in the Cambridge University Library, and guided by the advice of experts in each subject area, Cambridge University Press is using state-of-the-art scanning machines in its own Printing House to capture the content of each book selected for inclusion. The files are processed to give a consistently clear, crisp image, and the books finished to the high quality standard for which the Press is recognised around the world. The latest print-on-demand technology ensures that the books will remain available indefinitely, and that orders for single or multiple copies can quickly be supplied.

The Cambridge Library Collection will bring back to life books of enduring scholarly value (including out-of-copyright works originally issued by other publishers) across a wide range of disciplines in the humanities and social sciences and in science and technology.

NEW FRAGMENTS

PRINTED BY
SPOTTISWOODE AND CO., NEW-STREET SQUARE
LONDON

NEW FRAGMENTS

BY

JOHN TYNDALL, F.R.S.

LONDON
LONGMANS, GREEN, AND CO.
1892

CONTENTS

1880.

THE SABBATH.[1]

IN the opening words of a Lecture delivered in this city four years ago,[2] I spoke of the desire and tendency of the present age to connect itself organically with preceding ages. The expression of this desire is not limited to the connection of the material organisms of to-day with those of the geologic past, as set forth in the doctrines of Mr. Darwin. It is equally manifested in the domain of mind. To this source may be traced the philosophical writings of Mr. Herbert Spencer. To it we are indebted for the series of learned and laborious works on 'The Sources of Christianity,' by M. Renan. To it we owe the researches of Professor Max Müller in the domain of comparative philology and mythology, and the endeavour to found on these researches a 'science of religion.' In this relation, moreover, the recent work of Principal Caird[3] is highly characteristic of the tendencies of the age. He has no words of vituperation for the earlier and grosser religions of the world. Throughout the ages he discerns a purpose and a growth, wherein the earlier and more imperfect religions constitute the natural and necessary precursors of the later and more perfect ones. Even in the slough

[1] Presidential Address delivered before the Glasgow Sunday Society.
[2] *Fermentation : Fragments of Science*, vol. ii. p. 253.
[3] *Introduction to the Philosophy of Religion.*

of ancient paganism, Principal Caird detects a power ever tending towards amelioration, ever working towards the advent of a better state, and finally emerging in the purer life of Christianity.[1]

These changes in religious conceptions and practices correspond to the changes wrought by augmented experience in the texture and contents of the human mind. Acquainted as we now are with this immeasurable universe, and with the energies operant therein, the guises under which the sages of old presented the Maker and Builder thereof seem to us to belong to the utter infancy of things. To point to illustrations drawn from the heathen world would be superfluous. We may mount higher, and still find our assertion true. When, for example, Moses and Aaron, Nadab and Abihu, and seventy Elders of Israel are represented as climbing Mount Sinai, and actually seeing there the God of Israel, we listen to language to which we can attach no significance. 'There is in all this,' says Principal Caird, 'much which, even when religious feeling is absorbing the latent nutriment contained in it, is perceived [by the philosophic Christian of to-day] to belong to the domain of materialistic and figurative conception.' The reason is that the Christian philosopher of to-day has larger capacities and fuller knowledge than the Israelite of the time of Moses. What the one accepted as literal truth the other cannot accept save as a myth or figure. The children of Israel received without idealisation the statements of their great lawgiver. To them the tables of the law were true tablets of stone, prepared, engraved, broken, and re-engraved; while the graving tool which thus inscribed the law

[1] In Prof. Max Müller's *Introduction to the Science of Religion* some fine passages occur, embodying the above view of the continuity of religious development.

was held undoubtingly to be the finger of God. To us
such conceptions are impossible. We may by habit use
the words, but we attach to them no definite meaning.
' As the religious education of the world advances,'
says Principal Caird, ' it becomes impossible to attach
any literal meaning to those representations of God
and his relations to mankind, which ascribe to Him
human senses, appetites, passions, and the actions and
experiences proper to man's lower and finite nature.'
To Principal Caird, nevertheless, this imaging of the
Unseen is of inestimable value. It furnishes an objective counterpart to religious emotion, permanent but
plastic—capable of indefinite change and purification
in response to the changing thoughts and aspirations of
mankind.

It is, moreover, solely on this mutable element that
Principal Caird fixes his attention in estimating the
religious character of individuals, or the point of progress which has at any time been attained by nations
or races in the religious history of the world. ' Here,'
he says, 'the fundamental inquiry is as to the objective
character of their religious ideas or beliefs. The first
question is, not how they feel, but what they think and
believe ; not whether their religion manifests itself in
emotions more or less vehement or enthusiastic, but
what are the conceptions of God and divine things by
which these emotions are called forth?' These conceptions 'of God and divine things' were, it is admitted,
once ' materialistic and figurative,' and therefore objec-
tively untrue. Nor is their purer essence yet distilled ;
for the religious education of the world still ' advances,'
and is, therefore, incomplete. Hence the essentially
fluxional character of that objective counterpart to
religious emotion to which Principal Caird attaches
most importance. He, moreover, assumes that the

emotion is called forth by the conception. There is
doubtless action and reaction here ; but it may be
questioned whether the conception, which is a construction of the human understanding, could be at all put
together without materials drawn from the experience
of the human heart.[1]

The changes of conception here adverted to have
not always been peacefully brought about. The 'trans-
mutation ' of the old beliefs was often accompanied by
conflict and suffering. It was conspicuously so during
the passage from paganism to Christianity. Some of
the Roman emperors treated the Christians with fairness. Adrian was one of these. ' If anybody,' he says,
writing to the proconsul of Asia, ' appear as accuser,
and can prove that the Christians have broken the laws,
let punishment be inflicted in proportion to the gravity
of the offence. But, by Hercules ! if any should de-
nounce a Christian slanderously, you must punish the
slanderer still more severely.' This seems a very honest
line for a pagan emperor to pursue. Some of his successors followed his example, but others did not. During
the reign of Nero the cruelties inflicted on the Christians
at Rome can hardly be mentioned without a freezing of
the blood. According to Renan, the Antichrist of the
Apocalypse was the Emperor Nero ; he being raised to
this bad eminence by reason of his atrocities against the
new religion. The mystic number 666, which Protestants have so often fastened upon the Pope, answers
accurately to Nero's name and title. The numerical

---

[1] While reading the volume of Principal Caird I was reminded
more than once of the following passage in Renan's *Antéchrist* : ' Et
d'ailleurs, quel est l'homme vraiment religieux qui répudie complète-
ment l'enseignement traditionnel à l'ombre duquel il sentit d'abord
l'idéal, qui ne cherche pas les conciliations, souvent impossibles,
entre sa vieille foi et celle à laquelle il est arrivé par le progrès de
sa pensée ? '

values of the Hebrew letters added together make up
this number.

In his work entitled 'L'Eglise Chrétienne,' Renan
describes the sufferings of a group of Christians at
Smyrna which may be taken as typical. The victims
were cut up by the lash till the inner tissues of their
bodies were laid bare. They were dragged naked over
pointed shells. They were torn by lions; and finally,
while still alive, were committed to the flames. But all
these tortures failed to extract from them a murmur or
a cry. A youth named Germanicus, on this occasion,
gave his companions in agony an example of superhuman courage. His conflict with the lions called forth
such admiration that the proconsul entreated him to
have mercy on his own youth. Mercy was to be obtained
by recanting; but, instead of yielding, the youth provoked and excited the beasts, anxious to be torn to
pieces, and thus removed from so perverse a world.
His heroism simply exasperated his brutal persecutors,
who, when he was despatched, demanded another
victim. The Christians were called Atheists—a name
then and long afterwards of terrible import. 'Death
to the Atheists! let us seek Polycarp!' shouted the
maddened crowd. Polycarp, the friend of St. John,
and the principal personage in the Churches of Asia,
was then resident at Smyrna. They sought, found, and
arrested him. Those in power tried at first to coax him
into apostasy, but threats and entreaties proved equally
vain. 'Insult Christ!' exclaimed Statius Quadratus.
Polycarp replied: 'For eighty and six years have I
served Him, and He has never wronged me—I am a
Christian!' The grand old man felt a profound disdain
for the roaring crowd around him. 'Give me a day,'
said he to Quadratus, 'and I will show you what it is
to be a Christian.' 'Persuade the people,' retorted

Quadratus. 'I will reason with *you*,' replied Polycarp, 'because our precepts oblige us to show respect to those in authority; but I refuse to plead my cause before a mob.' His resolution was made known to the crowd, who shouted for the lions. They were informed that for that day the beasts had finished their work. 'To the flames, then!' cried the people; and the aged man was led to the stake. There he publicly thanked God for admitting him amongst those who had suffered death for his name. The fate of Polycarp reminds one of that of the Jew Eleazar, described in the sixth chapter of the Second Book of Maccabees. The Apocrypha, I would remark, ought to be bound up with all your Bibles; it contains much that is beautiful and wise, and there is in history nothing finer than the description of Eleazar's end.

The fortitude of the early Christians gained many converts to their cause; still, when the evidential value of fortitude is considered, it must not be forgotten that almost every faith can point to its rejoicing martyrs. Even the murderers of Polycarp had a faith of their own, the imperilling of which by Christianity spurred them on to murder. From faith they extracted the diabolical energy which animated them. The strength of faith is, therefore, no proof of the objective truth of faith. Indeed, at the very time here referred to we find two classes of Christians equally strong—Jewish Christians and Gentile Christians—who, while dying for the same Master, turned their backs upon each other, mutually declining all fellowship and communion. The forces which, acting on a large scale, had differentiated Christianity from paganism, soon made themselves manifest in details, producing disunion and opposition among those whose creeds and interests were in great part identical. Struggles for priority, moreover, were not uncommon. Jesus himself had to quell such contentions. His exhortations to humility were frequent. 'He that is least among you shall be greatest of all.' There were also conflicts upon points of doctrine. Among communities so diverse in temperament and antecedents differences were sure to arise. The point of difference which concerns us most had reference to the binding power of the Jewish law. Here dissensions arose among the apostles themselves. Nobody who reads with due attention the epistles of Paul can fail to see that this mighty propagandist had to carry on a lifelong struggle to maintain his authority as a preacher of Christ. There were not wanting those who denied him all vocation. James was the head of the Church at Jerusalem, and Judeo-Christians held that the ordination of James was alone valid. Paul, therefore, having no mission from James, was deemed by some a criminal intruder. The real fault of Paul was his love of freedom, and his uncompromising rejection, on behalf of his Gentile converts, of the chains of Judaism. He proudly calls himself 'the Apostle of the Gentiles.' He says to the Corinthians, 'I suppose I was not a whit behind the chiefest apostle. Are they Hebrews? So am I. Are they Israelites? So am I. Are they of the seed of Abraham? So am I. Are they ministers of Christ? I am more; in labours more abundant, in stripes above measure, in deaths oft.' He then establishes his right to the position which he claimed by recounting in detail the sufferings he had endured. I leave it to you to compare this Christian hero with some of the 'freethinkers' of our own day, who, 'more intolerant than the intolerance they deprecate,' flaunt in public their cheap and trumpery theories of the great Apostle and the Master whom he served.

Paul was too outspoken to escape assault. All in-
sincerity or double-facedness—all humbug, in short—
was hateful to him; and even among his colleagues
he found scope for this feeling. Judged by our standard
of manliness, Peter, in moral stature, fell far short of
Paul. In that supreme moment when his Master
required of him 'the durance of a granite ledge' Peter
proved 'unstable as water.' He ate with the Gentiles
when no Judeo-Christian was present to observe him;
but when such appeared he withdrew himself, fearing
those which were of the circumcision. Paul charged
him openly with dissimulation. But Paul's quarrel with
Peter was more than personal. Paul contended for a
principle, and was determined at all hazards to shield
his Gentile children in the Lord from the yoke which
their Jewish co-religionists would have imposed upon
them. ' If thou,' he says to Peter, ' being a Jew, livest
after the manner of the Gentiles, and not as do the Jews,
why compellest thou the Gentiles to live as the Jews?'
In the spirit of a liberal, not in name but in deed, he
overthrew the Judaic preferences for days, deferring at
the same time to the claims of conscience. ' Let him
who desires a Sabbath,' he virtually says, 'enjoy it; but
let him not impose it on his brother who does not.'
The rift thus revealed in the apostolic lute widened with
time, and Christian love was not the feeling which
long animated the respective followers of Peter and
Paul.

We who have been born into a settled state of things
can hardly realise the commotion out of which this tranquillity has emerged. We have, for example, the canon
of Scripture already arranged for us. But to sift and
select these writings from the mass of spurious documents afloat at the time of compilation was a work of
vast labour, difficulty, and responsibility. The age was

rife with forgeries. Even good men lent themselves to these pious frauds, believing that true Christian doctrine, which of course was *their* doctrine, would be thereby quickened and promoted. There were gospels and counter-gospels; epistles and counter-epistles—some frivolous, some dull, some speculative and romantic, and some so rich and penetrating, so saturated with the Master's spirit, that, though not included in the canon, they enjoyed an authority almost equal to that of the canonical books. When arguments or proofs were needed, whether on the side of the Jewish Christians or of the Gentile Christians, a document was discovered which met the case, and on which the name of an apostle, or of some authoritative contemporary of the apostles, was boldly inscribed. The end being held to sanctify the means, there was no lack of manufactured testimony. The Christian world seethed not only with apocryphal writings, but with hostile interpretations of writings not apocryphal. Then arose the sect of the Gnostics—men who *know*—who laid claim to the possession of a perfect science, and who, if they were to be believed, had discovered the true formula for what philosophers called 'the Absolute.' But these speculative Gnostics were rejected by the conservative and orthodox Christians of their day as fiercely as their successors the Agnostics—men who *don't know*—are rejected by the orthodox in our own. The good Polycarp one day met Marcion, an ultra-Paulite, and a celebrated member of the Gnostic sect. On being asked by Marcion whether he, Polycarp, did not know him, Polycarp replied, 'Yes, I know you very well; you are the first-born of the devil.' This is a sample of the bitterness then common. It was a time of travail—of throes and whirlwinds. Men at length began to yearn for peace

L'Eglise Chrétienne, p. 450.

and unity, and out of the embroilment was slowly consolidated that great organisation the Church of Rome.
The Church of Rome had its precursor in the Church
at Rome. But Rome was then the capital of the world ;
and, in the end, that famous city gave the Christian
Church, established in her midst, such a decided preponderance that it eventually laid claim to the proud
title of ' Mother and Matrix of all other Churches.'

With terrible jolts and oscillations the religious life
of the world has run down ' the ringing grooves of change.'
A smoother route may have been undiscoverable. At
all events it was undiscovered. Some years ago I found
myself in discussion with a friend who entertained the
notion that the general tendency of things in this world
is towards equilibrium, the result of which would be
peace and blessedness to the human race. My notion
was that equilibrium meant not peace and blessedness,
but death. No motive power is to be got from heat, save
during its *fall* from a higher to a lower temperature, as
no power is to be got from water save during its descent
from a higher to a lower level. Thus also life consists, not
in equilibrium, but in the passage towards equilibrium.
In man it is the leap from the potential through the
actual to repose. The passage often involves a fight.
Every natural growth is more or less of a struggle with
other growths, in which the fittest survive. In times
of strife and commotion we may long for peace ; but
knowledge and progress are the fruits of action. Some
are, and must be, wiser than the rest; and the enunciation of a thought in advance of the moment provokes
dissent or evokes approval, and thus promotes action.
The thought may be unwise ; but it is only by discussion, checked by experience, that its value can be
determined. Discussion, therefore, is one of the motive
powers of life, and, as such, is not to be deprecated.

Still one can hardly look without despair on the passions excited, and the energies wasted, over questions which, after ages of strife, are shown to be mere fatuity and foolishness. Thus the theses which shook the world during the first centuries of the Christian era have, for the most part, shrunk into nothingness. It may, however, be that the human mind could not become fitted to pronounce judgment on a controversy otherwise than by wading through it. We get clear of the jungle by traversing it. Thus even the errors, conflicts, and sufferings of bygone times may have been necessary factors in the education of the world. Let nobody, however, say that it has not been a hard education. The yoke of religion has not always been easy, nor its burden light—a result arising, in part from the ignorance of the world at large, but more especially from the mistakes of those who had the charge and guidance of a great spiritual force, and who guided it blindly. Looking over the literature of the Sabbath question, as catalogued and illustrated in the laborious, able, and temperate work of the late Mr. Robert Cox, we can hardly repress a sigh in thinking of the gifts and labours of intellect which this question has absorbed, and the amount of bad blood which it has generated. Further reflection, however, reconciles us to the fact that waste in intellect may be as much an incident of growth as waste in nature.

When the various passages of the Pentateuch which relate to the observance of the Sabbath are brought together, as they are in the excellent work of Mr. Cox, and when we pass from them to the similarly collected utterances of the New Testament, we are immediately exhilarated by a freer atmosphere and a vaster sky. Christ found the religions of the world oppressed almost to suffocation by the load of formulas piled upon them by the priesthood. He removed the load, and rendered

respiration free. He cared little for forms and ceremonies, which had ceased to be the raiment of man's
spiritual life. To that life he looked, and it he sought
to restore. It was remarked by Martin Luther that
Jesus broke the Sabbath deliberately, and even ostenta-
tiously, for a purpose. He walked in the fields; he
plucked, shelled, and ate the corn ; he treated the sick,
and his spirit may be detected in the alleged imposition
upon the restored cripple of the labour of carrying his
bed on the Sabbath day. He crowned his protest against
a sterile formalism by the enunciation of a principle
which applies to us to-day as much as to the world in
the time of Christ. 'The Sabbath was made for man,
and not man for the Sabbath.' No priestly power, he
virtually declares, shall henceforth interfere with man's
freedom to decide how the Sabbath is to be spent.

Though the Jews, to their detriment, kept themselves as a nation intellectually isolated, the minds of
individuals were frequently coloured by Greek thought
and culture. The learned and celebrated Philo, who
was contemporary with Josephus, was thus influenced.
Philo expanded the uses of the seventh day by including
in its proper observance studies which might be called
secular. 'Moreover,' he says, 'the seventh day is also
an example from which you may learn the propriety of
studying philosophy. As on that day it is said God
beheld the works that He had made, so you also may
yourself contemplate the works of Nature.' Permission
to do this is exactly what the members of the Sunday
Society humbly claim. The Jew, Philo, would grant
them this permission, but our straiter Christians will
not. Where shall we find such samples of those works
of Nature which Philo commended to the Sunday contemplation of his countrymen, as in the British Museum?
Within those walls we have, as it were, epochs disen-

tombed—ages of divine energy illustrated. But the efficient authorities—among whom I would include a short-sighted portion of the public—resolutely close the doors, and exclude from the contemplation of these things the multitudes who have only Sunday to devote to them. Are the authorities logical in doing so? Do they who thus stand between them and the public really believe those treasures to be the work of God? Do they or do they not hold, with Paul, that 'the eternal power and Godhead' may be clearly seen from 'the things that are made'? If they do—and they dare not affirm that they do not—I fear that Paul, with his customary plainness of language, would pronounce their conduct to be 'without excuse' [1]

Science, which is the logic of nature, demands proportion between the house and its foundation. Theology sometimes builds weighty structures on a doubtful base. The tenet of Sabbath observance is an illustration. With regard to the time when the obligation to keep the Sabbath was imposed, and the reasons for its imposition, there are grave differences of opinion between learned and pious men. Some affirm that it was instituted at the Creation in remembrance of the rest of God. Others allege that it was imposed after the departure of the Israelites from Egypt, and in memory of that departure. The Bible countenances both interpretations. In Exodus we find the origin of the Sabbath described with unmistakable clearness, thus : 'For in six days the Lord made heaven and earth, the sea and all that in them is. *Wherefore* the Lord blessed the seventh day, and

---

[1] I refer, of course, to those who object to the opening of the museums on religious grounds. The administrative difficulty stands on a different footing. But surely *it* ought to vanish in presence of the benefits to tens of thousands which in all probability would accrue.

hallowed it.' In Deuteronomy this reason is suppressed
and another is assigned. Israel being a servant in
Egypt, God, it is stated, brought them out of it with a
mighty hand and by an outstretched arm. ' *Therefore*
the Lord thy God commandeth thee to keep the
Sabbath day.' After repeating the Ten Commandments,
and assigning the foregoing origin to the Sabbath, the
writer in Deuteronomy proceeds thus : 'These words
the Lord spake unto all your assembly in the mount,
out of the midst of the fire, of the cloud and the thick
darkness, with a loud voice ; and he added no more.'
But in Exodus God not only added more, but something
entirely different. This has been a difficulty with
commentators—not formidable, if the Bible be treated
as any other ancient book, but extremely formidable on
the theory of plenary inspiration. I remember in the
days of my youth being shocked and perplexed by an
admission made by Bishop Watson in his celebrated
' Apology for the Bible,' written in answer to Tom Paine.
' You have,' says the bishop, 'disclosed a few weeds
which good men would have covered up from view.'
That there were 'weeds' in the Bible requiring to be
kept out of sight was to me, at that time, a new revelation. I take little pleasure in dwelling upon the errors
and blemishes of a book rendered venerable to me by
intrinsic wisdom and imperishable associations. But
when that book is wrested to our detriment, when its
passages are invoked to justify the imposition of a yoke,
irksome because unnatural, we are driven in self-defence
to be critical. In self-defence, therefore, we plead these
two discordant accounts of the origin of the Sabbath,
one of which makes it a purely Jewish institution, while
the other, unless regarded as a mere myth and figure,
is in irreconcilable antagonism to the facts of geology.

With regard to the alleged 'proofs' that Sunday

was introduced as a substitute for Saturday, and that
its observance is as binding upon Christians as their
Sabbath was upon the Jews, I can only say that those
which I have seen are of the flimsiest and vaguest
character. ' If,' says Milton, ' on the plea of a divine
command, they impose upon us the observances of a
particular day, how do they presume, without the
authority of a divine command, to substitute another
day in its place ? ' Outside the bounds of theology no
one would think of applying the term 'proofs' to the
evidence adduced for the change ; and yet on this
pivot, it has been alleged, turns the eternal fate of
human souls.[1] Were such a doctrine not actual it
would be incredible. It has been truly said that the
man who accepts it sinks, in doing so, to the lowest
depth of Atheism. It is perfectly reasonable for a
religious community to set apart one day in seven for
rest and devotion. Most of those who object to the
Judaic observance of the Sabbath recognise not only
the wisdom but the necessity of some such institution,
not on the ground of a divine edict, but of common
sense.[2] They contend, however, that it ought to be as
far as possible a day of cheerful renovation both of

---

[1] In 1785 the first mail-coach reached Edinburgh from London,
and in 1788 it was continued to Glasgow. The innovation was de-
nounced by a minister of the Secession Church of Scotland as
' contrary to the laws both of Church and State ; contrary to the
laws of God ; contrary to the most conclusive and constraining
reasons assigned by God ; and calculated not only to promote the
hurt and ruin of the nation, but also the eternal damnation of multitudes.'—Cox, vol. ii. p. 248. Even in our day there are clergymen
foolish enough to indulge in this dealing out of damnation.

[2] ' That public worship,' says Milton, ' is commended and in-
culcated as a voluntary duty, even under the Gospel, I allow ; but
that it is a matter of compulsory enactment, binding on believers
from the authority of this commandment, or of any Sinaitical
precept whatever, I deny.'

body and spirit, and not a day of penal gloom. There
is nothing that I should withstand more strenuously
than the conversion of the first day of the week into a
common working day. Quite as strenuously, however,
do I oppose its being employed as a day for the exercise
of sacerdotal rigour.

The early reformers emphatically asserted the freedom of Christians from Sabbatical bonds; indeed Puritan writers have reproached them with dimness of
vision regarding the observance of the Lord's Day.
'The fourth Commandment,' says Luther, 'literally
understood, does not apply to us Christians; for it is
entirely outward, like other ordinances of the Old
Testament, all of which are now left free by Christ.
If a preacher,' he continues, 'wishes to force you back
to Moses, ask him whether you were brought by Moses
out of Egypt. If he says no, then say, How, then,
does Moses concern me, since he speaks to the people
that have been brought out of Egypt? In the New
Testament Moses comes to an end, and his laws lose
their force. He must bow in the presence of Christ.'
'The Scripture,' says Melanchthon, 'allows that we are
not bound to keep the Sabbath; for it teaches that the
ceremonies of the law of Moses are not necessary after
the revelation of the Gospel. And yet,' he adds,
'because it was requisite to appoint a certain day that
the people might know when to assemble together, it
appeared that the Church appointed for this purpose
the Lord's Day.' I am glad to find my grand old
namesake on the side of freedom in this matter. 'As
for the Sabbath,' says the martyr Tyndale, 'we are
lords over it, and may yet change it into Monday, or
into any other day, as we see need; or may make every
tenth day holy day, only if we see cause why. Neither
need we any holy day at all if the people might be

taught without it.' Calvin repudiated 'the frivolities
of false prophets who, in later times, have instilled
Jewish ideas into the people. Those,' he continues,
' who thus adhere to the Jewish institution go thrice as
far as the Jews themselves in the gross and carnal
superstition of Sabbatism.' Even John Knox, who
has had so much Puritan strictness unjustly laid to his
charge, knew how to fulfil on the Lord's Day the duties
of a generous, hospitable host. His Master feasted on
the Sabbath day, and he did not fear to do the same on
Sunday. ' There be two parts of the Sabbath day '
says Cranmer : ' one is the outward bodily rest from
all manner of labour and work ; this is mere cere-
monial, and was taken away with other sacrifices and
ceremonies by Christ at the preaching of the gospel.
The other part of the Sabbath day is the inward rest or
ceasing from sin.' This higher symbolism, as regards
the Sabbath, is frequently employed by the Reformers.
It is the natural recoil of the living spirit from the
mechanical routine of a worn-out hierarchy.

Towards the end of the sixteenth century, demands
for a stricter observance of the Sabbath began to be
made—probably in the first instance with some reason,
and certainly with good intent. The manners of the
time were coarse, and Sunday was often chosen for their
offensive exhibition. But if there was coarseness on
the one side, there was ignorance both of Nature and
human nature on the other. Contemporaneously with
the demands for stricter Sabbath rules, God's judgments on Sabbath-breakers began to be pointed out.
Then and afterwards ' God's Judgments ' were much in
vogue, and man, their interpreter, frequently behaved
as a fiend in the supposed execution of them. But of
this subsequently. A Suffolk clergyman named Bownd,
who, according to Cox, was the first to set forth at large

the views afterwards embodied in the Westminster Confession, adduces many such judgments. One was the case of a nobleman, ' who for hunting on the holy day was punished by having a child with a head like a dog's.' Though he cites this instance, Bownd, in the matter of Sabbath observance, was very lenient towards noblemen. ' Concerning the feasts of noblemen and great personages or their ordinary diet upon this day (which in comparison may be called feasts), because they represent,' says the doctor, ' in some measure the majesty of God on the earth, in carrying the image as it were of the magnificence and puissance of the Lord, much is to be granted to them.'

Imagination once directed towards this question was sure to be prolific. Instances accordingly grew apace in number and magnitude. Memorable examples of God's judgments upon Sabbath-breakers, and other like libertines, in their unlawful sports happening within this realm of England, were collected. Innumerable cases of drowning while bathing on Sunday were adduced, without the slightest attention to the logical requirements of the question. Week-day drownings were not dwelt upon, and nobody knew or cared how the question of proportion stood between the two classes of bathers. The Civil War was regarded as a punishment for Sunday desecration. The fire of London, and a subsequent great fire in Edinburgh, were ascribed to this cause ; while the fishermen of Berwick lost their trade through catching salmon on Sunday. Their profanation was thus nipped by a miracle in the bud, and they were brought to repentance. A Nonconformist minister named John Wells, whose huge volume is described by Cox as ' the most tedious of all the Puritan productions about the Sabbath,' is specially copious in illustration. A drunken pedlar, ' fraught with commodities' on Sunday, drops into a river: God's retributive justice is seen in the fact. Wells travelled far in search of instances. One Utrich Schrœtor, a Swiss, while playing at dice on the Lord's Day, lost heavily, and apparently to gain the devil to his side broke out into this horrid blasphemy : ' If fortune deceive me now I will thrust my dagger in the body of God.' Whereupon he threw the dagger upwards. It disappeared, and five drops of blood, which afterwards proved indelible, fell upon the gaming table. The devil then appeared, and with a hideous noise carried off the vile blasphemer. His two companions fared no better. One was struck dead and turned into worms, the other was executed. A vintner who on the Lord's Day tempted the passers-by with a pot of wine was carried into the air by a whirlwind and never seen more. ' Let us read and tremble,' adds Mr. Wells. At Tidworth a man broke his leg on Sunday while playing at football. By a secret judgment of the Lord the wound turned into a gangrene, and in pain and terror the criminal gave up the ghost.

You may smile at these recitals, but is there not a survival of John Wells still extant among you ? Are there not people in your midst so well informed as to ' the secret judgments of the Lord ' as to be able to tell you their exact value and import, from the damaging of the share market through the running of Sunday trains to the calamitous overthrow of a railway bridge ? Alphonso of Castile boasted that if he had been consulted at the beginning of things he could have saved the Creator some worlds of trouble. It would not be difficult to give the God of our more rigid Sabbatarians a lesson in justice and mercy ; for his alleged judgments savour but little of either. How are calamities to be classified ? Almost within earshot of those who

note these Sunday judgments, the poor miners of Blan-
tyre are blown to pieces, while engaged in their sinless
week-day toil. A little further off the bodies of two
hundred and sixty workers, equally innocent of Sabbath-
breaking, are entombed at Abercarne. Dinas holds its
sixty bodies, while the present year has furnished a
fearful tale of similar disasters. Whence comes the
vision which differentiates the Sunday calamity from
the week-day calamity, seeing in the one a judgment
of heaven, and in the other a natural event? We may
wink at the ignorance of John Wells, for he lived in a
prescientific age ; but it is not pleasant to see his
features reproduced, on however small a scale, before
an educated nation in the latter half of the nineteenth
century.

Notwithstanding their strictness about the Sabbath,
which possibly carried with it the usual excess of a reaction, some of the straitest of the Puritan sect saw
clearly that unremitting attention to business, whether
religious or secular, was unhealthy. These considered
recreation to be as necessary to health as daily food ;
and hence exhorted parents and masters, if they would
avoid the desecration of the Sabbath, to allow to children and servants time for honest recreation on other
days. They might have done well to inquire whether
even Sunday devotions might not, without ' moral cul-
pability ' on their part, keep the minds of children and
servants too long upon the stretch. I fear many of the
good men who insisted, and insist, on a Judaic observance
of the Sabbath, and who dwell upon the peace and
blessedness to be derived from a proper use of the
Lord's Day, generalise beyond their data, applying the
experience of the individual to the case of mankind.
What is a conscious joy and blessing to themselves they
cannot dream of as being a possible misery, or even a

curse, to others. It is right that your most spiritually-minded men—men who, to use a devotional phrase, enjoy the closest walk with God—should be your pastors. But they ought also to be practical men, able to look not only on their personal feelings, but on the capacities of humanity at large, and willing to make their rules and teachings square with these capacities. There is in some minds a natural bias towards religion, as there is in others towards poetry, art, or mathematics; but the poet, artist, or mathematician who would seek to impose upon others, not possessing his tastes, the studies which give him delight, would be deemed an intolerable despot. The philosopher Fichte was wont to contrast his mode of rising into the atmosphere of faith with the experience of others. In his case the process, he said, was purely intellectual. Through reason he reached religion; while in the case of many whom he knew this process was both unnecessary and unused, the bias of their minds sufficing to render faith, without logic, clear and strong. In making rules for the Community these natural differences must be taken into account. The yoke which is easy to the few may be intolerable to the many, not only defeating its own immediate purpose, but frequently introducing recklessness or hypocrisy into minds which a franker and more liberal treatment would have kept free from both.[1]

The moods of the times—the ' climates of opinion,'

[1] ' When our Puritan friends,' says Mr. Frederick Robertson, 'talk of the blessings of the Sabbath, we may ask them to remember some of its curses.' Other and more serious evils than those recounted by Mr. Robertson may, I fear, be traced to the system of Sabbath observance pursued in many of our schools. At the risk of shocking some worthy persons, I would say that the invention of an invigorating game for fine Sunday afternoons, and healthy indoor amusement for wet ones, would prove infinitely more effectual as an aid to moral purity than most of our plans of religious meditation.

as Glanvil calls them—have also to be considered in
imposing disciplines which affect the public. For the
ages, like the individual, have their periods of mirth
and earnestness, of cheerfulness and gloom. From this
point of view a better case might be made out for the
early Sabbatarians than for their survivals at the present day. They were more in accord with the needs
and spirit of their age. Sunday sports were barbarous;
bull- and bear-baiting, interludes, and bowling were
reckoned amongst them, and the more earnest spirits
longed not only to promote edification but to curb excess. Sabbatarianism, therefore, though opposed, made
rapid progress. Its opponents were not always wise.
They did what religious parties, when in power, always
do—exercised that power tyrannically. They invoked
the arm of the flesh to suppress or change conviction.
In 1618 James I. published a declaration, known afterwards as ' The Book of Sports,' because it had reference
to Sunday recreations. It seems to have been, in itself,
a reasonable book. Puritan magistrates had interfered
with the innocent amusements of the people, and the
king wished to insure their being permitted, after divine
service, to those who desired them; but not enjoined
upon those who did not. Coarser sports, and sports
tending to immorality, were prohibited. Charles I.
renewed the declaration of his father. Not content,
however, with expressing his royal pleasure—not content with restraining the arbitrary civil magistrate—the
king decreed that the declaration should be published
' through all the parish churches,' the bishops in their
respective dioceses being made the vehicles of the royal
command. Defensible in itself, the declaration thus
became an instrument of oppression. The High Church
party, headed by Archbishop Laud, forced the reading
of the documents on men whose consciences recoiled

from the act. 'The precise clergy,' as Hallam calls
them, refused in general to comply, and were suspended
or deprived in consequence. 'But,' adds Hallam,
'mankind loves sport as little as prayer by compulsion ;
and the immediate effect of the king's declaration was
to produce a far more scrupulous abstinence from diversions on Sundays than had been practised before.'

The Puritans, when they came into power, followed
the evil example of their predecessors. They, the
champions of religious freedom, showed that they could,
in their turn, deprive their antagonists of their benefices,
fine them, burn their books by the common hangman,
and compel them to read from the pulpit things of
which they disapproved. On this point Bishop Heber
makes some excellent remarks. 'Much,' he says, 'as
each religious party in its turn had suffered from persecution, and loudly and bitterly as each had, in its own
particular instance, complained of the severities exercised against its members, no party had yet been found
to perceive the great wickedness of persecution in the
abstract, or the moral unfitness of temporal punishment
as an engine of religious controversy.' In a very different strain writes the Dr. Bownd who has been already
referred to as a precursor of Puritanism. He is so sure
of his ' doxy ' that he will unflinchingly make others
bow to it. ' It behoveth,' he says, ' all kings, princes,
and rulers, that profess the true religion to enact such
laws and to see them diligently executed, whereby the
honour of God in hallowing these days might be maintained. And, indeed, this is the chiefest end of all
government, that men might not profess what religion
they list, and serve God after what manner it pleaseth
them best, but that the parts of God's true worship
[Bowndean worship] might be set up everywhere, and
all men compelled to stoop unto it.'

There is, it must be admitted, a sad logical consistency in the mode of action deprecated by Bishop Heber. As long as men hold that there is a hell to be shunned, they seem logically warranted in treating lightly the claims of religious liberty upon earth. They dare not tolerate a freedom whose end they believe to be eternal perdition. Cruel they may be for the moment, but a passing pang vanishes when compared with an eternity of pain. Unreligious men might call it hallucination, but if I accept undoubtingly the doctrine of eternal punishment, then, whatever society may think of my act, I am self-justified not only in ' letting ' but in destroying that which I hold dearest, if I believe it to be thereby stopped in its progress to the fires of hell. Hence, granting the assumptions common to both, the persecution of Puritans by High Churchmen, and of High Churchmen by Puritans, was not without a basis in reason. I do not think the question can be decided on à priori grounds, as Bishop Heber seemed to suppose. It is not the abstract wickedness of persecution so much as our experience of its results that causes us to set our faces against it. It has been tried, and found the most ghastly of failures. This experimental fact overwhelms the plausibilities of logic, and renders persecution, save in its meaner and stealthier aspects, in our day impossible.

The combat over Sunday continued, the Sabbatarians continually gaining ground. In 1643 the divines who drew up the famous document known as the Westminster Confession began their sittings in Henry VII.'s Chapel. Milton thought lightly of these divines, who, he said, were sometimes chosen by the whim of members of Parliament; but the famous Puritan, Baxter, extolled them for their learning, godliness, and ministerial abilities. A journal of their earlier proceedings was

kept by Lightfoot, one of their members. On November
13, 1644, he records the occurrence of 'a large debate'
on the sanctification of the Lord's Day. After fixing
the introductory phraseology, the assembly proceeded to
consider the second proposition : 'To abstain from all
unnecessary labours, worldly sports, and recreations.' It
was debated whether ' worldly thoughts' should not be
added. ' This was scrupulous,' says the naive journalist,
' whether we should not be a scorn to go about to bind
men's thoughts, but at last it was concluded upon to be
added, both for the more piety and for the Fourth
Commandment.' The question of Sunday cookery was
then discussed and settled ; and, as regards public
worship, it was decreed ' that all the people meet so
timely that the whole congregation be present at the
beginning, and not depart until after the blessing.
That what time is vacant between or after the solemn
meetings of the congregation be spent in reading, meditation, repetition of sermons,' &c. These holy men
were full of that strength already referred to as imparted
by faith. They needed no natural joy to brighten their
lives, mirth being displaced by religious exaltation.
They erred, however, in making themselves a measure
for the world at large, and insured the overthrow of
their cause by drawing too heavily upon average human
nature. 'This much,' says Hallam, ' is certain, that
when the Puritan party employed their authority in
proscribing all diversions, and enforcing all the Jewish
rigour about the Sabbath, they rendered their own yoke
intolerable to the young and gay ; nor did any other
cause, perhaps, so materially contribute to bring about
the Restoration.'

From the records of the Town Council of Edinburgh,
Mr. Cox makes certain extracts which amusingly illustrate both the character of Sabbath discipline and the

difficulty of enforcing it. In 1560 it was, among other things, decreed that on Sundays 'all persons be astricted to be present at the ordinary sermons, as well after noon as before noon, and that from the last jow of the bell to the said sermons to the final end.' In 1581 the Council ordained that 'proclamation be made through this burgh, discharging all kinds of games and plays now commonly used the said day, such as bowling in yards, dancing, playing, running through the high street of hussies, bairns, and boys, with all manner of dissolution of behaviour.' The people obeyed and went to church, but it seems they chose their own preachers. This galavanting among the kirks was, however, quickly put an end to; for in 1584 it was ordained ' that all freemen and freemen's wives in times coming be found in their own parish kirk every Sunday, as also at the time of the Communions, under the pain of payment of an unlaw for every person being found absent.' In 1586 the Council ' finds it expedient that a bailie ilk Sunday his week about, visit the street taverns and other common places in time of sermon, and pones all offenders according to the town statutes.' Vaging (strolling) in the High Gate was also forbidden.

These restrictions, applying at first to the time of divine service only, were afterwards extended to the entire Sunday; but sabbath profanation resembled hydraulic pressure, and broke forth whenever it found a weak point in the municipal dam. The repairing and strengthening of the dam were incessant. Proclamation followed proclamation, forbidding the practice of buying and selling, the opening of eating- and coffee-houses, and prohibiting such sports as golf, archery, row-bowles, penny-stone, and kaitch-pullis. The gates of the city were ordered to be closed on Saturday night and not to be opened before four o'clock on Monday morning. At

the time these edicts were published the Provost complained of the little obedience hitherto given to the
manifold acts of council for keeping the Sabbath. A
decree on January 14, 1659, runs thus :—

' Whereas many both young and old persons walk,
or sit and play on the Castle hill, and upon the streets
and other places on the Sabbath day after sermons, so
that it is manifest that family worship is neglected by
such, the Council appoint that there be several pairs of
stocks provided to stand in several public places of the
city, that whosoever is needlessly walking or sitting
idly in the streets shall either pay eighteen-pence sterling penalty or be put in the stocks.'

The parents of children found playing are fined 6d.
a head. 'And if any children be found on the Castle
hill after supper to pay 18d. penalty or to be put in
the stocks.' Even this drastic treatment did not cure
the evil, for thirty years later the edict against
'vaging' on the Castle hill had to be renewed. At
the same time it was ordered that the public wells be
closed on Sunday from 8 A.M. till noon; then to open
till 1 P.M., and afterwards from 5 P.M. None to bring
any greater vessels to the wells for the carrying of
water than a pint stoup or a pint bottle on the Lord's
Day. Our present sanitary notions were evidently not
prevalent in Edinburgh in 1689. Mr. Cox remarks
that 'these ordinances were usually enacted at the instance of the clergy.' It would have been well had
the evils which the clergy inflicted on the world at the
time here referred to been limited to the stern manipulation of Sabbath laws.[1]

---

[1] In Massachusetts it was attempted to make Sabbath-breaking
a capital offence, but Governor Winthrop had the humanity and good
sense to erase it from the list of acts punishable with death   In
the laws of the colony of New Plymouth, presumptuous Sabbath-

In 1646 the ' Confession,' after ' endless janglings,'
being agreed upon, it was presented to Parliament,
which, in 1648, accepted and published its doctrinal
portion, thus securing uniformity of doctrine as far as
it could be secured by legislation. There was no lack
of definiteness in the Assembly's statements. They
spoke as confidently of the divine enactments as if
each member had been personally privy to the counsels
of the Most High. When Luther in the Castle of
Marburg had had enough of the arguments of Zuin-
glius on the ' real presence,' he is said to have ended the
controversy by taking up a bit of chalk and writing
firmly and finally upon the table ' Hoc est corpus
meum.' Equally downright and definite were the
divines at Westminster. They were modest in offering
their conclusions to Parliament as ' humble advice,'
but there was no flicker of doubt either in their theology or their cosmology. ' From the beginning of the
world,' they say, ' to the Resurrection of Christ the last
day of the week was kept holy as a Sabbath ; ' while
from the Resurrection it ' was changed into the first
day of the week, which in Scripture is called the Lord's
Day, and is to be continued to the end of the world as
the Christian Sabbath.' The notions of the divines,
regarding the ' beginning and the end ' of the world,
were primitive, but decided. An ancient philosopher
was once mobbed for venturing the extravagant opinion
that the sun, which appeared to be a circle less than a
yard in diameter, might really be as large as the whole
country of Greece. Imagine a man with the knowledge of a modern geologist lifting up his voice among
these Westminster divines ! ' It pleased God,' they
continue, ' at the beginning, to create, or make of

breaking was either followed by death or 'grievously punished at
the judgment of the court.'

nothing, the world and all things therein, whether visible or invisible, in the space of six days, and all very good.' Judged from our present scientific standpoint this, of course, is mere nonsense. But the calling of it by this name does not exhaust the question. The real point of interest to me, I confess, is not the cosmological errors of the Assembly, but the hold which theology has taken of the human mind, and which enables it to survive the ruin of what was long deemed essential to its stability. On this question of ' essentials ' the gravest mistakes are constantly made. Save as a passing form no part of objective religion is essential. It is, as already shown, in its nature fluxional. Posterity will refuse to subscribe to the Nicene creed. Religion lives not by the force and aid of dogma, but because it is ingrained in the nature of man. To draw a metaphor from metallurgy, the moulds have been broken and reconstructed over and over again, but the molten ore abides in the ladle of humanity. An influence so deep and permanent is not likely soon to disappear; but of the future form of religion little can be predicted. Its main concern may possibly be to purify, elevate, and brighten the life that now is, instead of treating it as the more or less dismal vestibule of a life that is to come.

The term ' nonsense,' which has been just applied to the views of creation enunciated by the Westminster Assembly, is used, as already stated, in reference to our present knowledge and not to the knowledge of three or four centuries ago. To most people the earth was at that time all in all; the sun and moon and stars being set in heaven merely to furnish lamplight to our planet. But though in relation to the heavenly bodies the earth's position and importance were thus exaggerated, very inadequate and erroneous notions were entertained regarding the shape and magnitude of the earth itself. Theologians were horrified when first informed that our planet was a sphere. The question of antipodes exercised them for a long time, most of them pouring ridicule on the idea that men could exist with their feet turned towards us, and with their heads pointing downwards. I think it was Sir George Airy who referred to the case of an over-curious individual, asking what we should see if we went to the edge of the world and looked over. That the earth was a flat surface on which the sky rested was the belief entertained by the founders of all our great religious systems. The growth of the Copernican theory in public favour filled even liberal Protestant theologians with apprehension. They stigmatised it as being ' built on fallible phenomena and advanced by many arbitrary assumptions against evident testimonies of Scripture.' [1] Newton finally placed his intellectual crowbar beneath these ancient notions, and heaved them into irretrievable ruin.

Then it was that penetrating minds among the theologians, seeing the nature of the change wrought by the new astronomy in our conceptions of the universe, also discerned the difficulty, if not the impossibility, of accepting literally the Mosaic account of creation. With characteristic tenacity they clung to that account, but they assigned to it a meaning entirely new. Dr. Samuel Clarke, who was the personal friend of Newton and a supporter of his theory, threw out the idea that ' possibly the six days of creation might be a typical representation of some greater periods.' Clarke's contemporary, Dr. Thomas Burnet, wrote with greater decision in the same strain. The Sabbath being

---

[1] Such was the view of Dr. John Owen, who is described by Cox as ' the most eminent of the Independent divines.'

regarded as a shadow or type of that heavenly repose which the righteous will enjoy when this world has passed away, ' so these six days of creation are so many periods or millenniums for which the world and the toils and labours of our present state are destined to endure.' [1] The Mosaic account was thus reduced to a poetic myth—a view which afterwards found expression in the vast reveries of Hugh Miller. But if this symbolic interpretation, which is now generally accepted, be the true one, what becomes of the Sabbath day? It is absolutely without ecclesiastical meaning. The man who was executed for gathering sticks on that day must therefore be regarded as the victim of a rude legal rendering of a religious epic.

There were many minor offshoots of discussion from the great central controversy. Bishop Horsley had defined a day ' as consisting of one evening and one morning, or, as the Hebrew words literally import, of the decay of light and the return of it.' But what then, it was asked, becomes of the Sabbath in the Arctic regions, where light takes six months to ' decay,' and as long to ' return '? Differences of longitude, moreover, render the observance of the Sabbath at the same hours impossible. To some people such questions might appear trifling ; to others they were of the gravest import. Whether the Sabbath should stretch from sunset to sunset, or from midnight to midnight, was also a subject of discussion. ' If it should begin at midnight,' says one writer, ' what man of a thousand can readily tell the certain time when it begins, that so they may in a holy manner begin the Sabbath with God ? All men have not the midnight clocks and bells to awaken them, nor can the crowing of cocks herein

---

[1] Cox, vol. ii. p. 211, note.

give a certain sound.   A poor Christian man had need
to be a good and watchful mathematician that holds
this opinion, or else I see not how he will know when
midnight is come.'   In 1590 the Presbytery of Glasgow
enjoined that the Sabbath should be ' from sun to sun.'
In 1640 the Sabbath was declared to extend from midnight to midnight.   Uncertainty reigned, and innocent
people were prosecuted for beginning to work immediately after sunset.   Already, prior to the date last
mentioned, voices were heard refusing to acknowledge
the propriety of the change from Saturday to Sunday,
and the doctrine of Seventh Day observance was afterwards represented by a sect.[1]   The earth's sphericity
and rotation, which had at first been received with
such affright, came eventually to the aid of those
afflicted with qualms and difficulties regarding the
respective claims of Saturday and Sunday.   The sun
moves apparently from east to west.   Suppose then we
start on a voyage round the world in a westerly direction.   In doing so we sail away, as it were, from the
sun, which follows and periodically overtakes us, reaching the meridian of our ship each succeeding day
somewhat later than if we stood still.   For every 15°
of longitude traversed by the vessel the sun will be
exactly an hour late; and after the ship has traversed

[1] Theophilus Brabourne, a sturdy Puritan minister of Norfolk,
whom Cox regards as the founder of this sect, thus argued the question in 1628 : ' And now let me propound unto your choice these two
days : the Sabbath-day on Saturday or the Lord's Day on Sunday ;
and keep whether of the twain you shall in conscience find the more
safe.   If you keep the Lord's Day, but profane the Sabbath Day, you
walk in great danger and peril (to say the least) of transgressing
one of God's eternal and inviolable laws—the Fourth Commandment.
But, on the other side, if you keep the Sabbath Day, though you
profane the Lord's Day, you are out of all gun-shot and danger, for
so you transgress no law at all, since neither Christ nor his apostles
did ever leave any law for it.'

twenty-four times 15°, or 360°, that is to say, the entire circle of the eaith, the sun will be exactly a day behind. Here, then, is the expedient suggested by Dr. Wallis, F.R.S., Savilian Professor of Geometry in the University of Oxford, to quiet the minds of those in doubt regarding Saturday observance. He recommends them to make a voyage round the world, as Sir Francis Drake did, ' going out of the Atlantic Ocean westward by the Straits of Magellan to the East Indies, and then from the east returning by the Cape of Good Hope homeward, and let them keep their Saturday-Sabbath all the way. When they come home to England they will find their Saturday to fall upon our Sunday, and they may thenceforth continue to observe their Saturday-Sabbath on the same day with us!'

Large and liberal minds were drawn into this Sabbatarian conflict, but they were not the majority Between the booming of the bigger guns we have an incessant clatter of small arms. We ought not to judge superior men without reference to the spirit of their age. This is an influence from which they cannot escape, and so far as it extenuates their errors it ought to be pleaded in their favour. Even the atrocities of the individual excite less abhorrence when they are seen to be the outgrowth of his time. But the most fatal error that could be committed by the leaders of religious thought is the attempt to force into their own age conceptions which have lived their life, and come to their natural end in preceding ages. History is the record of a vast experimental investigation—of a search by man after tne best conditions of existence. The Puritan attempt was a grand experiment. It had to be made. Sooner or later the question must have forced itself upon earnest believers possessed of power:—Is it not possible to rule the world in accordance with

III. D

the wishes of God as revealed in the Bible?—Is it not possible to make human life the copy of a divine pattern? The question could only have occurred in the first instance to the more exalted minds. But instead of working upon the inner forces and convictions of men, legislation presented itself as a speedier way to the attainment of the desired end. To legislation, therefore, the Puritans resorted. Instead of guiding, they repressed, and thus pitted themselves against the unconquerable impulses of human nature. Believing that nature to be depraved, they felt themselves logically warranted in putting it in irons. But they failed; and their failure ought to be a warning to their successors.

Another error, of a far graver character than that just noticed, may receive a passing mention here. At the time when the Sabbath controversy was hottest, and the arm of the law enforcing the claims of the Sabbath strongest and most unsparing, another subject profoundly stirred the religious mind of Scotland. A grave and serious nation, believing intensely in its Bible, found therein recorded the edicts of the Almighty against witches, wizards, and familiar spirits, and were taught by their clergy that such edicts still held good. The same belief had overspread the rest of Christendom, but in Scotland it was intensified by the rule of Puritanism and the natural earnestness of the people. I have given you a sample of the devilish cruelties practised in the time of Polycarp on the Christians at Smyrna. These tortures were far less shocking than those inflicted upon witches in Scotland. I say less shocking because the victims at Smyrna courted martyrdom. They counted the sufferings of this present time as not worthy to be compared with the glory to be revealed; while the sufferers for witchcraft, in the midst of all their agonies, felt themselves God-forsaken, and saw before them instead of the glories of heaven the infinite tortures of hell. Not to the fall of Sarmatia, but to the treatment of witches in the seventeenth century, ought to be applied the words of your poet Campbell :—

Oh ! bloodiest picture in the book of time !

The mind sits in sackcloth and ashes while contemplating the scenes so powerfully described by Mr. Lecky in his chapter on Magic and Witchcraft. But I will dwell no further upon these tragedies than to point out how terrible are the errors which our clergy may commit after they have once subscribed to the creed and laws of Judaism, and constituted themselves the legal exponents and interpreters of those laws.[1]

Turning over the leaves of the Pentateuch, where God's alleged dealings with the Israelites are recorded, it strikes one with amazement that such writings should be considered for a moment as binding upon us. The overmastering strength of habit, the power of early education—possibly a defiance of the claims of reason involved in the very constitution of the mental organ—are forcibly illustrated by the fact that learned men are still to be found willing to devote their time and endowments to these writings under the assumption that they are not human but divine. Claiming the same origin as other books, the Old Testament is without a rival, but its unnatural exaltation as a court of appeal provokes recoil and rejection. Leviticus, for example, when read in the light of its own age, is full of interest

---

[1] The sufferings of reputed witches in the seventeenth century as well as those of the early Christians, might be traced to panics and passions similar in kind to those which produced the atrocities of the Reign of Terror in France.

and instruction. We see there described the efforts of
the best men then existing to civilise the rude society
around them. Violence is restrained by violence medi-
cinally applied. Passion is checked, truth and justice
are extolled, and all in a manner suited to the needs of
a barbarian host. But read in the light of our age,
its conceptions of the deity are seen to be shockingly
mean, and many of its ordinances brutal. Foolishness
is far too weak a word to apply to any attempt to force
upon a scientific age the edicts of a Jewish lawgiver.
The doom of such an attempt is sure, and if the destruction of things really precious should be involved
in its failure, the blame will justly be ascribed to those
who obstinately persisted in the attempt. Let us then
cherish our Sunday as an inheritance derived from the
wisdom of the past, but let it be understood that we
cherish it because it is in principle reasonable and in
practice salutary. Let us uphold it, because it commends itself to that 'light of nature' which, despite
the catastrophe in Eden, the most famous theologians
mention with respect, and not because it is enjoined by
the thunders of Sinai. We have surely heard enough
of divine sanctions founded upon myths which, however
beautiful and touching when regarded from the proper
point of view, are seen, when cited for our guidance as
matters of fact, to offer warrant and condonation for
the greatest crimes, or to sink to the level of the most
palpable absurdities.[1]

Melanchthon writes finely thus : 'Wherefore our decision is
this : that those precepts which learned men have committed to
writing, transcribing them from the common reason and common
feelings of human nature are to be accounted as no less divine than
those contained in the tables of Moses.' (Dugald Stewart's translation.) Hengstenberg quotes from the same reformer as follows :
'The law of Moses is not binding upon us, though some things
which the law contains are binding, because they coincide with the

In this, as in all other theological discussions, it is interesting to note how character colours religious feeling and conduct. The reception into Christ's kingdom has been emphatically described as being born again. A certain likeness of feature among Christians ought, one would think, to result from a common spiritual parentage. But the likeness is not observed. Men professing to be born of the same spirit, prove to be as diverse as those who claim no such origin. Christian communities embrace some of the loftiest and many of the lowest of mankind. It may be urged that the lofty ones only are truly religious. To this it is to be replied that the others are often as religious as their natures permit them to be. *Character* is here the overmastering force. That religion should influence life in a high way implies the pre-existence of natural dignity. This is the mordant which fixes the religious dye. He who is capable of feeling the finer glow of religion would possess a substratum available for all the relations of life, even if his religion were taken away. Religion, on the other hand, cannot charm away malice, or make good defects of character. I have already spoken of persecution in its meaner forms. On the lower levels of theological warfare such are commonly resorted to. If you reject a dogma on intellectual grounds it is because there is a screw loose in your morality. Some personal sin besets and blinds you. The intellect is captive to a corrupt heart. Thus good men have been often calumniated by others who were not good; thus frequently have the noble become a target for the wicked and the mean. With the advance

law of nature.'—See Cox, vol i. p. 389. The Catechism of the Council of Trent expresses a similar view. There are, then, ' data of ethics ' over and above the revealed ones.

of public intelligence the day of such assailants is
happily drawing to a close.

These reflections, which connect themselves with
reminiscences outside the Sabbath controversy, have
been more immediately prompted by the aspersions
cast by certain Sabbatarians upon those who differ from
them. Mr. Cox notices and reproves some of these.
According to the Scottish Sabbath Alliance, for example, all who say that the Sabbath was an exclusively
Jewish institution, including, be it noted, such men as
Jeremy Taylor and Milton, 'clearly prove either their
dishonesty or ignorance, or inability to comprehend a
very plain and simple subject.' This becomes real
humour when we compare the speakers with the persons spoken of. A distinguished English dissenter, who
deals in a lustrous but rather cloudy logic, declares
that whoever asks demonstration of the divine appointment of the Christian Sabbath 'is blinded *by a moral
cause* to those exquisite pencillings, to those unob-
truded vestiges which furnish their clearest testimony
to this Institute.' A third writer charitably professes
his readiness ' to admit, in reference to this and many
other duties, that it is quite a possible thing for a mind
that is desirous of *evading the evidence* regarding it to
succeed in doing so.' A fourth luminary, whose knowledge obviously extends to the mind and methods of the
Almighty, exclaims, 'Is it not a principle of God's
Word in many cases to give enough and no more—to
satisfy the devout, not to overpower the *uncandid*?'
It is, of course, as easy as it is immoral to argue thus ;
but the day is fast approaching when the most atra-
bilious presbyter will not venture to use such language.
Let us contrast with it the utterance of a naturally
sweet and wholesome mind. 'Since all Jewish festivals,
new moons, and Sabbaths,' says the celebrated Dr. Isaac

Watts, 'are abolished by St. Paul's authority, since the
religious observation of days in the 14th chapter to the
Romans in general is represented as a matter of doubtful disputation, since the observation of the Lord's Day
is not built upon any express or plain institution by
Christ or his apostles in the New Testament, but rather
on examples and probable inferences, and on the reasons
and relations of things; I can never pronounce anything
hard or severe upon any fellow Christian who maintains
real piety in heart and life, though his opinion on this
subject may be very different from mine.' Thus
through the theologian radiates the gentleman.

Up to the end of the eighteenth century the catalogue of Mr. Cox embraces 320 volumes and publications. It is a monument of patient labour; while the
remarks of the writer, which are distributed throughout
the catalogue, illustrate both his intellectual penetration
and his reverent cast of mind. He wrought hard and
worthily with a pure and noble aim. I had the
pleasure of meeting Mr. Cox at Dundee in 1867, when
the British Association met there, and I could then
discern the earnestness with which he desired to see his
countrymen relieved from the Sabbath incubus, and at
the same time the moderation and care for the feelings
of others with which he advocated his views. He has
also given us a rapid 'Sketch of the Chief Controversies
about the Sabbath in the Nineteenth Century.' The
sketch is more compressed than the catalogue, and the
changes of thought in passing from author to author,
being more rapid, are more bewildering. It is, to a
great extent, what I have already called a clatter of
small arms mingled with the occasional thunder of
heavier guns. One thing is noticeable and regrettable
in these discussions, namely, the unwise and undis-
criminating way in which different Sunday occupations

are classed together and condemned. Bishop Bloomfield, for example, seriously injures his case when he
places drinking in gin-shops and sailing in steamboats
in the same category. I remember some years ago
standing by the Thames at Putney with my lamented
friend Dr. Bence Jones, when a steamboat on the river
with its living freight passed us. Practically acquainted
with the moral and physical influence of pure oxygen,
my friend exclaimed, ' What a blessing for these people
to be able thus to escape from London into the fresh
air of the country!' I hold the physician to have been
right and, with all respect, the Bishop to have been
wrong.

Bishop Bloomfield also condemns resorting to tea-
gardens on Sunday. But we may be sure that it is not
the tea-gardens, but the minds which the people bring
to them, which produce disorder. These minds already
possess the culture of the city, to which the Bishop
seems disposed to confine them. Wisely and soberly
conducted—and it is perfectly possible to conduct them
wisely and soberly—such gardens might be converted
into aids towards a life which the Bishop would commend. Purification and improvement are often possible,
where extinction is neither possible nor desirable. I
have spent many a Sunday afternoon in the tea-gardens
of the little university town of Marburg, in the company
of intellectual men and cultivated women, without
observing a single occurrence which, as regards morality,
might not be permitted in the Bishop's drawing-room.
I will add to this another observation made at Dresden
on a Sunday, immediately after the suppression of the
insurrection by the Prussian soldiery in 1849. The
victorious troops were encamped in some meadows on
the banks of the Elbe, and I went among them and saw
how they occupied themselves. Some were engaged in

physical games and exercises which in England would
be considered innocent in the extreme, some were conversing sociably, some singing the songs of Uhland,
while others, from elevated platforms, recited to listening
groups poems and passages from Goethe and Schiller.
Through this crowd of military men passed and repassed
the girls of the city, linked together with their arms
round each other's necks. During hours of observation,
I heard no word which was unfit for a modest ear ; while
from beginning to end I failed to notice a single case
of intoxication.[1]

It may appear uncivil and inappropriate for a person
invited to come amongst you as I have been to seek to
establish contrasts with other countries unfavourable to
your own ; but let me take an extract from an account
of Scotland written by a Scot, a short time prior to the
date of my visit to Dresden. 'A tree,' says this writer,
' is best known by its fruits. What are these in the
present instance ? The protracted effort to enforce a
stern Sabbatical observance per fas et nefas has no
doubt evoked an exceedingly decorous state of affairs
on Sunday ; but in a great measure only so far as
external appearances are concerned. Puritanism with
its uncompromising demands has had a sway of three
centuries in Scotland ; and yet at this moment, in proportion to the population, the amount of crime, vice, and
intemperance is as great, if not in some details greater,
than it is in England. But the most frightful feature
of Scotland is the loathsome squalor and heathenism of
its large towns. The combination of brutal iniquity,
filth, absence of self-respect, and intemperance visible

---

[1] The late Mr. Joseph Kay, as Travelling Bachelor of the University of Cambridge, has borne strong and earnest testimony to the
' humanising and civilising influence ' of the Sunday recreations of
the German people.

daily in the meaner class of streets of Edinburgh and Glasgow fills every traveller with surprise and horror.'

Here indeed we touch the core of the whole matter —the appeal to experience. Sabbatical rigour has been tried, and the question is: Have its results been so beneficent—so conducive to good morals and national happiness—as to render criminal every attempt to modify it? The advances made in all kinds of knowledge in this our age by special cultivators are known to be enormous, and the public desire for instruction, which the intellectual triumphs of the time naturally and inevitably arouse, is commensurate with the growth of knowledge. Must this desire, which is the motive power of all real and healthy progress, be quenched or left unsatisfied lest Sunday observances, unknown to the early Christians, repudiated by the heroes of the reformation, and insisted upon for the first time during a period of national gloom and suffering in the seventeenth century, should be interfered with? To justify this position the demonstration of the success of Sabbatarianism must be complete. Is it so? Are we so much better than other nations who have neglected to adopt our rules, that we can point to the working of these rules in the past as a conclusive reason for maintaining them immovable in the future? The answer must be, No! Within the range of my recollection no German man would have ventured to assert of Berlin or Dresden that its brutal iniquity, filth, and intemperance filled every traveller with surprise and horror. The statement would have been immediately branded as a flagrant untruth. And yet this is the language which, thirty years ago, when the Sabbath was observed more strictly than it is now, was used by a Scot in reference to the towns of Scotland. My Sabbatarian friends, you have no ground to stand upon. I say

friends, for I would far rather have you as friends than
as enemies—far rather see you converted than anni-
hilated. You possess a strength and earnestness with
which the world cannot dispense; but to be productive
of anything permanently good, that strength and that
earnestness must build upon the sure foundation of
human nature. This is that law of the universe spoken
of so frequently by your illustrious countryman, Mr.
Carlyle, to quarrel with which is to provoke and precipitate ruin. Join with us then in our endeavours to
turn our Sundays to better account. Back with your
support the moderate and considerate demands of the
Sunday Society, which scrupulously avoids interfering
with the hours devoted by common consent to public
worship. Offer the museum, the picture gallery, and
the public garden as competitors to the public-house.
By so doing you will fall in with the spirit of your time,
and row with, instead of against, the resistless current
along which man is borne to his destiny.

Most of you here are Liberals; perhaps Radicals,
perhaps even Republicans. In the proper sense of the
term, I am a Conservative. Madness or folly can de-
molish : it requires wisdom to conserve. But let us
understand each other. The first requisite of a true
conservatism is foresight. Humanity grows, and foresight secures room for future expansion. In your walks
in the country you sometimes see a wall built round a
growing tree. So much the worse for the wall, which is
sure to be rent and ruined by the energy it opposes. We
have here represented not a true, but a false and ignorant conservatism. The true conservative looks ahead
and prepares for the inevitable. He forestalls revolution
by securing, in due time, sufficient amplitude for the
national vibrations. He is a wrong-headed statesman
who imposes his notions, however right in the abstract,

on a nation unprepared for them. He is no statesman
at all who, without seeking to interpret and guide it in
advance, merely waits for the more or less coarse expression of the popular will, and then constitutes himself
its vehicle. *Untimeliness* is sure to be the characteristic
of the work of such a statesman. In virtue of the
position which he occupies, his knowledge and insight
ought to be in advance of the public knowledge and
insight ; and his action, in like degree, ought to precede
and inform public action. This is what I want my
Sabbatarian friends to bear in mind. If they look
abroad from the vantage-ground which they occupy,
they can hardly fail to discern that the intellect of this
country is gradually ranging itself upon our side.
Whether they hear or whether they forbear, we are sure
to unlock, for the public benefit, the doors of the
museums and galleries which we have purchased, and
for the maintenance of which we pay. But I would
have them not only to prepare for the coming change,
but to aid and further it by anticipation. They will
thus, in a new fashion, 'dish the Whigs,' prove themselves men of foresight and common sense, and obtain
a fresh lease of the respect of the community.

As the years roll by, the term 'materialist' will lose
more and more of its evil connotation ; for it will be
more and more seen and acknowledged that the true
spiritual nature of man is bound up with his material
condition. Wholesome food, pure air, cleanliness—hard
work if you will, but also fair rest and recreation—these
are necessary not only to physical but to spiritual well-
being. A clogged and disordered body implies a more
or less disordered mind. The seed of the spirit is cast
in vain amid stones and thorns, and thus your best
utterances become idle words when addressed to the
acclimatised inhabitants of our slums and alleys.

Drunkenness ruins the substratum of resolution. The physics of the drunkard's brain are incompatible with moral strength. Here your first care ought to be to cleanse and improve the organ. Break the sot's associations; change his environment; alter his nutrition; displace his base imaginations by thoughts drawn from the purer sources which we seek to render accessible to him. Such is the treatment of which the denizen of our slums stands in most immediate need—such the discipline requisite for the development of a force of will, able to resist the fascinations of the gin-shop. If you could establish Sunday tramways between these dens of filth and iniquity and the nearest green fields, you would, in so doing, be preaching a true Gospel. And not only the denizens of our slums, but the proprietors of our factories and counting-houses might, perhaps, be none the worse for an occasional excursion in the company of those whom they employ. A most blessed influence would also be shed upon the clergy if they were enabled from time to time to change their ' sloth urbane' for action on heath or mountain. Baxter was well aware of the soothing influence of fields, and countries, and walks and gardens, on a fretted brain Jeremy Taylor showed a profound knowledge of human nature when he wrote thus :—' It is certain that all which can innocently make a man cheerful, does also make him charitable. For grief, and age, and sickness, and weariness, these are peevish and troublesome ; but mirth and cheerfulness are content, and civil, and compliant, and communicative, and love to do good, and swell up to felicity only upon the wings of charity. Upon this account, here is pleasure enough for a Christian at present ; and if a facete discourse, and an amicable friendly mirth, can refresh the spirit and take it off from the vile temptation of peevish, despairing,

uncomplying melancholy, it must needs be innocent
and commendable.' I do not know whether you ever
read Thomas Hood's 'Ode to Rae Wilson,' with an
extract from which I will close this address. Hood
was a humourist, and to some of our graver theologians
might appear a mere feather-head. But those who
have read his more serious works will have discerned in
him a vein of deep poetic pathos. I hardly know anything finer than the apostrophe with which he turns
from those

> That bid you baulk
> A Sunday walk,
> And shun God's work as you should shun your own ;
>
> . . . .
>
> Calling all sermons contrabands,
> In that great Temple that's not made with hands,

to the description of what Sunday might be, and is, to
him who is competent to enjoy it aright.

> Thrice blessed, rather, is the man, with whom
> The gracious prodigality of nature,
> The balm, the bliss, the beauty, and the bloom,
> The bounteous providence in ev'ry feature,
> Recall the good Creator to his creature,
> Making all earth a fane, all heav'n its dome !
> To *his* tuned spirit the wild heather-bells
> Ring Sabbath knells ;
> The jubilate of the soaring lark
> Is chant of clerk ;
> For choir, the thrush and the gregarious linnet ;
> The sod's a cushion for his pious want ;
> And, consecrated by the heav'n within it,
> The sky-blue pool, a font.
> Each cloud-capp'd mountain is a holy altar ;
> An organ breathes in every grove ;
> And the full heart's a Psalter,
> Rich in deep hymns of gratitude and love !

1880.

GOETHE'S 'FARBENLEHRE.' [1]

IN the days of my youth, when life was strong and aspiration high, I found myself standing one fine summer evening beside a statue of Goethe in a German city. Following the current of thought and feeling started by the associations of the place, I eventually came to the conclusion that, judging even from a purely utilitarian point of view, a truly noble work of art was the most suitable memorial for a great man. Such a work appeared to me capable of exciting a motive force within the mind which no purely material influence could generate. There was then labour before me of the most arduous kind. There were formidable practical difficulties to be overcome, and very small means wherewith to overcome them, and yet I felt that no material means could, as regards the task I had undertaken, plant within me a resolve comparable with that which the contemplation of this statue of Goethe was able to arouse.

My reverence for the poet had been awakened by the writings of Mr. Carlyle, and it was afterwards confirmed and consolidated by the writings of Goethe himself. There was, however, one of the poet's works which, though it lay directly in the line of my own studies, remained for a long time only imperfectly known to me. My opinion of that work was not formed on hearsay. I

---

[1] A Friday evening discourse in the Royal Institution.

dipped into it so far as to make myself acquainted
with its style, its logic, and its general aim ; but having
done this I laid it aside as something which jarred upon
my conception of Goethe's grandeur. The mind willingly rounds off the image which it venerates, and only
acknowledges with reluctance that it is on any side incomplete ; and believing that Goethe in the 'Farben-.
lehre' was wrong in his intellectual, and perverse in his
moral judgments—seeing above all things that he had
forsaken the lofty impersonal calm which was his chief
characteristic, and which had entered into my conception of the god-like in literature—I abandoned the
'Farbenlehre,' and looked up to Goethe on that side
where his greatness was uncontested and supreme.

   But in the month of May 1878 Mr. Carlyle did me
the honour of calling upon me twice ; and not being
at home at the time, I visited him in Chelsea soon
afterwards. He was then in his eighty-third year, and
looking in his solemn fashion towards that portal to
which we are all so rapidly hastening, he remembered
his friends. He then presented to me, as 'a farewell
gift,' the two octavo volumes of letterpress, and the
single folio volume, consisting in great part of coloured
diagrams, which are here before you. Exactly half a
century ago these volumes were sent by Goethe to Mr.
Carlyle. They embrace the 'Farbenlehre'—a title
which may be translated, though not well translated,
'Theory of Colours'—and they are accompanied by a
long letter, or rather catalogue, from Goethe himself,
dated June 14, 1830, a little less than two years before
his death. My illustrious friend wished me to examine
the book, with a view of setting forth what it really
contained. This year for the first time I have been able
to comply with the desire of Mr. Carlyle ; and as I knew
that your wish would coincide with his, as to the propriety of making some attempt to weigh the merits of

a work which exerted so great an influence in its day,[1] I have not shrunk from the labour of such a review.

The average reading of the late Mr. Buckle is said to have amounted to three volumes a day. They could not have been volumes like those of the 'Farbenlehre.' For the necessity of halting and pondering over its statements is so frequent, and the difficulty of coming to any undoubted conclusion regarding Goethe's real conceptions is often so great, as to invoke the expenditure of an inordinate amount of time. I cannot even now say with confidence that I fully realise all the thoughts of Goethe. Many of them are strange to the scientific man. They demand for their interpretation a sympathy beyond that required, or even tolerated, in severe physical research. Two factors, the one external and the other internal, go to the production of every intellectual result. There is the evidence without, and there is the mind within on which that evidence impinges. Change either factor and the result will cease to be the same. In the region of politics, where mere opinion comes so much into play, it is only natural that the same external evidence should produce different convictions in different minds. But in the region of science, where demonstration instead of opinion is paramount, such differences ought hardly to be expected. That they nevertheless occur is strikingly exemplified by the case before us; for the very experimental facts which had previously converted the world to Newton's views, on appealing to the mind of Goethe, produced a theory of light and colours in violent antagonism to that of Newton.

[1] The late Sir Charles Eastlake translated a portion of the *Farbenlehre*; while the late Mr. Lewes, in his *Life of Goethe*, has given a brief but very clever account of the work. It is also dealt with by Dove and, in connection with Goethe's other scientific labours, by Helmholtz.

Goethe prized the 'Farbenlehre' as the most important of his works. 'In what I have done as a poet,' he
says to Eckermann, 'I take no pride, but I am proud
of the fact that I am the only person in this century
who is acquainted with the difficult science of colours.'
If the importance of a work were to be measured by the
amount of conscious labour expended in its production,
Goethe's estimate of the 'Farbenlehre' would probably
be correct. The observations and experiments there
recorded astonish us by their variety and number. The
amount of reading which he accomplished was obviously
vast. He pursued the history of optics not only along
its main streams, but on to its remotest rills. He was
animated by the zeal of an apostle, for he believed that
a giant imposture was to be overthrown, and that he
was the man to accomplish the holy work of destruction.
He was also a lover of art, and held that the enunciation
of the true principles of colour would, in relation to
painting, be of lasting importance. Thus positively
and negatively he was stimulated to bring all the
strength he could command to bear upon this question.
The greater part of the first volume is taken up with
Goethe's own experiments, which are described in 920
paragraphs duly numbered. It is not a consecutive
argument, but rather a series of jets of fact and logic
emitted at various intervals. I picture the poet in that
troublous war-time, walking up and down his Weimar
garden, with his hands behind his back, pondering his
subject, throwing his experiments and reflections into
these terse paragraphs, and turning occasionally into
his garden house to write them down. This first portion of the work embraces three parts, which deal, respectively, with Physiological or Subjective Colours,
with Physical or Prismatic Colours, and with Chemical
Colours and Pigments. To these are added a fourth
part, bearing the German title, 'Allgemeine Ansichten

nach innen;' a fifth part, entitled 'Nachbarliche Ver-
hältnisse,' neighbouring relations; and a sixth part,
entitled 'Sinnlich-sittliche Wirkung der Farbe,' sen-
suously-moral effect of colours. It is hardly necessary to
remark that some of these titles, though doubtless pregnant with meaning to the poet himself, are not likely to
commend themselves to the more exacting man of
science.

The main divisions of Goethe's book are subdivided
into short sections, bearing titles more or less shadowy
from a scientific point of view—Origin of white; Origin
of black; Excitement of colour; Heightening; Culmina-
tion; Balancing; Reversion; Fixation; Mixture real;
Mixture apparent; Communication actual; Communication apparent. He describes the colours of minerals,
plants, worms, insects, fishes, birds, mammals, and
men. Hair on the surface of the human body he considers indicative rather of weakness than of strength.
The disquisition is continued under the headings—How
easily colour arises; How energetic colour may be;
Heightening to red; Completeness of manifold phenomena; Agreement of complete phenomena; How easily
colour disappears; How durable colour remains; Relation
to philosophy; Relation to mathematics; Relation to
physiology and pathology; Relation to natural history;
Relation to general physics; Relation to tones. Then
follows a series of sections dealing with the primary
colours and their mixtures. These sections relate less
to science than to art. The writer treats, among other
things, of Æsthetic effects; Fear of the Theoretical;
Grounds and Pigments; Allegorical, Symbolical, and
Mystical use of colours. The headings alone indicate
the enormous industry of the poet; showing at the same
time an absence of that scientific definition which he
stigmatised as 'pedantry' in the case of Newton.

In connection with his subject, Goethe charged himself with all kinds of kindred knowledge. He refers to ocular spectra, quoting Boyle, Buffon, and Darwin; to the paralysis of the eye by light; to its extreme sensitiveness when it awakes in the morning; to irradiation—quoting Tycho Brahe on the comparative apparent size of the dark and the illuminated moon. He dwells upon the persistence of impressions upon the retina, and quotes various instances of abnormal duration. He possessed a full and exact knowledge of the phenomena of subjective colours, and described various modes of producing them. He copiously illustrates the production by red of subjective green, and by green of subjective red. Blue produces subjective yellow, and yellow subjective blue. He experimented upon shadows, coloured in contrast to surrounding light. The contrasting subjective colours he calls 'geforderte Farben,' colours 'demanded' by the eye. Goethe gives the following striking illustration of these subjective effects. 'I once,' he said, 'entered an inn towards evening, when a well-built maiden, with dazzlingly white face, black hair, and scarlet bodice and skirt came towards me. I looked at her sharply in the twilight, and when she moved away, saw upon the white wall opposite, a black face with a bright halo round it, while the clothing of the perfectly distinct figure appeared of a beautiful sea-green.' With the instinct of the poet, Goethe discerned in these antitheses an image of the general method of nature. Every action, he says, implies an opposite. Inhalation precedes expiration, and each systole has its corresponding diastole. Such is the eternal formula of life. Under the figure of systole and diastole the rhythm of nature is represented in other portions of his work.

Goethe handled the prism with great skill, and his

experiments with it are numberless. He places white rectangles on a black ground, black rectangles on a white ground, and shifts their apparent positions by prismatic refraction. He makes similar experiments with coloured rectangles and discs. The shifted image is sometimes projected on a screen, the experiment being then ' objective.' It is sometimes looked at directly through the prism, the experiment being then 'subjective.' In the production of chromatic effects, he dwells upon the absolute necessity of *boundaries*— ' Gränzen.' The sky may be looked at and shifted by a prism without the production of colour; and if the white rectangle on a black ground be only made wide enough, the centre remains white after refraction, the colours being confined to the edges. Goethe's earliest experiment, which led him so hastily to the conclusion that Newton's theory of colours was wrong, consisted in looking through a prism at the white wall of his own room. He expected to see the whole wall covered with colours, this being, he thought, implied in the theory of Newton. But to his astonishment it remained white, and only when he came to the boundary of a dark or a bright space did the colours reveal themselves. This question of ' boundaries ' is one of supreme importance to the author of the ' Farbenlehre ; ' the end and aim of his theory being to account for the coloured fringes produced at the edges of his refracted images.

Darkness, according to Goethe, had as much to do as light with the production of colour. Colour was really due to the commingling of both. Not only did his white rectangles upon a black ground yield the coloured fringes, but his black rectangles on a white ground did the same. The order of the colours seemed, however, different in the two cases. Let a visiting card, held in the hand between the eye and a window facing

the bright firmament, be looked at through a prism;
then supposing the image of the card to be shifted
upwards by refraction, a red fringe is seen above and a
blue one below. Let the back be turned to the window
and the card so held that the light shall fall upon it ;
on being looked at through the prism, blue is seen above
and red below. In the first case the fringes are due to
the decomposition of the light adjacent to the edge of
the card, which simply acts as an opaque body, and
might have been actually black. In the second case
the light decomposed is that coming from the surface
of the card itself. The first experiment corresponds to
that of Goethe with a black rectangle on a white
ground ; while the second experiment corresponds to
Goethe's white rectangle on a black ground. Both
these effects are immediately deducible from Newton's
theory of colours. But this, though explained to him
by physicists of great experience and reputation, Goethe
could never be brought to see, and he continued to
affirm to the end of his life that the results were utterly
irreconcilable with the theory of Newton.

In his own explanations Goethe began at the wrong
end, inverting the true order of thought, and trying to
make the outcome of theory its foundation. Apart
from theory, however, his observations are of great
interest and variety. He looked to the zenith at midnight, and found before him the blackness of space,
while in daylight he saw the blue firmament overhead ;
and he rightly adopted the conclusion that this colouring of the sky was due to the shining of the sun upon
a turbid medium with darkness behind. He by no
means understood the physical action of turbid media,
but he made a great variety of experiments bearing
upon this point. Water, for example, rendered turbid
by varnish, soap, or milk, and having a black ground

behind it, always appeared blue when shone upon by white light. When, instead of a black background, a bright one was placed behind, so that the light shone, not *on* but *through* the turbid liquid, the blue colour disappeared, and he had yellow in its place. Such experiments are capable of endless variation. To this class of effects belongs the painter's ' chill.' A cold bluish bloom, like that of a plum, is sometimes observed to cover the browns of a varnished picture. This is due to a want of optical continuity in the varnish. Instead of being a coherent layer it is broken up into particles of microscopic smallness, which virtually constitute a turbid medium and send blue light to the eye.

Goethe himself describes a most amusing illustration, or, to use his own language, 'a wonderful phenomenon,' due to the temporary action of a turbid medium on a picture. ' A portrait of an esteemed theologian was painted several years ago by an artist specially skilled in the treatment of colours. The man stood forth in his dignity clad in a beautiful black velvet coat, which attracted the eyes and awakened the admiration of the beholder almost more than the face itself. Through the action of humidity and dust, however, the picture had lost much of its original splendour. It was therefore handed over to a painter to be cleaned, and newly varnished. The painter began by carefully passing a wet sponge over the picture. But he had scarcely thus removed the coarser dirt, when to his astonishment the black velvet suddenly changed into a light blue plush : the reverend gentleman acquiring thereby a very worldly, if, at the same time, an old-fashioned appearance. The painter would not trust himself to wash further. He could by no means see how a bright blue could underlie a dark black, still less that he could have so rapidly

washed away a coating capable of converting a blue
like that before him into the black of the original
painting.'

Goethe inspected the picture, saw the phenomenon,
and explained it. To deepen the hue of the velvet coat
the painter had covered it with a special varnish, which,
by absorbing part of the water passed over it, was converted into a turbid medium, through which the black
behind instantly appeared as blue. To the great joy
of the painter, he found that a few hours' continuance
in a dry place restored the primitive black. By the
evaporation of the moisture the optical continuity of
the varnish (to which essential point Goethe does not
refer) was re-established, after which it ceased to act as
a turbid medium.

This question of turbid media took entire possession
of the poet's mind. It was ever present to his observation. It was illustrated by the azure of noonday, and
by the daffodil and crimson of the evening sky. The
inimitable lines written at Ilmenau—

> Ueber allen Gipfeln
> Ist Ruh',
> In allen Wipfeln
> Spürest Du
> Kaum einen Hauch—

suggest a stillness of the atmosphere which would allow
the columns of fine smoke from the foresters' cottages
to rise high into the air. He would thus have an
opportunity of seeing the upper portion of the column
projected against bright clouds, and the lower portion
against dark pines, the brownish yellow of the one, and
the blue of the other, being strikingly and at once
revealed. He was able to produce artificially at will
the colours which he had previously observed in nature.

He noticed that when certain bodies were incorporated with glass this substance also played a double part, appearing blue by reflected and yellow by transmitted light.[1]

The action of turbid media was to Goethe the ultimate fact—the *Urphänomen*—of the world of colours. ' We see on the one side Light and on the other side Darkness. We bring between both Turbidity, and from these opposites develop all colours.' As long as Goethe remains in the region of fact his observations are of permanent value. But by the coercion of a powerful imagination he forced his turbid media into regions to which they did not belong, and sought to overthrow by their agency the irrefragable demonstrations of Newton. Newton's theory, as known by everybody, is that white light is composed of a multitude of differently refrangible rays, whose coalescence produces the impression of white. By prismatic analysis these rays are separated from each other, the colour of each ray being strictly determined by its refrangibility. The experiments of Newton, whereby he sought to establish this theory, had long appealed with overmastering evidence to every mind trained in the severities of physical investigation. But they did not thus appeal to Goethe. Accepting for the most part the experiments of Newton, he rejected with indignation the conclusions drawn from them, and turned into utter ridicule the notion that white light possessed the composite character ascribed to it. Many of the naturalists of his time supported him, while among philosophers Schelling and Hegel shouted in acclamation over the supposed defeat of Newton. The physicists, however, gave the poet no countenance. Goethe met their scorn with scorn, and

---

[1] Beautiful and instructive samples of such glass are to be seen in the Venice Glass Company's shop, No. 30 St. James's Street.

under his lash these deniers of his theory, their Master
included, paid the penalty of their arrogance.

How, then, did he lay down the lines of his own
theory? How, out of such meagre elements as his
yellow, and his blue, and his turbid medium, did he
extract the amazing variety and richness of the Newtonian spectrum? Here we must walk circumspectly,
for the intellectual atmosphere with which Goethe surrounds himself is by no means free from turbidity. In
trying to account for his position, we must make ourselves acquainted with his salient facts, and endeavour to
place our minds in sympathy with his mode of regarding them. He found that he could intensify the yellow
of his transmitted light by making the turbidity of his
medium stronger. A single sheet of diaphanous parchment placed over a hole in his window-shutter appeared
whitish. Two sheets appeared yellow, which by the
addition of other sheets could be converted into red. It
is quite true that by simply sending it through a medium charged with extremely minute particles we can
extract from white light a ruby red. The red of the
London sun, of which we have had such fine and frequent examples during the late winter, is a case to
some extent in point. Goethe did not believe in Newton's differently refrangible rays. He refused to entertain the notion that the red light obtained by the
employment of several sheets of parchment was different
in quality from the yellow light obtained with two.
The red, according to him, was a mere intensification—
'Steigerung'—of the yellow. Colours in general consisted, according to Goethe, of light on its way to
darkness, and the only difference between yellow and
red consisted in the latter being nearer than the former
to its final goal.

But how in the production of the spectrum do tur-

bid media come into play ? If they exist, where are
they ? The poet's answer to this question is subtle in
the extreme. He wanders round the answer before he
touches it, indulging in various considerations regarding
penumbræ and double images, with the apparent aim of
breaking down the repugnance to his logic which the
mind of his reader is only too likely to entertain. If
you place a white card near the surface of a piece of
plate-glass, and look obliquely at the image of the
card reflected from the two surfaces, you observe two
images, which are hazy at the edges and more dense
and defined where they overlap. These hazy edges
Goethe pressed into his service as turbid media. He
fancied that they associated themselves indissolubly
with his refracted rectangles—that in every case the
image of the rectangle was accompanied by a secondary
hazy image, a little in advance of the principal one.
At one edge, he contended, the advanced secondary
image had black behind it, which was converted into
blue ; while at the other edge it had white behind it,
and appeared yellow. When the refracted rectangle is
made very narrow, the fringes approach each other and
finally overlap. Blue thus mingles with yellow, and
the green of the spectrum is the consequence. This,
in a nutshell, is the theory of colours developed in
the 'Farbenlehre.' Goethe obviously regarded the
narrowing of the rectangle, of the cylindrical beam, or
of the slit from which the light passed to the prism—
according to Newton the indispensable requisite for the
production of a pure spectrum—as an impure and
complicated mode of illustrating the phenomenon.
The elementary fact is, according to Goethe, obtained
when we operate with a wide rectangle the edges only
of which are coloured by refraction, while the centre
remains white. His experiments with the parchment

had made him acquainted with the passage of yellow
into red as he multiplied his layers ; but how this passage occurs in the spectrum he does not explain. That,
however, his hazy surfaces—his virtual turbid media—
produced, in some way or other, the observed passage
and intensification, Goethe held as firmly, and enunciated as confidently, as if his analysis of the phenomena
had been complete.

The fact is, that between double images and turbid
media there is no kinship whatever. Turbidity is due
to the diffusion, in a transparent medium, of minute
particles having a refractive index different from that
of the medium. But the act of reflection, which produced the penumbral surfaces, whose aid Goethe invoked, did not charge them with such discrete particles.
On various former occasions I have tried to set forth
the principles on which the chromatic action of turbid
media depends. When such media are to be seen blue,
the light scattered by the diffused particles, and that
only, ought to reach the eye. This feeble light may
be compared to a faint whisper which is easily rendered
inaudible by a louder noise. The scattered light of the
particles is accordingly overpowered, when a stronger
light comes, not from the particles, but from a bright
surface behind them. Here the light reaches the eye,
minus that scattered by the particles. It is therefore
the complementary light, or yellow. Both effects are
immediately deducible from the principles of the undulatory theory. As a stone in water throws back a
larger fraction of a ripple than of a larger wave, so do
the excessively minute particles which produce the
turbidity scatter more copiously the small waves of the
spectrum than the large ones. Light scattered by such
particles will therefore always contain a preponderance
of the waves which produce the sensation of blue.

During its transmission through the turbid medium the white light is more and more robbed of its blue constituents, the transmitted light which reaches the eye being therefore complementary to the blue.

Some of you are, no doubt, aware that it is possible to take matter in the gaseous condition, when its smallest parts are molecules, incapable of being either seen themselves or of scattering any sensible portion of light which impinges on them ; that it is possible to shake these molecules asunder by special light-waves, so that their liberated constituents shall coalesce anew and form, not *molecules*, but *particles*; that it is possible to cause these particles to grow, from a size bordering on the atomic, to a size which enables them to copiously scatter light. Some of you are aware that in the early stages of their growth, when they are still beyond the grasp of the microscope, such particles, no matter what the substance may be of which they are composed, shed forth a pure firmamental blue ; and that from them we can manufacture in the laboratory artificial skies which display all the phenomena, both of colour and polarisation, of the real firmament.

With regard to the production of the green of the spectrum by the overlapping of yellow and blue, Goethe, like a multitude of others, confounded the mixture of blue and yellow lights with that of blue and yellow pigments. This was an error shared by the world at large. But in Goethe's own day, Wünsch of Leipzig, who is ridiculed in the ' Farbenlehre,' had corrected the error, and proved the mixture of blue and yellow lights to produce white. Any doubt that might be entertained of Wünsch's experiments—and they are obviously the work of a careful and competent man—is entirely removed by the experiments of Helmholtz and others in our own day. Thus, to sum up, Goethe's theory, if such it may be called, proves incompetent to account even approximately for the Newtonian spectrum. He refers it to turbid media, but no such media come into play. He fails to account for the passage of yellow into red and of blue into violet ; while his attempt to deduce the green of the spectrum from the mixture of yellow and blue, is contradicted by facts which were extant in his own time.

One hole Goethe did find in Newton's armour, through which his lance incessantly worried the Englishman. Newton had committed himself to the doctrine that refraction without colour was impossible. He therefore thought that the object-glasses of telescopes must for ever remain imperfect, achromatism and refraction being incompatible. The inference of Newton was proved by Dollond to be wrong.[1] With the same mean refraction, flint glass produces a longer and richer spectrum than crown glass. By diminishing the refracting angle of the flint-glass prism, its spectrum may be made equal in length to that of the crown glass. Causing two such prisms to refract in opposite directions, the colours may be neutralised, while a considerable residue of refraction continues in favour of the crown. Similar combinations are possible in the case of lenses; and hence, as Dollond showed, the possibility of producing a compound achromatic lens. Here, as elsewhere, Goethe proves himself master of the experimental conditions. It is the power of interpretation that he lacks. He flaunts this error regarding achromatism incessantly in the face of Newton and his followers. But the error, which was a real one,

[1] Dollond was the son of a Huguenot. Up to 1752 he was a silk weaver at Spitalfields; he afterwards became an optician.

leaves Newton's theory of colours perfectly unimpaired.

Newton's account of his first experiment with the prism is for ever memorable. 'To perform my late promise to you,' he writes to Oldenburg, 'I shall without further ceremony acquaint you, that in the year 1666 (at which time I applied myself to the grinding of optick-glasses of other figures than spherical) I procured me a triangular glass prism, to try therewith the celebrated phenomena of colours. And in order thereto, having darkened my chamber, and made a small hole in my window-shuts, to let in a convenient quantity of the sun's light, I placed my prism at its entrance, that it might be thereby refracted to the opposite wall. It was at first a very pleasing divertisement, to view the vivid and intense colours produced thereby; but after a while applying myself to consider them more circumspectly, I became surprised to see them in an oblong form, which according to the received laws of refractions, I expected should have been circular. They were terminated at the sides with straight lines, but at the ends the decay of light was so gradual, that it was difficult to determine justly what was their figure, yet they seemed semi-circular.

'Comparing the length of this coloured *spectrum* with its breadth, I found it about five times greater; a disproportion so extravagant, that it excited me to a more than ordinary curiosity of examining from whence it might proceed.' This curiosity Newton gratified by instituting a series of experimental questions, the answers to which left no doubt upon his mind that the elongation of his spectrum was due to the fact 'that *light* is not similar or homogeneal, but consists of *difform rays, some of which are more refrangible than others*; so that, without any difference in their incidence on the same medium, some shall be more *refracted* than others; and therefore that, according to their *particular degrees of refrangibility*, they were transmitted through the prism to divers parts of the opposite wall. When,' continues Newton, ' I understood this, I left off my aforesaid glass works; for I saw that the perfection of telescopes was hitherto limited, not so much for want of glasses truly figured according to the prescriptions of optick authors, as because that *light* itself is an heterogeneous mixture of *differently refrangible rays*; so that were a glass so exactly figured as to collect any one sort of rays into one point, it could not collect those also into the same point, which, having the same incidence upon the same medium, are apt to suffer a different refraction.'

Goethe harped on this string without cessation. ' The Newtonian doctrine,' he says, ' was really dead the moment achromatism was discovered. Gifted men— our own Klügel, for example—felt this, but expressed themselves in an undecided way. On the other hand, the school which had been long accustomed to support, patch up, and glue their intellects to the views of Newton, had surgeons at hand to embalm the corpse, so that even after death, in the manner of the Egyptians, it should preside at the banquets of the natural philosophers.'

In dealing with the chromatic aberration of lenses, Goethe proves himself to be less heedful than usual as an experimenter. With the clearest perception of principles, Newton had taken two pieces of cardboard, the one coloured a deep red, the other a deep blue. Around those cards he had wound fine black silk, so that the silk formed a series of separate fine dark lines upon the two coloured surfaces. He might have drawn black lines over the red and blue, but the silk lines were finer than any that he could draw. Illuminating

both surfaces, he placed a lens so as to cast an image of
the surfaces upon a white screen. The result was, that
when the dark lines were sharply defined upon the red,
they were undefined upon the blue ; and that when, by
moving the screen, they were rendered distinct upon
the blue, they were indistinct upon the red. A distance
of an inch and a half separated the focus of red rays
from the focus of blue rays, the latter being nearer to
the lens than the former. Goethe appears to have
attempted a repetition of this experiment ; at all events,
he flatly contradicts Newton, ascribing his result not
to the testimony of his bodily eyes, but to that of the
prejudiced eyes of his mind. Goethe always saw the
dark lines best defined upon the brighter colour. It
was to him purely a matter of contrast, and not of
different refrangibility. He argues caustically that
Newton proves too much ; for were he correct, not only
would a dioptric telescope be impossible, but when presented to our naked eyes, differently-coloured objects
must appear utterly confusing. Let a house, he says,
be supposed to stand in full sunshine ; let the roof-tiles
be red, the walls yellow, with blue curtains behind the
open windows, while a lady with a violet dress steps
out of the door. Let us look at the whole from a
point in front of the house. The tiles we will suppose
appear distinct, but on turning to the lady we should
find both the form and the folds of her dress undefined.
We must move forward to see her distinctly, and then
the red tiles would appear nebulous. And so with
regard to the other objects, we must move to and fro
in order to see them clearly, if Newton's pretended
second experiment were correct. Goethe seems to have
forgotten that the human eye is not a rigid lens, and
that it is able to adjust itself promptly and without
difficulty to differences of distance enormously greater

than that due to the different refrangibility of the
differently-coloured rays.

Newton's theory of colours, it may be remarked, is
really less a 'theory' than a direct presentation of
facts. Given the accepted definition of refraction, it is
a matter of fact, and not of theoretic inference, that
white light is not 'homogeneal,' but composed of differently refrangible rays. The demonstration is ocular
and complete. Having palpably decomposed the white
light into its constituent colours, Newton recom-
pounded these colours to white light. Both the analysis
and the synthesis are matters of fact. The so-called
'theory of light and colours' is in this respect very
different from the corpuscular theory of light. Newton's explanation of colour stands where it is, whether
we accept the corpuscular or the undulatory theory;
and it stands because it is at bottom, not a theory, but
a body of fact, to which theory must bow or disappear.
Newton himself pointed out that his views of colours
were entirely independent of his belief in the 'corporeity' of light.

After refraction-colours, Goethe turns to those produced by diffraction, and, as far as the phenomena are
concerned, he deals very exhaustively with the colours
of thin plates. He studies the colours of Newton's
rings both by reflected and transmitted light. He
states the conditions under which this class of colours
is produced, and illustrates the conditions by special
cases. He presses together flat surfaces of glass, observes the flaws in crystals and in ice, refers to the
iridescences of oil on water, to those of soap-bubbles,
and to the varying colours of tempered steel. He is
always rich in facts. But when he comes to deal with
physical theory, the poverty and confusion of his otherwise transcendent mind become conspicuous. His turbid

media entangle him everywhere, leading him captive and committing him to almost incredible delusions. The colours of tempered steel, he says, and kindred phenomena, may perhaps be *quite conveniently deduced* from the action of turbid media. Polished steel powerfully reflects light, and the colouring produced by heating may be regarded as a feeble turbidity, which, acted upon by the polished surface behind, produces a bright yellow. As the turbidity augments, this colour becomes dense, until finally it exhibits an intense ruby-red. Supposing this colour to reach its greatest proximity to darkness, the turbidity continuing to augment as before, we shall have behind the turbid medium a dark background, against which we have first violet, then dark blue, and finally light blue, thus completing the cycle of the phenomena. The mind that could offer such an explanation as this must be qualitatively different from that of the natural philosopher.

The words 'quite conveniently deduced,' which I have italicised in the last paragraph, are also used by Goethe in another place. When the results of his experiments on prismatic colours had to be condensed into one commanding inference, he enunciated it thus:—
'Und so lassen sich die Farben bei Gelegenheit der Refraction aus der Lehre von den trüben Mitteln gar bequem ableiten.' This is the crown of his edifice, and it seems a feeble ending to so much preparation. Kingsley once suggested to Lewes that Goethe might have had a vague feeling that his conclusions were not sound, and that he felt the jealousy incident to imperfect conviction. The ring of conscious demonstration, as it is understood by the man of science, is hardly to be found in the words, 'gar bequem ableiten.' They fall flaccid upon the ear in comparison with the mind-compelling Q.E.D. of Newton.

Throughout the first 350 pages of his work, wherein he develops and expounds his own theory, Goethe restrains himself with due dignity. Here and there there is a rumble of discontent against Newton, but there is no sustained ill-temper or denunciation. After, however, unfolding his own views, he comes to what he calls the 'unmasking of the theory of Newton.' Here Goethe deliberately forsakes the path of calm, objective research, and delivers himself over to the guidance of his emotions. He immediately accuses Newton of misusing, as an advocate, his method of exposition. He goes over the propositions in Newton's optics one by one, and makes even the individual words of the propositions the objects of his criticism. He passes on to Newton's experimental proofs, invoking, as he does so, the complete attention of his readers, if they would be freed to all eternity from the slavery of a doctrine which has imposed upon the world for a hundred years. It might be thought that Goethe had given himself but little trouble to understand the theorems of Newton and the experiments on which they were based. But it would be unjust to charge the poet with any want of diligence in this respect. He repeated Newton's experiments, and in almost every case obtained his results. But he complained of their incompleteness and lack of logical force. What appears to us as the very perfection of Newton's art, and absolutely essential to the purity of the experiments, was regarded by Goethe as needless complication and mere torturing of the light. He spared no pains in making himself master of Newton's data, but he lacked the power of penetrating either their particular significance, or of estimating the force and value of experimental evidence generally.

He will not, he says, shock his readers at the outset by the utterance of a paradox, but he cannot withhold the assertion that by experiment nothing can really be proved. Phenomena may be observed and classified; experiments may be accurately executed, and made thus to represent a certain circle of human knowledge; but deductions must be drawn by every man for himself. Opinions of things belong to the individual, and we know only too well that conviction does not depend upon insight but upon will—that man can only assimilate that which is in accordance with his nature, and to which he can yield assent. In knowledge, as in action, says Goethe, prejudice decides all, and prejudice, as its name indicates, is judgment prior to investigation. It is an affirmation or a negation of what corresponds or is opposed to our own nature. It is the cheerful activity of our living being in its pursuit of truth or of falsehood, as the case may be—of all, in short, with which we feel ourselves to be in harmony.

There can be no doubt that Goethe, in thus philosophising, dipped his bucket into the well of profound self-knowledge. He was obviously stung to the quick by the neglect of the physicists. He had been the idol of the world, and accustomed as he was to the incense of praise, he felt sorely that any class of men should treat what he thought important with indifference or contempt. He had, it must be admitted, some ground for scepticism as to the rectitude of scientific judgments, seeing that his researches on morphology met at first no response, though they were afterwards lauded by scientific men. His anger against Newton incorporates itself in sharp and bitter sarcasm. Through the whole of Newton's experiments, he says, there runs a display of pedantic accuracy; but how the matter really stands with Newton's gift of observation, and with his experimental aptitudes, every man possessing eyes and senses may make himself aware. Where, it may be boldly

asked, can the man be found, possessing the extraordinary gifts of Newton, who could suffer himself to
be deluded by such a *hocus pocus*, if he had not in the
first instance wilfully deceived himself? Only those
who know the strength of self-deception, and the extent
to which it sometimes trenches on dishonesty, are in a
condition to explain the conduct of Newton and of
Newton's school. ' To support his unnatural theory,' he
continues, ' Newton heaps experiment on experiment,
fiction upon fiction, seeking to dazzle where he cannot
convince.'

It may be that Goethe is correct in affirming that
the will and prejudice of the individual are all-influential.
We must, however, add the qualifying words, ' as far as
the individual is concerned.' For in science there exists,
apart from the individual, objective truth ; and the fate
of Goethe's own theory, though commended to us by
so great a name, illustrates how, in the progress of
humanity, the individual, if he err, is left stranded and
forgotten—truth, independent of the individual, being
more and more grafted on to that tree of knowledge
which is the property of the human race.

The imagined ruin of Newton's theory did not
satisfy Goethe's desire for completeness. He would
explore the ground of Newton's error, and show how it
was that one so highly gifted could employ his gifts for
the enunciation and diffusion of such unmitigated
nonsense. It was impossible to solve the riddle on
purely intellectual grounds. Scientific enigmas, he
says, are often only capable of ethical solution, and
with this maxim in his mind he applies himself, in the
second volume of the ' Farbenlehre,' to the examination
of ' Newton's Persönlichkeit.' He seeks to connect him
with, or rather to detach him from, the general character
of the English nation—that sturdy and competent race,

which prizes above all things the freedom of individual
action. Newton was born in a storm-tossed time—none
indeed more pregnant in the history of the world. He
was a year old when Charles I. was beheaded, and he
lived to see the first George upon the throne. The
shock of parties was in his ears; changes of Ministries,
Parliaments, and armies were occurring before his eyes,
while the throne itself, instead of passing on by inheritance, was taken possession of by a stranger. What,
asks Goethe, are we to think of a man who could put
aside the claims, seductions, and passions incident to
such a time, for the purpose of tranquilly following out
his bias as an investigator ?

So singular a character arrests the poet's attention.
Goethe had laid down his theory of colours, he must add
to it a theory of Newton. The great German is here
at home, and Newton could probably no more have gone
into these disquisitions regarding character, than Goethe
could have developed the physical theories of Newton.
He prefaces his sketch of his rival's character by reflections and considerations regarding character in general.
Every living thing, down to the worm that wriggles
when trod upon, has a character of its own. In this
sense even the weak and cowardly have characters, for
they will give up the honour and fame which most men
prize highest, so that they may vegetate in safety and
comfort. But the word character is usually applied to
the case of an individual with great qualities, who
pursues his object undeviatingly, and without permitting either difficulty or danger to deflect him from his
course.

'Although here, as in other cases,' says Goethe, ' it is
the Exuberant (*Ueberschwängliche*) that impresses the
imagination, it must not be imagined that this attribute
has anything to do with moral feeling. The main

foundation of the moral law is a *good* will [1] which, in
accordance with its own nature, is anxious only for the
right. The main foundation of character is a *strong*
will, without reference to right or wrong, good or bad,
truth or error. It is that quality which every Party
prizes in its members. A good will cherishes freedom,
it has reference to the inner man and to ethical aims.
The strong will belongs to Nature and has reference to
the outer world—to action. And inasmuch as the
strong will in this world is swayed and limited by the
conditions of life, it may almost be assumed as certain
that it is only by accident that the exercise of a strong
will and of moral rectitude find themselves in harmony
with each other.' In determining Newton's position in
the series of human characters, Goethe helps himself to
images borrowed from the physical cohesion of matter.
Thus, he says, we have strong, firm, compact, elastic,
flexible, rigid or obstinate, and viscous characters.
Newton's character he places under the head of rigid or
obstinate, and his theory of colours Goethe pronounces
to be a petrified *aperçu.*

Newton's assertion of his theory, and his unwavering
adherence to it to the end of his life, Goethe ascribes
straight off to moral obliquity on Newton's part. In
the heat of our discussion, he says, we have even ascribed
to him a certain dishonesty. Man is subject to error,
but when errors form a series, which is followed pertina-
ciously, the erring individual becomes false to himself
and to others. Nevertheless reason and conscience
will not yield their rights. We may belie them, but
they are not deceived. It is not too much to say that
the more moral and rational a man is, the greater

---

[1] I have rendered Goethe's 'gute Wille' by *good* will; his
'Wollen,' which he contrasts with ' Wille,' I have rendered by *strong*
will.

will be his tendency to lie when he falls into error, and
the vaster will be that error when he makes up his mind
to persist in it.

This is all intended to throw light upon Newton.
When Goethe passes from Newton himself to his followers,
the small amount of reserve which he exhibited when
dealing with the master entirely disappears. He mocks
their blunders as having not even the merit of originality.
He heaps scorn on Newton's imitators. The expression
of even a truth, he says, loses grace in repetition, while
the repetition of a blunder is impertinent and ridiculous.
To liberate oneself from an error is difficult, sometimes
indeed impossible for even the strongest and most gifted
minds. But to take up the error of another, and persist
in it with stiffnecked obstinacy, is a proof of poor
qualities. The obstinacy of a man of originality when
he errs may make us angry, but the stupidity of the
copyist irritates and renders us miserable. And if in
our strife with Newton we have sometimes passed the
bounds of moderation, the whole blame is to be laid
upon the school of which Newton was the head, whose
incompetence is proportional to its arrogance, whose
laziness is proportional to its self-sufficiency, and whose
virulence and love of persecution hold each other in
perfect equilibrium.

There is a great deal more invective of this kind,
but you will probably, and not without sadness, consider this enough. Invective may be a sharp weapon,
but over-use blunts its edge. Even when the denunciation is just and true, it is an error of art to indulge in
it too long. We not only incur the risk of becoming
vapid, but of actually inverting the force of reproba-
tion which we seek to arouse, and of bringing it back
by recoil upon ourselves. At suitable intervals, separated from each other by periods of dignified reserve,

invective may become a real power of the tongue or
pen.   But indulged in constantly it degenerates into
scolding, and then, instead of being regarded as a proof
of strength, it is accepted, even in the case of a Goethe,
as an evidence of weakness and lack of self-control.

If it were possible to receive upon a mirror Goethe's
ethical image of Newton and to reflect it back upon its
author, then, as regards vehement persistence in wrong
thinking, the image would accurately coincide with
Goethe himself.   It may be said that we can only solve
the character of another by the observation of our own.
This is true, but in the portraiture of character we are
not at liberty to mix together subject and object as
Goethe mixed himself with Newton.   So much for the
purely ethical picture.   On the scientific side something more is to be said.   I do not know whether
psychologists have sufficiently taken into account that,
as regards intellectual endowment, vast wealth may co-
exist with extreme poverty.   I do not mean to give
utterance here to the truism that the field of culture is
so large that the most gifted can master only a portion
of it.   This would be the case supposing the individual
at starting to be, as regards natural capacity and poten-
tiality, rounded like a sphere.   Something more radical
is here referred to.   There are individuals who at
starting are not spheres, but hemispheres ; or, at least,
spheres with a segment sliced away—full-orbed on one
side, but flat upon the other.   Such incompleteness of
the mental organisation no education can repair.   Now
the field of science is sufficiently large, and its studies
sufficiently varied, to bring to light in the same individual antitheses of endowment like that here indicated.

So far as science is a work of ordering and classifi-
cation, so far as it consists in the discovery of analogies
and resemblances which escape the common eye—of

the fundamental identity which often exists among apparently diverse and unrelated things—so far, in short, as it is observational, descriptive, and imaginative, Goethe, had he chosen to make his culture exclusively scientific, might have been without a master, perhaps even without a rival. The instincts and capacities of the poet lend themselves freely to the natural-history sciences. But when we have to deal with stringently physical and mechanical conceptions, such instincts and capacities are out of place. It was in this region of mechanical conceptions that Goethe failed. It was on this side that his sphere of endowment was sliced away. He probably was not the only great man who possessed a spirit thus antithetically mixed. Aristotle himself was a mighty classifier, but not a stringent physical reasoner. And had Newton attempted to produce a Faust, the poverty of his intellect on the poetic and dramatic side might have been rendered equally manifest. But here, if not always, Newton abstained from attempting that for which he had no gift, while the exuberance of Goethe's nature caused him to undertake a task for which he had neither ordination nor vocation, and in the attempted execution of which his deficiencies became revealed.

One task among many—one defeat amid a hundred triumphs. But any recognition on my part of Goethe's achievements in other realms of intellectual action would justly be regarded as impertinent. You remember the story of the first Napoleon when the Austrian plenipotentiary, in arranging a treaty of peace, began by formally recognising the French Republic. 'Efface that,' said the First Consul; 'the French Republic is like the sun; he is blind who fails to recognise it.' And were I to speak of recognising Goethe's merits, my effacement would be equally well deserved.

'Goethe's life,' says Carlyle, 'if we examine it, is well represented in that emblem of a solar day. Beautifully rose our summer sun, gorgeous in the red, fervid east, scattering the spectres and sickly damps, of both of which there were enough to scatter; strong, benignant, in his noonday clearness, walking triumphant through the upper realms—and now mark also how he sets! "So stirbt ein Held;" so dies a hero!'

Two grander illustrations of the aphorism 'To err is human' can hardly be pointed out in history than Newton and Goethe. For Newton went astray, not only as regards the question of achromatism, but also as regards vastly larger questions touching the nature of light. But though as errors they fall into the same category, the mistake of Newton was qualitatively different from that of Goethe. Newton erred in adopting a wrong mechanical conception in his theory of light, but in doing so he never for a moment quitted the ground of strict scientific method. Goethe erred in seeking to engraft in his 'Farbenlehre' methods altogether foreign to physics or to the treatment of a purely physical theme.

We frequently hear protests made against the cold mechanical mode of dealing with æsthetic phenomena employed by scientific men. The dissection by Newton of the light to which the world owes all its visible splendour seemed to Goethe a desecration. We find, even in our own day, the endeavour of Helmholtz to arrive at the principles of harmony and discord in music resented as an intrusion of the scientific intellect into a region which ought to be sacred to the human heart. But all this opposition and antagonism has for its essential cause the incompleteness of those with whom it originates. The feelings and aims with which Newton and Goethe respectively ap-

proached Nature were radically different, but they
had an equal warrant in the constitution of man. As
regards our tastes and tendencies, our pleasures and
pains, physical and mental, our action and passion, our
sorrows, sympathies, and joys, we are the heirs of all
the ages that preceded us; and of the human nature
thus handed down poetry is an element just as much
as science. The emotions of man are older than his
understanding, and the poet who brightens, purifies,
and exalts these emotions may claim a position in the
world at least as high and as well assured as that of
the man of science. They minister to different but to
equally permanent needs of human nature ; and the
incompleteness of which I complain consists in the
endeavour on the part of either to exclude the other.
There is no fear that the man of science can ever
destroy the glory of the lilies of the field ; there is no
hope that the poet can ever successfully contend against
our right to examine, in accordance with scientific
method, the agent to which the lily owes its glory.
There is no necessary encroachment of the one field
upon the other. Nature embraces them both, and
man, when he is complete, will exhibit as large a
toleration.

1882.

ATOMS, MOLECULES, AND ETHER WAVES.[1]

MAN is prone to idealisation. He cannot accept as final the phenomena of the sensible world, but looks behind that world into another which rules the sensible one. From this tendency of the human mind systems of mythology and scientific theories have equally sprung. By the former the experiences of volition, passion, power, and design, manifested among ourselves, were transplanted, with the necessary modifications, into an unseen universe, from which the sway and potency of these magnified human qualities were exerted. ' In the roar of thunder and in the violence of the storm was felt the presence of a shouter and furious strikers, and out of the rain was created an Indra or giver of rain.' It is substantially the same with science, the principal force of which is expended in endeavouring to rend the veil which separates the sensible world from an ultra-sensible one. In both cases our materials, drawn from the world of the senses, are modified by the imagination to suit intellectual needs. The 'first beginnings' of Lucretius were not objects of sense, but they were suggested and illustrated by objects of sense. The idea of atoms proved an early want on the part of minds in pursuit of the knowledge of Nature. It has

[1] Written at Alp Lusgen for the first number of *Longman's Magazine*.

never been relinquished, and in our own day it is growing steadily in power and precision.

The union of bodies in fixed and multiple proportions constitutes the basis of modern atomic theory.
The same compound retains, for ever, the same elements,
in an unalterable ratio. We cannot produce pure
water containing one part, by weight, of hydrogen and
nine of oxygen ; nor can we produce it when the ratio
is one to ten ; but we can produce it from the ratio of
one to eight, and from no other. So, also, when water
is decomposed by the electric current, the proportion,
as regards volumes, is as fixed as in the case of weights.
Two volumes of hydrogen and one of oxygen invariably
go to the formation of water. Number and harmony,
as in the Pythagorean system, are everywhere dominant
in this under-world.

Following the discovery of fixed proportions we
have that of *multiple* proportions. For the same compound, as above stated, the elementary factors are constant ; but one elementary body often unites with
another so as to form different compounds. Water,
for example, is an oxide of hydrogen ; but a peroxide
of that substance also exists, containing exactly double
the quantity of oxygen. Nitrogen also unites with
oxygen in various ratios, but not in all. The union
takes place, not gradually and uniformly, but by steps,
a definite weight of matter being added at each step.
The larger combining quantities of oxygen are thus
multiples of the smaller ones. It is the same with
other combinations.

We remain thus far in the region of fact : why not
rest there ? It might as well be asked why we do not,
like our poor relations of the woods and forests, rest
content with the facts of the sensible world. In virtue
of our mental idiosyncrasy, we demand *why* bodies

should combine in multiple proportions, and the outcome and answer of this question is the atomic theory.
The definite weights of matter above referred to
represent the weights of atoms, indivisible by any force
which chemistry has hitherto brought to bear upon
them. If matter were a *continuum*—if it were not
rounded off, so to say, into these discrete atomic masses
—the impassable breaches of continuity which the law
of multiple proportions reveals could not be accounted
for. These atoms are what Maxwell finely calls 'the
foundation-stones of the material universe,' which, amid
the wreck of composite matter, 'remain unbroken and
unworn.'

A group of atoms drawn and held together by what
chemists term affinity, is called a molecule. The
ultimate parts of all compound bodies are molecules.
A molecule of water, for example, consists of two atoms
of hydrogen, which grasp and are grasped by one atom
of oxygen. When water is converted into steam, the
distances between the molecules are greatly augmented,
but the molecules themselves continue intact. We
must not, however, picture the constituent atoms of
any molecule as held so rigidly together as to render
intestine motion impossible. The interlocked atoms
have still liberty of vibration, which may, under certain
circumstances, become so intense as to shake the
molecule asunder. Most molecules—probably all—are
wrecked by intense heat, or in other words by intense
vibratory motion ; and many are wrecked by a very
moderate heat of the proper quality. Indeed, a weak
force which bears a suitable relation to the constitution
of the molecule can, by timely savings and accumula-
tions, accomplish what a strong force out of such relation fails to achieve.

We have here a glimpse of the world in which the

physical philosopher for the most part resides. Science has been defined as ' organised common sense,' by whom I have forgotten ; but, unless we stretch unduly the definition of common sense, I think it is hardly applicable to this world of molecules. I should be inclined to ascribe the creation of that world to inspiration, rather than to what is currently known as common sense. For the natural-history sciences the definition may stand—but hardly for the physical and mathematical sciences.

The sensation of light is produced by a succession of waves which strike the retina in periodic intervals ; and such waves, impinging on the molecules of bodies, agitate their constituent atoms. These atoms are so small and, when grouped to molecules, are so tightly clasped together, that they are capable of tremors equal in rapidity to those of light and radiant heat. To a mind coming freshly to these subjects, the numbers with which scientific men here habitually deal must appear utterly fantastical ; and yet, to minds trained in the logic of science, they express most sober and certain truth. The constituent atoms of molecules can vibrate to and fro millions of millions of times in a second. The waves of light and of radiant heat follow each other at similar rates through the luminiferous ether. Further, the atoms of different molecules are held together with varying degrees of tightness—they are tuned, as it were, to notes of different pitch. Suppose, then, light-waves, or heat-waves, to impinge upon an assemblage of such molecules, what may be expected to occur? The same as what occurs when a piano is opened and sung into. The waves of sound select the strings which respectively respond to them—the strings, that is to say, whose rates of vibration are the same as their own. Of the whole series of strings these only sound. The vibratory motion of the voice, imparted first to the air, is taken up by the strings. It may be regarded as *absorbed*, each string constituting itself thereby a new centre of motion. Thus also as regards the tightly-locked atoms of molecules on which waves of light or radiant heat impinge. Like the waves of sound just adverted to, the waves of ether select those atoms whose periods of vibration synchronise with their own periods of recurrence, and to such atoms they deliver up their motion. It is thus that light and radiant heat are absorbed.

And here the statement, though elementary, must not be omitted, that the colours of the prismatic spectrum, which are presented in an impure form in the rainbow, are due to different rates of atomic vibration in their source, the sun. From the extreme red to the extreme violet, between which are embraced all colours visible to the human eye, the rapidity of vibration steadily increases, the length of the waves of ether produced by these vibrations diminishing in the same proportion. I say ' visible to the human eye,' because there may be eyes capable of receiving visual impression from waves which do not affect ours. There is a vast store of rays, or more correctly waves, beyond the red, and also beyond the violet, which are incompetent to excite our vision ; so that could the whole length of the spectrum, visible and invisible, be seen by the same eye, its length would be vastly augmented.

I have spoken of molecules being wrecked by a moderate amount of heat of the proper quality : let us examine this point for a moment. There is a liquid called nitrite of amyl—frequently administered to patients suffering from heart disease. The liquid is volatile, and its vapour is usually inhaled by the patient. Let a quantity of this vapour be introduced into a wide glass tube, and let a concentrated beam of solar light be sent through the tube along its axis. Prior to the entry of the beam, the vapour is as invisible as the purest air. When the light enters, a bright cloud is immediately precipitated on the beam. This is entirely due to the waves of light, which wreck the nitrite of amyl molecules, the products of decomposition forming innumerable liquid particles, which constitute the cloud. Many other gases and vapours are acted upon in a similar manner. Now the waves that produce this decomposition are by no means the most powerful of those emitted by the sun. It is, for example, possible to gather up the ultra-red waves into a concentrated beam, and to send it through the vapour, like a beam of light. But though possessing vastly greater energy than the light waves, they fail to produce decomposition. Hence the justification of the statement already made, that a suitable relation must subsist between the molecules and the waves of ether to render the latter effectual.

A very impressive illustration of the decomposing power of the waves of light is here purposely chosen ; but the processes of photography illustrate the same principle. The photographer, without fear, illuminates his developing-room with light transmitted through red or yellow glass; but he dares not use blue glass, for blue light would decompose his chemicals. And yet the waves of red light, measured by the amount of energy which they carry, are immensely more powerful than the waves of blue. The blue rays are usually called chemical rays—a misleading term ; for, as Draper and others have taught us, the rays that produce the grandest chemical effects in Nature, by decomposing the carbonic acid and water which form the nutriment of plants, are not the blue ones. In regard, however, to the salts of silver, and many other compounds, the blue rays are the most effectual. How is it, then, that weak waves can produce effects which strong waves are incompetent to produce ? This is a feature characteristic of periodic motion. In the experiment of singing into an open piano already referred to, it is the *accord* subsisting between the vibrations of the voice and those of the string that causes the latter to sound. Were this accord absent, the intensity of the voice might be quintupled, without producing any response. But when voice and string are identical in pitch, the successive impulses add themselves together, and this addition renders them in the aggregate powerful, though individually they may be weak. In some such fashion the periodic strokes of the smaller ether waves accumulate, till the atoms on which their timed impulses impinge are jerked asunder, and what we call chemical decomposition ensues.

Savart was the first to show the influence of musical sounds upon liquid jets, and I have now to describe an experiment belonging to this class, which bears upon the present question. From a screw-tap in my little Alpine kitchen I permitted, an hour ago, a vein of water to descend into a trough, so arranging the flow that the jet was steady and continuous from top to bottom. A slight diminution of the orifice caused the continuous portion of the vein to shorten, the part further down resolving itself into drops. In my experiment however the vein, before it broke, was intersected by the bottom of the trough. Shouting near the descending jet produced no sensible effect upon it. The higher notes of the voice, however powerful, were also ineffectual. But when the voice was lowered to about 130 vibrations a second, the feeblest utterance of this note sufficed to shorten, by one-half, the continuous portion of the jet. The responsive drops ran along the vein, pattered against the trough, and scattered a copious spray round their place of impact. When the note ceased, the continuity and steadiness of the vein were immediately restored. The formation of the drops was here periodic; and when the vibrations of the note accurately synchronised with the periods of the drops, the waves of sound aided what the illustrious Plateau proved to be the natural tendency of the liquid cylinder to resolve itself into spherules, and virtually decomposed the vein.

I have stated, without proof, that where absorption occurs the motion of the ether-waves is taken up by the constituent atoms of molecules. It is conceivable that the ether waves, in passing through an assemblage of molecules, might deliver up their motion to each molecule as a whole, leaving the relative positions of the constituent atoms unchanged. But the long series of reactions represented by the deportment of nitrite of amyl vapour does not favour this conception ; for, were the atoms animated solely by a common motion, the molecules would not be decomposed. The fact of decomposition, then, goes to prove the atoms to be the seat of the absorption. They, in great part, take up the energy of the ether waves, whereby their union is severed, and the building-materials of the molecules are scattered abroad.

Molecules differ in stability ; some of them, though hit by waves of considerable force, and taking up the motions of these waves, nevertheless hold their own with a tenacity which defies decomposition. And here, in passing, I may say that it would give me extreme pleasure to be able to point to my researches in confirmation of the solar theory recently enunciated by my friend the President of the British Association. But though the experiments which I have made on the decomposition of vapours by light might be numbered by

the thousand, I have, to my regret, encountered no fact
which proves that free aqueous vapour is decomposed by
the solar rays, or that the sun is nourished by the re-
combination of gases, in the severance of which it had
previously sacrificed its heat.

The memorable investigations of Leslie and Rumford, and the subsequent classical researches of Melloni
and Knoblauch, dealt, in the main, with the properties
of radiant heat ; while, in my investigations, radiant
heat, instead of being regarded as an end, was employed
as a means of exploring molecular condition. On this
score little could be said until the gaseous form of
matter was brought under the dominion of experiment.
This was first effected in 1859, when it was proved that
gases and vapours, notwithstanding the open door which
the distances between their molecules might be supposed
to offer to the heat waves, were, in many cases, able effectually to bar their passage. It was then proved that
while the elementary gases and their mixtures, including
among the latter the earth's atmosphere, were almost as
pervious as a vacuum to ordinary radiant heat, the
compound gases were one and all absorbers, some of
them indeed taking up with intense avidity the motion
of the ether waves.

A single illustration will here suffice. Let a mixture of hydrogen and nitrogen in the proportion of three
to fourteen by weight be enclosed in a space through
which are passing the heat-rays from an ordinary stove.
The gaseous mixture offers no measurable impediment
to the rays of heat. Let the hydrogen and nitrogen
now unite to form the compound ammonia. A magical
change instantly occurs. The number of atoms present
remains unchanged. The transparency of the compound is quite equal to that of the mixture prior to

combination. No change is perceptible to the eye, but the keen vision of experiment soon detects the fact that the perfectly transparent and highly attenuated ammonia resembles pitch or lampblack in its behaviour to the rays of obscure heat.

There is probably boldness, if not rashness, in the attempt to make these ultra-sensible actions generally intelligible, and I may have already transgressed the limits beyond which the writer of a familiar article cannot profitably go. There may, however, be a remnant of readers willing to accompany me, and for their sakes I proceed. A hundred compounds might be named which, like the ammonia, are transparent to light but more or less opaque—often, indeed, intensely opaque—to the rays of heat from obscure sources. Now the difference between these latter rays and the light-rays is purely a difference of period of vibration. The vibrations in the case of light are more rapid, and the ether waves which they produce are shorter, than in the case of obscure heat. Why then should the ultra-red waves be intercepted by bodies like ammonia, while the more rapidly recurrent waves of the whole visible spectrum are allowed free transmission? The answer I hold to be that, by the act of chemical combination, the vibrations of the constituent atoms of the molecules are rendered so sluggish as to synchronise with the motions of the longer waves. They resemble loaded piano-strings, or slowly descending water-jets, requiring notes of low pitch to set them in motion.

The influence of synchronism between the 'radiant' and the 'absorbent' is well shown by the behaviour of carbonic acid gas. To the complex emission from our heated stove, carbonic acid would be one of the most transparent of gases. For such waves olefiant gas, for example, would vastly transcend it in absorbing power.

But when we select a radiant with whose waves the atoms of carbonic acid are in accord, the case is entirely altered. Such a radiant is found in a carbonic oxide flame, where the radiating body is really hot carbonic acid. To this special radiation carbonic acid is the most opaque of gases.

And here we find ourselves face to face with a question of great delicacy and importance. Both as a radiator and as an absorber carbonic acid is, in general, a feeble gas. It is beaten in this respect by chloride of methyl, ethylene, ammonia, sulphurous acid, nitrous oxide, and marsh gas. Compared with some of these gases, its behaviour in fact approaches that of elementary bodies. May it not help to explain their neutrality ? The doctrine is now very generally accepted that atoms of the same kind may, like atoms of different kinds, group themselves to molecules. Affinity exists between hydrogen and hydrogen, and between chlorine and chlorine, as well as between hydrogen and chlorine. We have thus homogeneous molecules as well as heterogeneous molecules, and the neutrality so strikingly exhibited by the elements may be due to a quality of which carbonic acid furnishes a partial illustration. The paired atoms of the elementary molecules may be so out of accord with the periods of the ultra-red waves —the vibrating periods of these atoms may, for example, be so rapid—as to disqualify them both from emitting those waves, and from accepting their energy. This would practically destroy their power, both as radiators and absorbers. I have reason to know that by a distinguished authority this hypothesis has for some time been entertained.

We must, however, refresh ourselves by occasional contact with the solid ground of experiment, and an interesting problem now lies before us awaiting experi-

mental solution. Suppose 200 men to be scattered
equably throughout the length of Pall Mall. By timely
swerving now and then a runner from St. James's
Palace to the Athenæum Club might be able to get
through such a crowd without much hindrance. But
supposing the men to close up so as to form a dense
file crossing Pall Mall from north to south : such a
barrier might seriously impede, or entirely stop, the
runner. Instead of a crowd of men, let us imagine a
column of molecules under small pressure, thus resembling the sparsely-distributed crowd. Let us suppose
the column to shorten, without change in the quantity
of matter, until the molecules are so squeezed together
as to resemble the closed file across Pall Mall. During
these changes of density, would the action of the molecules upon a beam of heat passing among them resemble the action of the crowd upon the runner ?

We must answer this question by direct experiment.
To form our molecular crowd we place, in the first instance, a gas or vapour in a tube 38 inches long, the
ends of which are closed with circular windows, air-
tight, but formed of a substance which offers little or
no obstruction to the calorific waves. Calling the
measured value of a heat-beam passing through this
tube, when empty, 100, we carefully determine the proportionate part of this total absorbed when the molecules
are in the tube. We then gather precisely the same
number of molecules into a column 10·8 inches long, the
one column being thus three and a half times the other.
In this case also we determine the quantity of radiant
heat absorbed. By the depression of a barometric
column, we can easily and exactly measure out the proper quantities of the gaseous body. It is obvious that
1 mercury inch of vapour, in the long tube, would
represent precisely the same amount of matter—or in

other words the same number of molecules—as $3\frac{1}{2}$ inches in the short one ; while 2 inches of vapour in the long tube would be equivalent to 7 inches in the short one.

The experiments have been made with the vapours of two very volatile liquids, namely, sulphuric ether and hydride of amyl. The sources of radiant heat were, in some cases, an incandescent lime cylinder, and in others a spiral of platinum wire heated to bright redness by an electric current. One or two of the measurements will suffice for the purposes of illustration. First, then, as regards the lime-light :—For 1 inch of pressure in the long tube, the absorption was 18·4 per cent. of the total beam ; while for 3·5 inches of pressure in the short tube, the absorption was 18·8 per cent., or almost exactly the same as the former. For 2 inches pressure, moreover, in the long tube, the absorption was 25·7 per cent.; while for 7 inches, in the short tube, it was 25·6 per cent. of the total beam. Thus closely do the absorptions in the two cases run together—thus emphatically do the molecules assert their individuality. As long as their number is unaltered, their action on radiant heat is unchanged. Passing from the lime-light to the incandescent spiral, the absorptions of the smaller equivalent quantities in the two tubes were 23·5 and 23·4 per cent.; while the absorptions of the larger equivalent quantities were 32·1 and 32·6 per cent. respectively. This constancy of absorption, when the density of a gas or vapour is varied, I have called ' the conservation of molecular action.'

But it may be urged that the change of density in these experiments has not been carried far enough to justify the enunciation of a law of molecular physics. The condensation into less than one-third of the space does not, it may be said, quite represent the close file

of men across Pall Mall. Let us therefore push matters
to extremes, and continue the condensation till the
vapour has been squeezed into a liquid. To the pure
change of density we shall then have added the change
in the state of aggregation. The experiments here
are more easily described than executed ; nevertheless,
by sufficient training, scrupulous accuracy, and minute
attention to details, success may be ensured. Knowing
the respective specific gravities, it is easy, by calculation, to determine the condensation requisite to reduce a column of vapour of definite density and length
to a layer of liquid of definite thickness. Let the
vapour, for example, be that of sulphuric ether, and let
it be introduced into our 38-inch tube till a pressure of
7·2 inches of mercury is obtained. Or let it be hydride
of amyl of the same length, at a pressure of 6·6
inches. Supposing the column to shorten, the vapour
would become proportionally denser, and would, in
each case, end in the production of a layer of liquid
exactly 1 millimeter in thickness.[1] Conversely, a layer
of liquid ether, or of hydride of amyl, of this thickness,
were its molecules freed from the thrall of cohesion,
would form a column of vapour 38 inches long, at a
pressure of 7·2 inches in the one case, and of 6·6 inches
in the other. In passing through the liquid layer, a
beam of heat encounters the same number of molecules
as in passing through the vapour layer ; and our problem is to decide, by experiment, whether, in both cases,
the molecule is not the dominant factor, or whether its
power is augumented, diminished, or otherwise over-
ridden by the state of aggregation.

Using the sources of heat before-mentioned, and
employing diathermanous lenses, or silvered mirrors, to
render the rays from those sources parallel, the absorp-

[1] The millimeter is $\frac{1}{25}$th of an inch.

tion of radiant heat was determined, first for the liquid layer, and then for its equivalent vaporous layer. As before, a representative experiment or two will suffice for illustration. When the substance was sulphuric ether, and the source of radiant heat an incandescent platinum spiral, the absorption by the column of vapour was found to be 66·7 per cent. of the total beam. The absorption of the equivalent liquid layer was next determined, and found to be 67·2 per cent. Liquid and vapour, therefore, differed from each other only 0·5 per cent. :—In other words, they were practically identical in their action. The radiation from the lime-light has a greater power of penetration through transparent substances than that from the spiral. In the emission from both of these sources we have a mixture of obscure and luminous rays; but the ratio of the latter to the former, in the lime-light, is greater than in the spiral; and, as the very meaning of transparency is perviousness to the luminous rays, the emission in which these rays are predominant must pass most freely through transparent substances. Increased transmission implies diminished absorption; and, accordingly, the respective absorptions of ether vapour and liquid ether when the lime-light was used, instead of being 66·7 and 67·2 per cent., were found to be—

Vapour .   .   .   . 33·3 per cent.
Liquid .   .   .   . 33·3 „

no difference whatever being observed between the two states of aggregation. The same was found true of hydride of amyl.

This constancy and continuity of the action exerted on the waves of heat when the state of aggregation is changed I have called 'the thermal continuity of liquids and vapours.' It is, I think, the strongest illustration hitherto adduced of the conservation of molecular action.

Thus, by new methods of search, we reach a result which was long ago enunciated on other grounds. Water is well known to be one of the most opaque of liquids to the waves of obscure heat. But if the relation of liquids to their vapours be that here shadowed forth—if in both cases the molecule asserts itself to be the dominant factor—then the dispersion of the water of our seas and rivers as invisible aqueous vapour in our atmosphere, does not annul the action of the molecules on solar and terrestrial heat. But as aqueous vapour is transparent—which, as before explained, means pervious to the luminous rays—and as the emission from the sun abounds in such rays while from the earth's emission they are wholly absent, the vapour screen offers a far greater hindrance to the outflow of heat from the earth towards space than to the inflow from the sun towards the earth. The elevation of our planet's temperature is therefore a direct consequence of the existence of aqueous vapour in our air. Were that garment removed, terrestrial life would probably perish through the consequent refrigeration.

I have thus endeavoured to give some account of recent incursions into that ultra-sensible world—the world of the ' scientific imagination '—mentioned at the outset of this paper. Invited by my publishers, with whom I have so long worked in harmony, to send some contribution to the first number of their new Magazine, I could not refuse them this proof of my good-will.

NOTE.—The researches glanced at in the foregoing brief article have been published *in extenso* in the ' Philosophical Transactions.' I would invite to them and others, in their correct historical sequence, the attention of my friend Professor Von Hofmann.

1883.

COUNT RUMFORD.[1]

ON a bright calm day in the autumn of 1872—that portion of the year called, I believe, in America the Indian summer—I made a pilgrimage to the modest birthplace of Count Rumford. My guide on the occasion was Dr. George Ellis of Boston, and a more competent guide I could not possibly have had. To Dr. Ellis the American Academy of Arts and Sciences had committed the task of writing a life of Rumford, and this labour of love had been accomplished in 1871, a year prior to my visit to the United States. In regard to Rumford's personal life, Dr. Ellis's elaborate volume constitutes, if I may so speak, the quarry out of which the building-materials of these lectures are drawn. The life of such a man, however, cannot be duly taken in without reference to his work, and the publication by the American Academy of Sciences of four large volumes of Rumford's essays renders the task of dealing with his labours lighter than it would have been had his writings been suffered to remain scattered in the magazines, journals, and transactions of learned societies in which they originally appeared.

The name of Count Rumford was Benjamin Thompson. For thirty years he was the contemporary of

[1] From a short course of lectures delivered in the Royal Institution.

another Benjamin, who reached a level of fame as high as his own. Benjamin Franklin and Benjamin Thompson were born within twelve miles of each other, and for six of the thirty years just referred to the one lived in England and the other in France. Still, there is nothing to show that they ever saw each other, or were in any way acquainted with each other, or, indeed, felt the least interest in each other. As regards posthumous fame, Rumford has fared worse than Franklin. For ten, or perhaps a hundred, people in this country who know something of the career of the one, hardly a unit is to be found acquainted with the career of the other. Among scientific men, however, the figure of Rumford presents itself with singular impressiveness at the present day—a result mainly due to the establishment of the grand scientific generalisation known as the Mechanical Theory of Heat. Boyle, and Hooke, and Locke, and Leibnitz, had already distinctly ranged themselves on the side of this theory. But by experiments conducted on a scale unexampled at the time, and by reasonings, founded on these experiments, of singular force and penetration, Rumford has made himself a conspicuous landmark in the history of the theory. His inference from his experiments was that heat is a form of motion.

The town of Woburn, connected in my memory with a cultivated companion, with genial sunshine and the bright colouring of American trees, is nine miles distant from the city of Boston. In North Woburn, a little way off, on March 26, 1753, Rumford was born. He came of people who had to labour for their livelihood, who tilled their own fields, cut their own timber and fuel, worked at their varied trades, and thus maintained the independence of New England yeomen. Thompson's

father died when he was two years old. His mother
married again, and had children by her second husband,
but the affection between her and her firstborn remained
strong and unbroken. The arrangements made for the
maintenance of mother and son throw some light upon
their position. She was to have the use of one-half of
a garden ; the privilege of land to raise beans for sauce;
to receive within a specified time 80 ' weight' of beef,
8 bushels of rye, 2 bushels of malt, and 2 barrels of
cider. Finally, she had the right of gathering apples
to bake, and a further allowance of three bushels of
apples every year.

The fatherless boy had been placed under the care
of a guardian, from whom his stepfather, Josiah Pierce,
received a weekly allowance of two shillings and five-
pence for the child's maintenance. Young Thompson
received his first education from Mr. John Fowle, a
graduate of Harvard College, described by Dr. Ellis as
' an accomplished and faithful man.' He also went to
a school at Byfield, kept by a relation of his own. At
the age of eleven he was placed for a time under the
tuition of a Mr. Hill, ' an able teacher in Medford,' adjoining Woburn. The lad's mind was ever active, and
his invention incessantly exercised, but for the most
part on subjects apart from his daily work. In relation
to that work he came to be regarded as ' indolent,
flighty, and unpromising.' His guardian at length
thinking it advisable to change his vocation, apprenticed
him in October 1766 to Mr. John Appleton, of Salem,
an importer of British goods. Here, however, instead
of wooing customers to his master's counter, he occupied
himself with tools and implements hidden beneath it.
He is reported to have been a skilful musician, passionately fond of music of every kind ; and during his stay
with Mr. Appleton, whenever he could do so without

being heard, he solaced his leisure with performances on the violin.

By the Rev. Thomas Barnard, minister of Salem, and his son, young Thompson was taught algebra, geometry, and astronomy. By self-practice he became an able and accurate draughtsman. He did not escape that last infirmity of ingenious minds—the desire to construct a perpetual motion. He experimented with fireworks, and was once seriously burnt by the unexpected ignition of his materials. His inquisitiveness is illustrated by the questions put to his friend Mr. Baldwin in 1769. He wishes to be told the direction pursued by the rays of light under certain conditions ; he desires to know the cause of the change of colour which fire produces in clay. ' Please,' he adds, ' to give the nature, essence, beginning of existence, and rise of the wind in general, with the whole theory thereof, so as to be able to answer all questions relative thereto.' One might suppose him to be preparing for a competitive examination. He grew expert in drawing caricatures, a spirited group of which has been reproduced by Dr. Ellis. It is called a Council of State, and embraces a jackass with twelve human heads. These sketches were found in a mutilated scrap-book, which also contained a kind of journal of his proceedings in 1769. He mentions a French class which he attended in the evenings, records the purchase of a certain measure of black cloth, states his debt to his uncle, Hiram Thompson, for part of the rent of a pew. The liabilities thus incurred he met by cutting and carting firewood. Mixed with entries such as these are ' directions for the backsword,' in which the postures of the combatants are defined and illustrated by sketches. The scrap-book also contained an account of the expense ' towards getting an electrical machine.'

Soon afterwards he began the study of medicine nuder
Dr. John Hay, of Woburn.

Thompson kept a strict account of his debts to Dr.
Hay, crediting him with such things as leather gloves, and
Mrs. Hay with knitting him a pair of stockings. These
items he tacks on to the more serious cost of his board,
from December 1770 to June 1772, at forty shillings,
old currency, per week, amounting to 156*l.* The specie
payments of Thompson were infinitesimal, eight of them
amounting in the aggregate to 2*l.* His further forms
of payment illustrate the habits of the community in
which he dwelt. Want of money caused them to fall
back upon barter. He debited Dr. Hay with the following items, the value of which no doubt had been previously agreed upon between them:—' To ivory for
smoke machine ; parcels of butter, coffee, sugar, and
tea ; parcels of various drugs, camphor, gum benzoine,
arsenic, calomel, and rhubarb ; one-half of white sheepskin leather ; brass wire ; white oak timber ; to sundry
lots of wood ; to other lots delivered while I was at
Wilmington, and left by me when I was at Wilmington
the last time ; to a blue Huzza cloak, bought of Zebe-
diah Wyman, and paid for by fifteen and a half cords
of wood ; a pair of knee buckles ; a chirurgical knife ;
to a cittern, and to the time I have been absent from
your house, nineteen weeks at forty shillings ; and for
the time my mother washed for me.' To help him,
moreover, to eke out the funds necessary for the prosecution of his studies, Thompson tried his hand from
time to time at school-teaching.

At this early age—for he was not more than seven-
teen—he had learnt the importance of order in the distribution of his time. The four-and-twenty hours of a
single day are thus spaced out :—' From eleven to six,
sleep. Get up at six o'clock and wash my hands and

face. From six to eight, exercise one-half, and study one-half. From eight to ten, breakfast, attend prayers, &c. From ten to twelve, study all the time. From twelve to one, dine, &c. From one to four, study constantly. From four to five, relieve my mind by some diversion or exercise. From five till bedtime, follow what my inclination leads me to ; whether it be to go abroad, or stay at home and read either Anatomy, Physic, or Chemistry, or any other book I want to peruse.'

In 1771 he managed, by walking daily from Woburn to Cambridge and back, a distance of some sixteen miles, to attend the lectures on natural philosophy delivered by Professor Winthrop in Harvard College. This privilege was secured to him by his friend Mr. Baldwin. Thompson had taught school for a short time at Wilmington, and afterwards for six weeks and three days at Bradford, where his repute rose so high that he received a call to Concord, the capital of New Hampshire. The Indian name of Concord was Penacook. In 1733 it had been incorporated as a town in Essex County, Massachusetts. Some of the early settlers had come from the English Essex ; and, as regards pronunciation, they carried with them the name of the English Essex town, Romford, of brewery celebrity. They, however, changed the first *o* into *u*, calling the American town Rumford. Strife had occurred as to the county or State to which Rumford belonged. But the matter was amicably settled at last ; and to denote the subsequent harmony, the name was changed from Rumford to Concord.[1] In later years, when hon-

[1] Not to be confounded with the Concord rendered famous as the dwelling-place of Emerson. In connection with the foregoing subject I have been favoured with the following interesting letter :—

'Addison Lodge, Barnes, S.W. : *August* 19.

' DEAR SIR,—I venture to proffer a remark upon a detail in your

ours fell thick upon him, Rumford was made a Count of the Holy Roman Empire. He chose for his title Count Rumford, in memory of his early association with Concord.

' When Benjamin Thompson went to Concord as a teacher, he was in the glory of his youth, not having yet reached manhood. His friend Baldwin describes him as of a fine manly make and figure, nearly six feet in height, of handsome features, bright blue eyes, and dark auburn hair. He had the manners and polish of a gentleman, with fascinating ways and an ability to interesting paper upon *Count Rumford*. My apology for so doing is that I am a Romford man, and that I think you may care for the mere crumb of information I possess bearing upon the spelling and pronunciation of the name of my native place.

' Romford is always pronounced Rumford by Essex folk. When I was a boy it was *spelled* almost indifferently, Romford and Rumford. I remember that the post-mark in my school-days (some forty years ago) was Rumford. Norden's map of Essex (1599) has Rumforde; and on Bowen's map (1775) the spelling is the same—Rumford. The registers in the vestry book, from 1665 until some fifty years ago, give *Rumford*. So that I think it safe to say that the traditional spelling and pronunciation with the Essex settlers at Concord must have been Rumford. I must, however, add—but I fear I am hardly justified in troubling you with so long a note—that the *o* occurs in two *Latin* entries in the Register :—

' "1564, Baptizata fuit Anna Baylie filia Hugonis Cissor, Romford."
' And in the same year there is an entry of a burial with "Romfordiae." I believe it was the *Latinising* of Rumford that modified the vowel, the alteration being prompted by the mistaken notion that the etymology of the place was Roman-ford. That the Rum is English (= broad) is, I think, hardly open to question. The nearest *ford* town is *Ilford*, with which the *roomy* ford contrasts. Of late the sluggish little river has come to be called the river Rom. This is quite a novel " notion," and is quite local.

' Thanking you for the pleasure and profit I have derived from reading your article,

' I remain, dear Sir,
' Yours very faithfully,
' HENRY ATTWELL.

'Professor Tyndall, F.R.S., &c., &c., &c.'

make himself agreeable.'[1] Thus writes his biographer. In Concord, at the time of Thompson's arrival, dwelt the widow of Colonel Rolfe with her infant son. Her husband had died in December 1771, leaving a large estate behind him. Thompson was indebted to Mrs. Rolfe's father, the Rev. Timothy Walker, minister of Concord, for counsel, and to her brother for civility and hospitality. The widow and the teacher met, and their meeting was a prelude to their marriage. Rumford, somewhat ungallantly, told his friend Pictet in after-years that she married him rather than he her. She was obviously a woman of decision. As soon as they were engaged an old curricle, left by her father, was fished up, and, therein mounted, she carried Thompson to Boston, and committed him to the care of the tailor and hairdresser. This journey involved a drive of sixty miles. On the return journey, it is said, they called at the house of Thompson's mother, who when she saw him exclaimed, ' Why, Ben, my son, how could you go and lay out all your winter's earnings in finery?' Thompson was nineteen when he married, his wife being thirty-three.

In 1772 he became acquainted with Governor Wentworth, then resident at Portsmouth. On the 13th of November there was a grand military review at Dover, New Hampshire, ten miles from Portsmouth, at which Thompson was present. On two critical occasions in the life of this extraordinary man his appearance on horseback apparently determined his career. As he rode among the soldiers at Dover, his figure attracted the attention of the governor, and on the day following he was the great man's guest. So impressed was Wentworth with his conversation that he at once made up his mind to attach Thompson to the public service.

[1] Ellis, p. 43.

To secure this wise end he adopted unwise means. ' A vacancy having occurred in a majorship in the Second Provincial Regiment of New Hampshire, Governor Wentworth at once commissioned Thompson to fill it.' Jealousy and enmity naturally followed the appointment of a man without name or fame in the army, over the heads of veterans with infinitely stronger claims. He rapidly became a favourite with the governor, and on his proposing, soon after his appointment, to make a survey of the White Mountains, Wentworth not only fell in with the idea, but promised, if his public duties permitted, to take part in the survey himself. It will be remembered that at this time Thompson was not quite twenty years old.

For a moment, in 1773, he appears in the character of a farmer, and invokes the aid of a friend to procure for him supplies of grass and garden seeds from England. But amid preoccupations of this kind his scientific bias emerges. After a brief reference to the seed procured for him by his friend Baldwin, he proposes to the latter the following question : ' A certain cistern has three brass cocks, one of which will empty it in fifteen minutes, one in thirty minutes, and the other in sixty minutes. Qu. How long would it take to empty the cistern if all three cocks were to be opened at once ? If you are fond of a correspondence of this kind, and will favour me with an easy question, arithmetical or algebraical, I will endeavour to give as good an account of it as possible. If you find out an answer to the above immediately, I hope you will not take as an affront my proposing anything which you may think so easy, for I must confess I scarce ever met with any little notion that puzzled me so much in my life.'

In 1774 the ferment of discontent with the legisla-

tion of the mother country had spread throughout the
colony. Clubs and committees were formed which often
compelled men to take sides before the requisite data
for forming a clear judgment had been obtained.
' Our candour,' says Dr. Ellis, ' must persuade us to
allow that there were reasons, or at least prejudices and
apprehensions, which might lead honest and right-
hearted men, lovers and friends of their birthland, to
oppose the rising spirit of independence, as inflamed by
demagogues, and as foreboding discomfiture and mischief.' Thompson became ' suspect,' though no record
of any unfriendly or unpatriotic act or speech on his
part is to be found. He was known to be on friendly
terms with Governor Wentworth; but the governor,
when he gave Thompson his commission, was highly
popular in the province. Prior to Wentworth's acces
sion to office he ' had strongly opposed every measure
of Great Britain which was regarded as encroaching
upon our liberties.' He thought himself, nevertheless,
in duty bound to stand by the royal authority when it
was openly defied. This rendered him obnoxious.

Thompson was a man of refractory temper, and the
circumstances of the time were only too well calculated
to bring that temper out. ' There was something,'
says Dr. Ellis, ' exceedingly humiliating and degrading
to a man of an independent and self-respecting spirit
in the conditions imposed at times by the " Sons of
Liberty," in the process of cleansing oneself from the
taint of Toryism. The Committees of Correspondence
and of Safety, whose services stand glorified to us
through their most efficient agency in a successful
struggle, delegated their authority to every witness or
agent who might be a self-constituted guardian of
patriotic interests, or a spy, or an eavesdropper, to
catch reports of suspected persons.' Human nature

is everywhere the same, and to protect a cherished cause these 'Sons of Liberty' sometimes adopted the tactics of the Papal Inquisition.

Public feeling grew day by day more exasperated against Thompson, and in the summer of 1774 he was summoned before a committee to answer to the charge of being unfriendly to the cause of liberty. ' He denied the charge, and challenged proof. The evidence, if any such was offered—and no trace of testimony, or even of imputation of that kind is on record—was not of a sort to warrant any proceeding against him, and he was discharged.' This, however, gave him but little relief, and extra-judicial plots were formed against him. The Concord mob resolved to take the matter into their own hands. One day they collected round his house, and with hoots and yells demanded that Thompson should be delivered up to them. Having got wind of the matter he escaped in time. In a letter addressed to his father-in-law at this time from Charlestown, near Boston, he gives his reasons for quitting home. ' To have tarried at Concord and have stood another trial at the bar of the populace would doubtless have been attended with unhappy consequences, as my innocence would have stood me in no stead against the prejudices of an enraged, infatuated multitude—and much less against the determined villainy of my inveterate enemies, who strive to raise their popularity on the ruins of my character.'

He returned to his mother's house in Woburn, where he was joined by his wife and child. While they were with him, shots were exchanged and blood was shed at Concord (Emerson's Concord) and Lexington. Thompson was at length arrested, and confined in Woburn. A ' Committee of Correspondence ' was formed to inquire into his conduct. They invited everyone who could give evidence in the affair to appear at the meeting-house on May 18. The committee met, but finding nothing against the accused, they adjourned the meeting. He then addressed a petition to the Committee of Safety for the colony of Massachusetts Bay, in which he begged for a full and searching trial, relying on an acquittal commensurate with the thoroughness of the examination. The petition was not attended to. On May 29, 1775, he was examined at Woburn, where he conducted his own defence. He was acquitted by the committee, who recommended him to the ' protection of all good people in this and the neighbouring provinces.' The committee, however, refused to make this acquittal a public one, lest, it was alleged, it should offend those who were opposed to Thompson.

Despair and disgust took possession of him more and more. In a long letter addressed to his father-in-law from Woburn, he defends his entire course of conduct. His principal offence was probably negative ; for silence at the time was deemed tantamount to antagonism. During his brief period of farming he had working for him some deserters from the British army in Boston. These he persuaded to go back, and this was urged as a crime against him. He defended himself with spirit, declaring after he had explained his motives that if his action were a crime, he gloried in being a criminal. He made up his mind to quit the country, expressing the devout wish ' that the happy time may soon come when I may return to my family in peace and safety, and when every individual in America may sit down under his own vine and under his own fig-tree, and have none to make him afraid.'

On this letter, and on the circumstances of the time, Dr. Ellis makes the following wise and pertinent re-marks : ' Major Thompson was not the only person in

those troubled times that had occasion to charge upon
those espousing the championship of public liberty a
tyrannical treatment of individuals who did not accord
with their schemes or views. Probably in our late war
of rebellion his case was paralleled by those of hundreds
in both sections of our country, who, with halting and
divided minds or unsatisfied judgments, were arrested
in the process of decision by treatment from others
which put them under the lead of passion. The choice
of a great many Royalists in our revolution would have
been wiser and more satisfactory to themselves, had
they been allowed to make it deliberately.' On October
13, 1775, Thompson quitted Woburn, reached the shore
of Narraganset Bay, and went on board a British frigate.
In this vessel he was conveyed to Boston, where he remained until the town was evacuated by the British
troops. The news of this catastrophe he carried to England.
Henceforward, till the close of the war, he was on the
English side. As a matter of course, he was proscribed
by his countrymen, and his property confiscated.

Thompson was not only a man of great capacity,
but, in early days, of a social pliancy and teachableness
which enabled him with extreme rapidity to learn the
manners and fall into the ways of great people. On
the English side the War of Independence was begun,
continued, and ended, in ignorance. Even now we can
hardly read the pages of 'The Virginians' which refer
to these times without exasperation. Blunder followed
blunder, and defeat followed defeat, until the knowledge
which ought to have been ready at the outset came too
late. Thompson for a time was the vehicle of such be-
lated knowledge. He was immediately attached to the
Colonial Office, then ruled over by Lord George Germain. Cuvier, in his ' Eloge,' thus describes his first
interview with that Minister : ' On this occasion, by the

clearness of his details and the gracefulness of his manners, he insinuated himself so far into the graces of Lord George Germain that he took him into his employment.' With Lord George he frequently breakfasted, dined, and supped, and was occasionally his guest in the country. But besides giving information useful to his chiefs, he occupied himself with other matters. He was a born experimentalist, handy, ingenious, full of devices to meet practical needs. He turned his attention to improvements in military matters; 'advised and procured the adoption of bayonets for the fusees of the Horse Guards, to be used in fighting on foot.' He had previously been engaged with experiments on gunpowder, which he now resumed. The results of these experiments he communicated to Sir Joseph Banks, then President of the Royal Society, with whom he soon became intimate. In 1779 he was elected a Fellow of the Royal Society.

When the war had become hopeless, many of the exiles who had been true to the Royalist cause came to England, where Thompson's official position imposed on him the duty of assuaging their miseries and adjusting their claims. In this connection, the testimony of Dr. Ellis regarding him is that, 'so far as the relations between these refugees and Mr. Thompson can be traced, I find no evidence that he failed to do in any case what duty and friendliness required of him.' Still he did not entirely escape the censure of his outlawed fellow-countrymen. One of them in particular had been a judge in Salem when Thompson was a shopboy in Appleton's store. Judge Curwen complained of Thompson's fair appearance and uncandid behaviour. He must have keenly felt the singular reversal in their relations. 'This young man,' says the judge, 'when a shop-lad to my next neighbour, ever appeared active, good-natured,

and sensible ; by a strange concurrence of events, he is
now Under-Secretary to the American Secretary of State,
Lord George Germain, a Secretary to Georgia, Inspector
of all the clothing sent to America, and Lieutenant-
Colonel Commandant of Horse Dragoons at New York ;
his income from these sources is, I have been told, near
7,000$l.$[1] a year—a sum infinitely beyond his most
sanguine expectations.'

As the prospects of the war darkened, Thompson's
patron in England became more and more the object of
attack. The people had been taxed in vain. England
was entangled in Continental war, and it became gradually recognised that the subjugation of the colony
was impossible. Burgoyne had surrendered, and the
issue of the war hung upon the fate of Cornwallis. On
October 19 he also was obliged to capitulate. The
effect of the disaster upon Lord North, who was then
Prime Minister, is thus described by Sir M. W. Wrax-
all :—' The First Minister's firmness, and even his presence of mind, gave way for a short time under this
awful disaster. I asked Lord George afterwards how
he took the communication. " As he would have
taken a ball in his breast," replied Lord George ; " he
opened his arms, exclaiming wildly, as he paced up and
down the apartment during a few minutes, ' O God ! it
is all over ! ' " '

To Thompson's credit be it recorded, that he showed
no tendency to desert the cause he had espoused when
he found it to be a failing one. In 1782 his chief was
driven from power. At this critical time he accepted
the commission of lieutenant-colonel in the British
army, and returned to America with a view of rallying
for a final stand such forces as he might find capable
of organisation. He took with him four pieces of

[1] This Dr. Ellis considers to be a delusion.

artillery, with which he made experiments during the
voyage. His destination was Long Island, New York,
but stress of weather carried him to Charleston, South
Carolina. ' Obliged,' says Pictet, ' to pass the winter
there, he was made commander of the remains of the
cavalry in the Royal army, which was then under the
orders of Lieutenant-General Leslie. This corps was
broken, but he promptly restored it, and won the confidence and attachment of the commander. He led them
often against the enemy, and was always successful in
his enterprises.'

About the middle of April Thompson reached New
York, and took command of the King's American
Dragoons. Colours were presented to the regiment on
August 1, a very vivid account of the ceremony being
given in Rivington's ' Royal Gazette ' of August 7, 1782.
Prince William Henry, afterwards King William IV.,
was there at the time. The regiment passed in review
before him, performing marching salutes. They then
returned, dismounted, and formed in a semicircle in front
of the canopy. After an address by their chaplain, the
whole regiment knelt down, laid their helmets and arms
on the ground, held up their right hands, and took a
most solemn oath of allegiance to their sovereign, and
fidelity to their standard. From Admiral Digby the
Prince received the colours, and presented them with his
own hands to Thompson, who passed them on to the
oldest cornets. ' On a given signal the whole regiment,
with all the numerous spectators, gave three shouts, the
music played " God save the King," the artillery fired
a royal salute, and the ceremony was ended.'

Many complaints have been made of the behaviour
of the troops during their stay at Long Island, New
York. But war is always horrible ; and it is pretty

clear from the account of Dr. Ellis, that the complaints
had no other foundation than events inseparable from
the carrying on of war.   In the statement of Thompson's
case his biographer, extenuating nothing, and setting
down naught in malice, winds up his third chapter with
these words : ' I may add to such praise as is due to him
as a good soldier, quick and true and bold in action,
and faithful to the Government which he served, the
higher tribute that, from the hour when the war closed,
he became, and ever continued to be, the constant
friend and generous benefactor of his native country.'

Early in April 1783 Thompson obtained leave to
return to England, but finding there no opportunity for
active service, he resolved to try his fortune on the Continent, intending to offer his services as a volunteer in
the Austrian army against the Turks. The historian
Gibbon crossed the Channel with him. In a letter dated
Dover, September 17, 1783, Gibbon writes thus :—' Last
night the wind was so high that the vessel could not
stir from the harbour ; this day it is brisk and fair.
We are flattered with the hope of making Calais Harbour by the same tide in three hours and a half ; but
any delay will leave the disagreeable option of a tottering
boat or a tossing night. What a cursed thing to live
in an island ! this step is more awkward than the whole
journey. The triumvirate of this memorable embarka-
tion will consist of the grand Gibbon ; Henry Laurens,
Esq., President of Congress; and Mr. Secretary, Colonel,
Admiral, Philosopher Thompson, attended by three
horses, who are not the most agreeable fellow-passengers.
If we survive, I will finish and seal my letter at Calais.
Our salvation shall be ascribed to the prayers of my
lady and aunt, for I do believe they both pray.' The
'grand Gibbon' is reported to have been terribly

frightened by the plunging of his fellow-passengers, the three blood-horses.

Thompson pushed on to Strasburg, where Prince Maximilian of Bavaria, then a field-marshal in the service of France, was in garrison. As on a former occasion in his native country, Thompson, mounted on one of his chargers, appeared on the parade-ground. He attracted the attention of the Prince, who spoke to him, and on learning that he had been serving in the American war, pointed to some of his officers, and remarked that they had been in the same war. An animated conversation immediately began, at the end of which Thompson was invited to dine with the Prince. After dinner, it is said, he produced a portfolio containing plans of the principal engagements, and a collection of excellent maps of the seat of war. Eager for information, the Prince again invited him for the next day, and when at length the traveller took leave, engaged him to pass through Munich, giving him a friendly letter to his uncle, the Elector of Bavaria.

Thompson carried with him wherever he went the stamp of power and the gift of address. The Elector, a sage ruler, saw in him immediately a man capable of rendering the State good service. He pressed his visitor to accept a post half military, half civil. The proposal was a welcome one to Thompson, and he came to England to obtain the King's permission to accept it. Not only was the permission granted, but on February 23, 1784, he was knighted by the King. Dr. Ellis publishes the 'grant of arms' to the new knight. In it he is described as 'Son of Benjamin Thompson, late of the Province of Massachusetts Bay in New England, Gent., deceased,' and as ' of one of the most antient Families in North America ; that an Island which belonged to

his Ancestors, at the entrance of Boston Harbour, where
the first New England Settlement was made, still bears
his name ; that his Ancestors have ever lived in reputable
Situations in that country where he was born, and have
hitherto used the Arms of the antient and respectable
Family of Thompson, of the county of York, from a
constant Tradition that they derived their Descent from
that Source.' The original parchment, perfect and
unsullied, with all its seals, is in the possession of Mrs.
James F. Baldwin, of Boston, widow of the executor
of Countess Sarah Rumford.[1] The knight himself,
observes his biographer, must have furnished the information written on that flowery and mythical parchment.
Thompson was fond of display, and he here gave rein to
his tendency. He returned to Munich, and on his
arrival the Elector appointed him colonel of a regiment
of cavalry and general aide-de-camp to himself. He
was lodged in a palace, which he shared with the
Russian Ambassador, and had a military staff and a
corps of servants. 'His imposing figure, his manly and
handsome countenance, his dignity of bearing, and his
courteous manners, not only to the great, but equally
to his subordinates and inferiors, made him exceedingly
popular.'

He soon acquired a mastery of the German and
French languages. He made himself minutely acquainted with everything concerning the dominions of
the Elector—their population and employments, their
resources and means of development, and their relations
to other powers. He found much that needed removal
and required reformation. Speaking of the Electorate,
Cuvier remarks that 'its sovereigns had encouraged
devotion, and made no stipulation in favour of industry.
There were more convents than manufactories in their

[1] Ellis.

States ; their army was almost a shadow, while ignorance
and idleness were conspicuous in every class of society.'
Thompson evoked no religious animosity. He avowed
himself a Protestant, but met with no opposition
on that score. Holding as he did the united offices of
Minister of War, Minister of Police, and Chamberlain
of the Elector, his influence and action extended to
all parts of the public service. Then, as now, the
armies of the Continent were maintained by conscrip-
tion. Drawn away from their normal occupations, the
peasants returned after their term of service lazy and
demoralised. This was a great difficulty ; and in dealing
with it patient caution had to be combined with administrative skill. Four years of observation were spent at
Munich before Thompson attempted anything practical.
The pay of the soldiers was miserable, their clothing bad,
their quarters dirty and mean ; the expense being out of
all proportion to the return. The officers, as a general
rule, regarded the soldiers as their slaves ; and here
special prudence was necessary in endeavouring to effect
a change. Thompson induced the more earnest among
the officers to co-operate with him, by making the proposed reforms to originate apparently with them. He
aimed at making soldiers citizens and citizens soldiers.
The situation of the soldier was to be rendered pleasant,
his pay was to be increased, his clothing rendered comfortable and even elegant, while all liberty consistent
with strict subordination was to be permitted him.
Within, the barracks were to be neat and clean ; and without, attractive. Reading, writing, and arithmetic were
to be taught, not only to the soldiers and their children,
but to the children of the neighbouring peasantry. The
paper used in the school would, it was urged, be
practically free of cost, *as it would serve afterwards
for cartridges.*

The marshes near Mannheim were dreary bogs, use-less for cultivation and ruinous to the health of the city. Thompson drained them, banked them in, and converted them into a garden for the use of the garrison. For the special purpose of introducing the culture of the potato, he extended the plan of military gardens to all other garrisons. The gardens were tilled, and their produce was owned by noncommissioned officers and privates, each of whom had a plot of 365 square feet allotted to him. Gravel walks divided the plots from each other. The plan proved completely successful. Indolent soldiers became industrious, while soldiers on furlough, spreading abroad their taste and knowledge, caused little gardens to spring up everywhere over the country. Having secured this end, he converted it into a means of suppressing the enormous evils of mendicity. Bavaria was infested with beggars, vagabonds, and thieves, native and foreign. These mendicant tramps were in the main stout, healthy, and able-bodied fellows, who found a life of thievish indolence pleasanter than a life of honest work. ' These detestable vermin had recourse to the most diabolical arts and the most horrid crimes in the prosecution of their infamous trade. They robbed, and maimed and exposed little children, so as to extract money from the tender-hearted. In the cities the beggars formed a distinct caste, with profes-sional rules to guide them. Their training was a training in robbery; the means they employed for extorting support being equivalent to direct plunder. Seeing no escape from the incubus, the public had come to bow to it as a necessity. The energy with which Thompson grappled with this evil may be inferred from the fact that out of a population of sixty thousand, two thousand six hundred beggars were impounded in a single week.

Four regiments of cavalry were so cantoned that every village in Bavaria and the adjoining provinces had a patrol party of four or five mounted soldiers 'daily coursing from one station to another.' The troopers were under strict discipline, extreme care being taken to avoid collision with the civil authorities. This disposition of the cavalry was antecedent to seizing, as a beginning, all the beggars in the capital. Aged and infirm mendicants were carefully distinguished from the sturdy and able-bodied. Voluntary contributions were essential, but the inhabitants, though groaning under the load of mendicancy, had been so often disappointed in their efforts to get rid of it that they now held back. Thompson resolved to give proof of success before asking for general aid. He interested persons of high rank in his scheme ; organised a bureau to relieve the needy and employ the idle. The members of his committee were presidents of the great offices of State, who worked without pay. The city was divided into sixteen districts, with a committee of charity for each ; while a respected citizen, assisted by a priest and a physician, serving gratuitously, looked after the worthy poor. He knew perfectly well that in Munich many bequests consecrated to charity were being abused and wasted, but he cautiously abstained from meddling with them.

The problem before him might well have daunted a courageous man. It was neither more nor less than to convert people bred up in lazy and dissolute habits into thrifty workers. Precepts, he knew, were unavailing, so his aim was to establish habits. Reversing the maxim that people must be virtuous to be happy, he made his beggars happy as a step towards making them virtuous. He affirmed that he had learnt the importance of cleanliness through observing the habits of birds and beasts.

Lawgivers and founders of religions never failed to recognise the influence of cleanliness on man's moral nature. ' Virtue,' he said, ' never dwelt long with filth and nastiness, nor do I believe there ever was a person scrupulously attentive to cleanliness who was a consummate villain.' He had to deal with wretches covered with filth and vermin, to cleanse them, teach them, and give them the pleasure and stimulus of earning money. He did not waste his means on fine buildings, but taking a deserted manufactory, he repaired it, enlarged it, adding to it kitchen, bakehouse, and workshops for mechanics. Halls were provided for the spinners of flax, cotton, and wool. Other halls were set up for weavers, clothiers, dyers, saddlers, wool-sorters, carders, combers, knitters, and seamstresses.

The next step was to get the edifice filled with suitable inmates. New Year's Day was the beggars' holiday, and their reformer chose that day to get hold of them. It was the 1st of January, 1790. In the prosecution of his despotic scheme all men seemed to fall under his lead. To relieve it of the odium which might accrue if it were effected wholly by the military, he associated with himself and his field officers the magistrates of Munich. They gave him willing sympathy and aid. On New Year's morning he and the chief magistrate walked out together. With extended hand a beggar immediately accosted them. Thompson, setting the example to his followers, laid his hand gently upon the shoulder of the vagabond, committed him to the charge of a sergeant, with orders to take him to the Town-hall, 'where he would be provided for in one way if he were really helpless, but in another way if he were not.' Thompson encouraged his associates, and with such alacrity was the work accomplished, that at the end of that day not a single beggar remained at

large. The name of every member of the motley crew
was inserted in prepared lists, and they were sent off to
their haunts with instructions to appear on the following
day at the military workhouse, where they would inhabit
comfortable warm rooms, enjoy a warm dinner daily, and
be provided with remunerative work. In the suburbs the
same measures were followed up successfully by patrols
of soldiers and police.

With his iron resolution was associated, in those
days, a plastic tact which enabled him to avoid
jealousies and collisions that a man of more hectoring
temper and less self-restraint would infallibly have
incurred. To the schools for poor students, the Sisters
of Charity, the hospital for lepers, and other institutions,
had been conceded the right of making periodic appeals
from house to house ; German apprentices had also been
permitted to beg upon their travels ; all of these had
their claims adjusted. After he had swept his swarm of
paupers into the quarters provided for them, Thompson's
hardest work began. Here the inflexible order which
characterised him through life came as a natural force
to his aid. ' He encouraged a spirit of industry, pride,
self-respect, and emulation, finding help even in trifling
distinctions of apparel.' His pauper workhouse was
self-supporting, while its inmates were happy. For
several years they made up all the clothing of the
Bavarian troops, realising sometimes a profit of 10,000
florins a year. Thompson himself constructed and
arranged a kitchen which provided daily a warm and
nutritious dinner for a thousand or fifteen hundred
persons, an incredibly small amount of fuel sufficing to
cook a dinner of this magnitude. The military workhouse was also remunerative. Its profits for six years
exceeded a hundred thousand dollars. The military
workhouse at Mannheim was unfortunately set on fire

and ruined during the siege of the city by Austrian
troops.

Thompson had the art of making himself loved and
honoured by the people whom he ruled in this arbitrary
way. Some very striking illustrations of this are given
in the ' Life and Essays.' He once, for example, broke
down at Munich under his self-imposed labours. It
was thought that he was dying, and one day while in
this condition, his attention was attracted by the confused noise of a passing multitude in the street. It
was the poor of Munich who were going in procession to
the church to offer public prayers for him. ' Public
prayers!' he exclaims, ' for me, a private person, a
stranger, a Protestant!' Four years afterwards, when
he was dangerously ill at Naples, the people of their
own accord set apart an hour each evening, after they
had finished their labours in the military workhouse, to
pray for his recovery.

Men find pleasure in exercising the powers they
possess, and Rumford possessed, in its highest and
strongest form, the power of organisation. The relief
of the poor, which occupied his attention for years, was
pursued by him as a scientific inquiry. He differentiated the people who had fair claims upon the State
from those whose infirmity and incapacity rendered continuous assistance necessary, but who could not be aided
by compulsory taxation. In this case the promptings
of humanity must be invoked. Persons of high rank
ought here to take the lead, combining with those immediately below them to secure efficient supervision
and relief. The expense thus incurred is small compared
with that incidental to beggary and its concomitant
thieving. Thompson's hope and confidence never for-
sook him. He faced, unquailing, problems from which
less daring spirits would have recoiled. He held, un-

doubtingly, that 'arrangement, method, provision for the minutest details, subordination, co-operation, and a careful system of statistics, will facilitate and make effective any undertaking, however burdensome and comprehensive.' Such a statement would surely have elicited a 'bravo!' from Carlyle. In Thompson, flexible wisdom formed an amalgam with despotic strength. With skill and resolution the objects of public benevolence must, he urged, be made to contribute as far as possible to their own support. The homeliest details did not escape him. He commended well-dressed vegetables as a cheap and wholesome aliment. He descanted on the potato, he gave rules for the construction of soup-kitchens, and determined the nutritive value of different kinds of food. During his boyhood at Woburn he had learnt the use of Indian corn, and at Munich he strongly recommended the dumplings, bread, and hasty pudding made from maize. Pure love of humanity would, at first sight, seem to have been the motive force of Thompson's action. Still, it has been affirmed by those who knew him that he did not really love his fellow-men. His work had for him the fascination of a problem above the capacities of most men, but which he felt himself able to solve. It was said to be the work of his intellect, not of his heart. In reference to him, Cuvier quotes what Fontenelle said of Dodard, who turned his rigid observance of the fasts of the Church into a scientific experiment on the effects of abstinence, thereby taking the path which led at once to heaven and into the French Academy. I should hesitate before accepting this as a complete account of Rumford's motives.

In the north-easterly environs of Munich a wild and neglected region of forest and marsh, which had formerly been the hunting-ground of the Elector, was

converted by Thompson into an 'English garden.'
Pleasure-grounds, parks, and fields were laid out, and
surrounded by a drive six miles long. Walks, pro-
menades, grottos, a Chinese pagoda, a racecourse, and
other attractions were introduced ; a lake was formed
and a mound raised ; while a refreshment-saloon, hand-
somely furnished, provided for the creature comforts
of the visitors. Here during Rumford's absence in
England in 1795, and without his knowledge, a monument was raised to commemorate his beneficent achievements. ' It stands within the garden, and is composed
of Bavarian freestone and marble. It is quadrangular,
its two opposite fronts being ornamented with basso-
relievos, and bearing inscriptions.' The wanderer on
one side is exhorted to halt, while thankfulness streng-
thens his enjoyment. 'A creative hint from Carl
Theodor, seized upon with spirit, feeling, and love by
the friend of man, Rumford, has ennobled into what
thou now seest this once desert region.' On the other
side of the monument is a dedication to ' Him who
eradicated the most scandalous of public evils, Idleness
and Mendicity ; who gave to the poor help, occupation,
and morals, and to the youth of the Fatherland so many
schools of culture. Go, wanderer ! try to emulate him
in thought and deed, and us in gratitude.'

Rumford's health, as already indicated, had given
way, and in 1793 he went to Italy to restore it. He
was absent for sixteen months, and during his absence
was seriously ill at Naples. Had he been less filled
with his projects, it might have been better for his
health. Had he known how to employ the sanative
power of Nature, he would have kept his vigorous frame
longer in working order. But the mountains of Maggiore were to him less attractive than the streets of

Verona, where he committed himself to the planning of soup-kitchens. He made similar plans for other cities, so that to call his absence a holiday would be a misnomer. He returned to Munich in August 1794, slowly recovering, but not able to resume the management of his various institutions. In September 1795 he returned to London, after an absence of eleven years. Dr. Ellis describes him as ' the victim of an outrage ' on his arrival, the meaning of which seems to be that the trunk containing his papers, which was carried behind his carriage, was appropriated by London thieves. ' By this cruel robbery,' he says, ' I have been deprived of the fruits of the labours of my whole life. . . . It is the more painful to me, as it has clouded my mind with suspicions that can never be cleared up.' What the suspicions were we do not know.

Soon afterwards he was invited by Lord Pelham, then Secretary of State for Ireland, to visit him in Dublin ; he went, and during his two months' stay there busied himself with improvements of warming, cooking, and ventilation, in the hospitals and workhouses of the city. He left behind him a number of models of useful mechanism. The Royal Irish Academy elected him a member. The grand jury of Dublin presented him with an address ; while the Viceroy and the Lord Mayor wrote to him officially to thank him for his services. Dr. Ellis has not been able to find these documents, but they were seen by Pictet, who describes them as ' filled with the most flattering expressions of esteem and gratitude.'

In Rumford's case the life of the intellect appeared to have interfered with the life of the affections. When he quitted America, he left his wife and infant daughter behind him, and whether any communication afterwards occurred between him and them is not known.

In 1793, in a letter to his friend Baldwin, he expressed the desire to visit his native country. He also wished exceedingly to be personally acquainted with his daughter, who was then nineteen. His affection for his mother, which appears to have been very real, also appears in this letter. With reference to the projected visit, he asks, ' Should I be kindly received ? Are the remains of party spirit and political persecution done away ? Would it be necessary to ask leave of the State ? ' A year prior to the date of this letter, Rumford's wife had died at the age of fifty-two. On January 29, 1796, his daughter sailed for London to see her father. She had a tedious passage, but soon after her arrival she writes to her friend Mrs. Baldwin, ' All fatigue and anxiety are now at an end, since my dear father is well and loves me.'

In a history of her life, written many years afterwards, she, however, describes the disappointment she experienced on first meeting her father. Her imagination had sketched a fancy picture of him. She ' had heard him spoken of as an officer, and had attached to this an idea of the warrior with a martial look, possibly the sword, if not the gun, by his side.' All this disappeared when she saw him. He did not strike her as handsome, or even agreeable—a result in part due to the fact that he had been ill, and was very thin and pale. She speaks, however, of his laughter ' quite from the heart,' while the expression of his mouth, with teeth described as ' the most finished pearls,' was sweetness itself. He did not seem to manage her very successfully. She had little knowledge of the world, and her purchases in London he thought both extravagant and extraordinary. After having, by due discipline, learnt how to make an English courtesy, to the horror of her father, almost the first use she made of her newly-acquired accomplishment was to courtesy to a house-keeper.

His labours in the production of cheap and nutritious food necessarily directed Rumford's attention to fire-places and chimney-flues. When he published his essay on this subject in London, he reported that he had not less than five hundred smoky chimneys on his hands. His aid and advice were always ready, and were given indiscriminately to all sorts and conditions of men. Devonshire House, Sir Joseph Banks's, the Earl of Bess-borough's, Countess Spencer's, Melbourne House, Lady Templeton's, Mrs. Montagu's, Lord Sudley's, the Marquis of Salisbury's, and a hundred and fifty other houses in London, were placed in his care. The saving of fuel, with gain instead of loss of warmth, varied in these cases from one-half to two-thirds. ' Giving very simple and intelligible information about the philosophical principles of combustion, ventilation, and draughts, he prepared careful diagrams to show the proper measure-ments and arrangements of all the parts of a fireplace and flue. He took out no patent for his inventions, but left them free to the public. In a poem published at this time by Thomas James Matthias we have the following reference to the labours of Rumford :—

> Nonsense, or sense, I'll bear in any shape—
> In gown, in lawn, in ermine, or in crape :
> What's a fine type, where truth exerts her rule ?
> Science is science, and a fool's a fool.
> Yet all shall read, and all that page approve,
> When public spirit meets with public love.
> Thus late, where poverty with rapine dwelt,
> Rumford's kind genius the Bavarian felt,
> Not by romantic charities beguiled,
> But calm in project, and in mercy mild ;
> Where'er his wisdom guided, none withstood,
> Content with peace and practicable good ;
> Round him the labourers throng, the nobles wait,
> Friend of the poor, and guardian of the State.

The pall of smoke which habitually hung over London, 'covering all its prominent edifices with a dingy and sooty mantle,' curiously and anxiously interested him. He 'saw in that smoke the unused material which was turned equally to waste and made a means of annoyance and insalubrity.' He would bind himself, if the opportunity were allowed him, 'to prove that from the heat, and the material of heat, which were thus wasted, he would cook all the food used in the city, warm every apartment, and perform all the mechanical work done by means of fire.' Under this wasted heat Rumford would doubtless comprise both the imperfectly-consumed gases, such as carbonic oxide, and the heated air and other gases discharged by the chimneys.

There is no doubt that the present age has entered largely into the labours of Rumford. Many of the devices and conveniences now employed in our kitchens owe their origin to him. The practical needs and mechanical ingenuity of his own countrymen have caused them to follow his lead with conspicuous success. We have, for example, in our modest little kitchen in the Alps, an American oven which, with the expenditure of an extremely small amount of firewood, heats our baths, cooks our meat, bakes our bread, boils our clothes, and contributes to the warmth and comfort of the house. This arrangement traces its pedigree to Rumford.

In 1796 he founded the historic medal which bears his name. On the 12th of July of that year he wrote thus to Sir Joseph Banks, then President of the Royal Society : 'I take the liberty to request that the Royal Society would do me the honour to accept of 1,000*l.* stock in the Funds of this country, which I have actually purchased, and which I beg leave to transfer to the President, Council, and Fellows of the Royal Society, to the end that the interest of the same may be by them, and by their successors, received from time to time for ever, and the amount of the same applied and given once every second year, as a premium to the author of the most important discovery or useful improvement which shall be made, or published by printing, or in any way made known to the public, in any part of Europe, during the preceding two years, on Heat or Light.'

He adds in a subsequent letter, as further defining his wishes, that the premium should be limited to new discoveries tending to improve theories of Fire, of Heat, of Light, and of Colours, and to new inventions and contrivances by which the generation, and preservation, and management of heat and of light may be facilitated. The device and inscriptions on the medal were determined by a committee. It was resolved ' that the diameter of the medal do not exceed three inches, and that Mr. Milton be employed in sinking the dies of the said medal.' Two medals are always given, one of gold, the other of silver, and a sum of about seventy pounds usually accompanies the medals. Rumford himself was the first recipient of the medal. The second was given to Sir John Leslie, the founder's celebrated rival in the domain of radiant heat. On the same date Rumford presented to the American Academy of Arts and Sciences the same sum for the promotion of the same object. In fact, the letters to Sir Joseph Banks and to the Honourable John Adams, then President of the American Academy, are identical in terms. For a long series of years the American Academy did not consider that the candidates for the medal had reached the level of merit which would justify its award. No award was therefore made ; and in 1829 the Rumford bequest had increased from five thousand to twenty thousand dollars. After some litigation the terms of the bequest were extended to embrace applications of it far beyond the design of the testator. Permission was obtained to apply the fund to the publication of books, or methods of discovery, bearing on the Count's favourite subjects of experiment; and to the aid and reward of scientific workers. Thus, in 1839, Dr. Hare, of Philadelphia, received from the Academy six hundred dollars for his invention of the compound blow-pipe, and his improvements in galvanic apparatus. In 1862 the Rumford medal was awarded to Mr. John B. Ericsson, for his caloric engine; while Mr. Alvan Clark, so celebrated for his improvements of the refracting telescope, and the eminent Dr. John Draper, of the University of New York, have been also numbered among the recipients.

Accompanied by his daughter, Rumford returned to Germany in 1796. 'Three weeks' constant travel; circuitous routes to avoid troops, bad roads, still worse accommodations—passing nights in the carriages for the want of an inn—scantiness of provisions, joined with great fatigue, rendered our journey by no means agreeable.' At Munich they were lodged in the splendid house allotted to the Count. France and Austria were then at war, while Bavaria sought to remain rigidly neutral. Eight days after Rumford's arrival, the Elector took refuge in Saxony. Moreau had crossed the Rhine and threatened Bavaria. After a defeat by the French, the Austrians withdrew to Munich, but found the gates of the city closed against them. They planted batteries on a height commanding the city. According to arrangement with the Elector, Rumford assumed the command of the Bavarian forces, and by his firmness and presence of mind prevented both French and Austrians from entering Munich. A foreigner acting thus was sure to excite jealousy and encounter opposition ; but, despite all this, he was eminently successful in realising his

aims. The consideration in which he was held by the
Elector is illustrated by the fact that he made Miss
Thompson a Countess of the Empire, conferring on her
a pension of 200*l.* a year, with liberty to enjoy it in any
country where she might wish to reside.

The following incident is worth recording. In March
1796, Rumford's daughter, wishing to celebrate his
birthday, chose out of his workhouse a dozen of the
most industrious little boys and girls, dressed them up
in the uniform of that establishment, and robing herself
in white, led them into his room and presented them to
him. He was so much touched by the incident, that
he made her a present of two thousand dollars (400*l.*)
on condition that she should, in her will, apply the
interest of the sum to the clothing every year for ever,
on her own birthday, of twelve meritorious children—
six girls and six boys—in the Munich uniform. The
poor children were to be chosen from her native town,
Concord. Habit must to some extent have blinded
Rumford's eyes to the objection which independent New
Englanders were likely to make to this fantastic apparel.
They bluntly stated their objections, but ' with grateful
hearts ' they nevertheless expressed their willingness to
accept the donation. Nothing further was done during
Rumford's lifetime.

The New England girl, brought up in Concord,
transplanted thence to London, and afterwards to
Munich, was subjected to a trying ordeal. After a
short period of initiation, she appears to have passed
through it creditably. Her writing does not exhibit
any marked qualities of intellect. She was bright,
gossipy, ' volatile,' and she throws manifold gleams
on the details of Rumford's life. He constantly kept
a box at the opera, though he hardly ever went there,
and hired by the year a doctor named Haubenal.
She amusingly describes a quintuple present made to

her by her father soon after her arrival in Munich. The first item was ' a little shaggy dog, as white as snow, excepting black eyes, ears, and nose ' ; the second was a lady named Veratzy, who was sent to teach her French and music; the third was a Catholic priest, named Dillis, who was to be her drawing-master ; the fourth was a teacher of Italian, named Alberti ; and the fifth, the before-mentioned Dr. Haubenal, who was to look after her health. She did not at all like the arrangement. She was particularly surprised and shocked at a doctor's offering his services before they were wanted. ' Said I to myself, Surrounded by people who speak French—and all genteel people speak it at Munich— and knowing considerable of the language already, where is the use of my fatiguing myself with masters ? Music the same.' In fact, the little dog ' Cora ' was the only welcome constituent of the gift.

She describes with considerable spirit a ball which was organised to celebrate her father's birthday. All united to do him honour. Wreaths surrounded his bust; his workhouse children, joined by some children of the nobility, all dressed in white, handed addresses to him, and sang in accompaniment to the swell of music, of which he was passionately fond. All this was arranged without his knowledge, and possibly not without an intention to give dramatic force to a revelation to be made at the time. It was observed that Rumford had singled out from the children a little girl of eight, who accompanied him when he walked, and took her place beside him when he sat. The little girl was his illegitimate child. Sarah, on learning this, threw herself into the dance she had previously declined, and thus whirled away her indignation. Her partner was the young Count Taxis, Rumford's aide-de-camp, between whom and Rumford's daughter a friendly intimacy was obviously growing up. Rumford noticed this, and disapproved of it. Being invited to dinner at the house of the Countess Lerchenfeld, with her father's consent Miss Thompson went. Count Taxis happened to be one of the party, and on hearing this Rumford jumped to the conclusion that a ladies' conspiracy was afoot to counteract his wishes. With a lowering look he taxed his daughter with what he supposed to be an intrigue. At first she could only stare at him in surprise. 'After which, on knowing what it meant, like many young people who laugh when there is nothing to laugh at, an irresistible inclination seized me to laugh.' She gave way to her inclination, 'and it ended in my father's boxing my ears.' She was stunned by the indignity, and 'quitted the room, without making an observation, or trying to appease him by saying I was innocent.'

The Elector put the seal to his esteem for Rumford by appointing him as Plenipotentiary from Bavaria to the Court of London. King George, however, declined to accept him in this capacity. Mr. Paget, the Minister at the Court of Bavaria, was desired ' to lose no time in apprising the Ministers of His Electoral Highness that such an appointment would be by no means agreeable to His Majesty, and that His Majesty relies therefore on the friendship and good understanding which have always hitherto subsisted between himself and the Elector of Bavaria, and that His Highness will have no hesitation in withdrawing it.' The King had made up his mind. ' Should there unexpectedly arise any difficulty about a compliance with the request, which His Majesty is so clearly warranted in making, I am to direct you, in the last resort, to state in distinct terms that His Majesty will by no means consent to receive Count Rumford in the character which has been assigned to him.' The fact of Rumford's being not only a British subject, but that he had actually filled a confidential situation under the British Government, was cited as rendering his appointment peculiarly objectionable. Some correspondence ensued between Lord Grenville and Rumford, but the appointment was not ratified.

Stung by the refusal of King George to accept him as Bavarian Minister, the thought, which had often occurred to him, of returning to his native country now revived. Mr. Rufus King was at that time American Ambassador in London ; and he, by Rumford's desire, wrote to Colonel Pickering, then Secretary of State for the United States, informing him that intrigues in Bavaria, and the refusal of the English king, had caused the Count to decide on establishing himself at, or near, Cambridge, Massachusetts. Mr. King described the Count's intention to live in the character of a German nobleman, renouncing all political action, and devoting himself to literary pursuits. He observed that Rumford had much experience of cannon foundries, and had made important improvements in the mounting of flying artillery. He was, moreover, the possessor of an extensive military library, and wished nothing more ardently than to be useful to his native country. Provision had been made for the institution of a military academy in the United States. This they offered to place under the superintendence of Rumford. 'I am authorised,' said Mr. King, ' to offer you, in addition to the superintendence of the military academy, the appointment of Inspector-General of the Artillery of the United States ; and we shall moreover be disposed to give to you such rank and emoluments as would be likely to afford you satisfaction, and to secure to us the advantage of your service.'

The hour for the final decision approached, but before it arrived another project had laid hold of Rum-

ford's imagination—a project which in its results has
proved of more importance to science, and probably of
more advantage to mankind, than any which this multi-
farious genius had previously undertaken. This project
was the foundation of the Royal Institution of Great
Britain. In answer to the American Ambassador, he says,
' Nothing could have afforded me so much satisfaction
as to have had it in my power to have given to my
liberal and generous countrymen such proof of my sentiments as would, in the most public and ostensible
manner, have evinced, not only my gratitude for the
kind attentions I have received from them, but also
the ardent desire I feel to assist in promoting the pro-
sperity of my native country; but engagements, which
great obligations have rendered sacred and inviolable,
put it out of my power to dispose of my time and services with that unreasoned freedom which would be
necessary in order to enable me to accept of those
generous offers which the Executive Government of
the United States has been pleased to propose to me.'

The climate of Europe, however, did not seem to
suit Rumford's daughter. Possibly also the simple tastes
and habits of her childhood were too deeply ingrained
in her constitution to permit of her deriving any real
enjoyment from the outsided, and apparently noisy life
which she was forced to lead in Munich and London.
Be this as it may, she returned to America, reaching
the port of Boston on October 10, 1799, ' being then
just twenty-five years of age.' Rumford himself
remained in England with the view of realising what
I have called the greatest project of his life—the founding of the Royal Institution.

His ideas on this subject took definite shape in 1799.
They were set forth in a pamphlet of fifty pages, bear-

ing the following lengthy title : ' Proposals for forming
by subscription, in the Metropolis of the British Empire,
a Public Institution for diffusing the Knowledge and
facilitating the general Introduction of Useful Mechanical Inventions and Improvements, and for teaching, by
courses of Philosophical Lectures and Experiments, the
Application of Science to the Common Purposes of Life.'
The introduction to this pamphlet is dated from Rumford's residence in Brompton Row, March 4, 1799.
His aim, he alleges, is to cause science and art to work
together ; to establish relations between philosophers
and workmen ; to bring their united efforts to bear in
the improvement of agriculture, manufactures, com-
merce, and the augmentation of domestic comforts.
He specially dwells on the management of fire, it
being, as he thinks, a subject of peculiar interest to
mankind.   Fuel, he asserted, costs the kingdom more
than ten millions sterling annually, which was much
more than twice what it ought to cost.   Rumford knew
human nature well, and for the greater portion of his
life knew how to appeal to it with effect.   In fact, the
knowledge never failed him, though towards the end
irritability, due to ill-health and crosses of various kinds,
rendered him less able to apply the knowledge than
when he was in the blossom of his prime.   As regards
the success of his new scheme, he urged upon those
with whom he acted the necessity of making the indolent and luxurious take an interest in it.   Such persons,
he says, ' must either be allured or shamed into action.'
Hence, he urges, the necessity of making benevolence
' fashionable.'

It ought to be mentioned that Rumford, at this
time, could count on the sympathy and active support
of a number of excellent men, who, in advance of him,
had founded a ' Society for Bettering the Condition and

Increasing the Comforts of the Poor.' The aid of the
committee of this society was now sought. It was agreed
on all hands that the proposed new Institution would
be too important to permit of its being made an ap-
pendage to any other. It was resolved that it should
stand alone. A committee consisting of eight members
of the above society was, however, appointed to confer
with Rumford regarding his plan. To each member of
this committee he submitted a statement of his views.
These are in part set forth in the title to his pamphlet
already quoted. The aim of the Institution, further-
more, was 'to excite a spirit of improvement among
all ranks of society, and to afford the most effectual
assistance to those who are engaged in the various pursuits of useful industry.' He begged, however, that His
Majesty's Ministers might be informed of the intention
of the founders of the Institution to accept his services.
This he deemed necessary because of his being, in the
first place, a subject of His Majesty, and also, by His
Majesty's special permission, the servant of a foreign
prince. The Government was to be fully informed, not
only as to the general aims, but also of the details of
the scheme. Not till then did he ask for the countenance and support necessary to carry it into execution.

The committee met and ratified Rumford's proposals. They agreed that subscribers of fifty guineas
each should be the perpetual proprietors of the Institution ; that a contribution of ten guineas should secure
the privileges of a life subscriber ; whilst a subscription
of two guineas should constitute an annual subscriber.
Besides other important rights, each proprietor was to
receive two transferable tickets, admitting him to every
part of the Institution, and to all the lectures and experiments. Each life subscriber was to receive one ticket,
not transferable, securing free admission to every part

of the establishment, and to all lectures and experiments. An annual subscriber had a single ticket for a single year, but might at any time become a life subscriber by the additional payment of eight guineas. The managers, nine in number, were to be chosen by ballot by the proprietors. The managers were to be unpaid, and, without any pecuniary advantage to themselves, they were held solemnly pledged to the faithful discharge of their duties. Three were to constitute a quorum, but in special cases six were required. A Committee of Visitors was also appointed, the same in number as the Committee of Managers, and holding office for the same number of years.

The managers were to devote the surplus funds of the Institution to the purchase of models of inventions and improvements in the mechanical arts, a room in the Institution being devoted to the reception of them. The room still exists, and, though diverted from its original purpose, is still called 'the Model Room.' A general meeting of the proprietors was held at the house of Sir Joseph Banks, in Soho Square, on March 7, 1799. Fifty-eight persons, comprising men of distinction in science, members of Parliament and of the nobility, including one bishop, were found to have qualified as proprietors by the subscription of fifty guineas each. The prelate was the Bishop of Durham. The Committee of Managers was chosen, and they held their first meeting at the house of Sir Joseph Banks on March 9, 1799. Mr. Thomas Bernard, one of the most active members of the society from whose committee the first managers were chosen, was appointed Secretary. To Rumford and Bernard was delegated the duty of preparing a draught of a charter; while Earls Morton and Spencer, Sir Joseph Banks, and Mr. Pelham, were requested 'to lay the proposals before His Majesty, the Royal Family, the Ministers, the great officers of State, the members of both Houses of Parliament, of the Privy Council, and before the twelve judges.'

On January 13, 1800, the Royal Seal was attached to the Charter of the Institution. In the same year was published, in quarto form, ' The Prospectus, Charter, Ordinance, and Bye-laws of the Royal Institution of Great Britain.' The King was its Patron, and the first officers of the Institution were appointed by him. The Earl of Winchilsea was President. Lord Morton, Lord Egremont, and Sir Joseph Banks were Vice-Presidents. The managers, chosen by sealed ballot by the proprietors, were divided into three classes of three each ; the first class serving for one, the second for two, and the third for three years. The Earls of Bessborough, Egremont, and Morton, respectively, headed the lists of the three classes. Rumford himself was appointed to serve for three years. The three lists of Visitors were headed by the Duke of Bridgewater, Viscount Palmerston, and Earl Spencer respectively. That Rumford possessed the power of persuasion, and the infection of enthusiasm, is sufficiently demonstrated by this powerful list. But neither persuasion nor enthusiasm might have been found availing had not his actual achievements in Bavaria occupied the background. The first Professor of Natural Philosophy and Chemistry was Dr. Thomas Garnett, while the first Treasurer was Mr. Thomas Bernard. But this was not enough. A home and foreign secretary, legal counsel, a solicitor and a clerk, were added to the list. One rule established at this time has been adhered to with great fidelity to the present day. No political subject was to be mentioned in the lectures.

In a somewhat florid style Rumford (for he was

obviously the writer) descants on the name and objects
of the new project. The word Institution is chosen
because it had been least used previously, and because
it best indicates the objects of the new society. The
influence of the mechanical arts on the progress of
civilisation and refinement is pointed out, and illustrated
by reference to nations, provinces, towns, and even villages, which thrive in proportion to the activity of their
industry. ' Exertion quickens the spirit of invention,
makes science flourish, and increases the moral and
physical powers of man.' The printing-press, navigation, gunpowder, the steam-engine, are referred to as
having changed the whole course of human affairs.
The slowness with which improvements make their
way among workmen is ascribed to the influence of
habit, prejudice, suspicion, jealousy, dislike of change,
and the narrowing effect of the subdivision of work
into many petty occupations. But slowness is also due
to the greed for wealth, the desire for monopoly, the
spirit of secret intrigue exhibited among manufacturers.
Between these two the philosopher steps in, whose business it is ' to examine every operation of Nature and
art, and to establish general theories for the direction
and conducting of future processes.' But philosophers
may become dreamers, and they have therefore habitually to be called back to the study of practical questions
which bear upon the ordinary pursuits of life. Science
and practice are in short to interact, to the advantage
of both. This object may be promoted by the offering
of premiums, after the manner of the Society of Arts,[1]
by the granting of patents ; and, finally, by the method
of the new Institution—the diffusion of the knowledge
of useful mechanical inventions, and their introduction
into life.

[1] Founded in 1753.

One of the first practical steps taken towards the realisation of these ideas was the purchase of the house, or rather houses, in Albemarle Street in which we are now assembled, and their modification to suit the objects in view. Rumford's obvious intention was to found an Institute of Technology and Engineering. Mere description was not sufficient. He demanded something visible and tangible, and therefore proposed that the Institution should be made a repository for models of all useful contrivances and improvements : cottage fireplaces and kitchen utensils ; kitchens for farm-houses and for the houses of gentlemen ; a laundry, including boilers, washing, ironing, and drying-rooms ; German, Swedish, and Russian stoves ; open chimney fireplaces, with ornamental grates ; ornamental stoves ; working models ' of that most curious and most useful machine, the steam-engine ' ; brewers' boilers ; distillers' coppers ; condensers ; large boilers for hospitals ; ventilating apparatus in hot-houses ; lime-kilns ; steam-boilers for preparing food for stall-fed cattle ; spinning-wheels ; looms ; agricultural implements ; bridges of various constructions ; human food ; clothing ; houses; towns ; fortresses ; harbours ; roads ; canals ; carriages ; ships ; tools ; weapons ; &c. Chemistry was to be applied to soils, tillage, and manures ; to the making of bread, beer, wine, spirits, starch, sugar, butter, and cheese ; to the processes of dyeing, calico-printing, bleaching, painting, and varnishing ; to the smelting of ores ; the formation of alloys ; to mortars, cements, bricks, pottery, glass, and enamels. Above all, ' the phenomena of *light* and *heat*—those great powers which give life and energy to the universe—powers which, by the wonderful process of combustion, are placed under the command of human beings—will engage a profound interest.'

In reference to the alleged size of the bed of Og, the

king of Basan, Bishop Watson proposed to Tom Paine
the problem to determine the bulk to which a human
body may be augmented before it will perish by its own
weight. As regards the projected Institution, Rumford
surely had passed this limit, and by the ponderosity of
his scheme had ensured either the necessity of change
or the certainty of death.   In such an establishment
Davy was sure to be an iconoclast.   He cared little for
models—not even for the apparatus with which his own
best discoveries were made, but incontinently broke it
up whenever he found it could be made subservient to
further ends.

The 'Journal of the Royal Institution' was established
at this time, and published under Rumford's direction.
No private advertisements were to appear in it, but it was
to be sold for threepence when its contents amounted
to eight pages, and for sixpence when they amounted to
sixteen.   The experiments and experimental lectures of
Davy were then attracting attention.   Rumours of the
young chemist reached Rumford through Mr. Under-
wood and Mr. James Thompson.   At Rumford's request
Davy came to London.   His life at the moment was
purely a land of promise, but Rumford had the sagacity
to see the promise, and the wisdom to act upon his insight.   Nor was his judgment rapidly formed; for several
interviews, doubtless meant to test the youth, preceded
his announcement to Davy, on February 16, 1801, of the
resolution of the Managers, ' That Mr. Humphry Davy
be engaged in the service of the Royal Institution, in
the capacity of Assistant Lecturer in Chemistry, Director of the Chemical Laboratory, and Assistant Editor of
the Journals of the Institution ; and that he be allowed
to occupy a room in the house, and be furnished with
coals and candles, and that he be paid a salary of 100
guineas per annum.'   Rumford, moreover, held out to

Davy the prospect, if he devoted himself entirely and permanently to the Institution, of becoming, in the course of two or three years, full Professor of Chemistry, with a salary of 300*l*. per annum, 'provided,' he adds, 'that within that period you shall have given proofs of your fitness to hold that distinguished situation.' This promise of the professorship in two or three years was ominous for Dr. Garnett, between whom and the Managers differences soon arose which led to his withdrawal from the Institution.

Davy began his duties on Wednesday, March 11, 1801. He was allowed the room adjoining that occupied by Dr. Garnett, to whom he was to refund the expenses incurred in furnishing the room. The committee of expenditure paid to Dr. Garnett 20*l*. 2*s*. 3*d*. for a new Brussels carpet, and 17*l*. 6*s*. for twelve chairs, the carpet and chairs being transferred to the room occupied by the Managers. 'Count Rumford reported further that he had purchased cheaper a second-hand carpet for Mr. Davy's room, together with such other articles as appeared to him necessary to render the room habitable, and among the rest a new sofa-bed, which, in order that it may serve as a model for imitation, has been made complete in all its parts.'

The name of a man who has no superior in its annals now appears for the first time in connection with the Institution. Here also the sagacity of Rumford was justified by events. At the suggestion of Sir Joseph Banks he had an interview with Dr. Thomas Young, destined to become so illustrious as the decipherer of the Egyptian hieroglyphics, and, by the discovery of Interference, the founder of the undulatory theory of light. It was proposed to him, by Rumford, to accept an engagement as Professor of Natural Philosophy in the Institution, as Editor of its Journals, and as super-

intendent of the house, at a salary of 300*l.* per annum.
Young accepted the appointment, and the Managers
confirmed it by resolution on August 3, 1801 :—
' Resolved, that the Managers approve of the measures
taken by Count Rumford ; and that the appointment of
Dr. Young be confirmed.'

Rumford's health fluctuated perpetually, and it was
said at the time that this was due in some measure to
the fanciful notions he entertained, and acted on, with
regard to diet and exercise. But Dr. Young affirms
that his habits in these respects were guided by his
physicians.

Many years ago, wishing to supplement my knowledge of the Turkish bath, I referred to a paper of
Rumford's which gave an account of a visit to Harrogate
and his experience there. According to the rules of the
place he had his bath in the evening, and went to bed
immediately afterwards. He found himself restless and
feverish ; the bath, indeed, seemed to do him more harm
than good. An observant fellow-lodger had made, and
had corrected the same experience. Acting on his
advice, Rumford took his bath two hours before dinner,
engaging afterwards in his usual work, or going out to
have a blow on the common. So far from suffering
chill from this exposure, he found himself invigorated
by it. My own experience, I may say, corroborates all
this. Rumford took the senses of man as he found
them, and tried to enhance the gratifications thence
derived :—' To increase the pleasure of a warm bath
he suggests the burning of sweet-scented woods and
aromatic gums and resins in small chafing-dishes in the
bathing-rooms, by which the air will be perfumed with
the most pleasant odours.' He spiritedly defends this
counsel :—' Effeminacy is no doubt very despicable,
especially in a person who aspires to the character and

virtues of a man. But I see no cause for calling anything effeminate which has no tendency to diminish
either the strength of the body, the dignity of the
sentiments, or the energy of the mind. I see no good
reason for considering those grateful aromatic perfumes,
which in all ages have been held in such high estimation, as a less elegant or less rational luxury than
smoking tobacco or stuffing the nose with snuff.'

Rumford, for a year or so, occupied rooms in the
Institution, but his private residence was in Brompton
Row, described by his friend Pictet as being about a mile
from London. Grass and trees grew in front of the
house. The windows had a double glazing, and outside
were placed vases of flowers and odorous shrubs. Pictet,
who was Rumford's guest in 1801, minutely describes
the whole arrangement of the house. Into Rumford's
working-room, which overlooked the country, the light
came through a set of windows arranged on the arc of a
circle. The window-sills were arranged with flowers
and shrubs, so that you might suppose yourself to be in
the country, close to a garden bordered by a park.
Pictet goes on to describe the various strokes of ingenuity shown in the management of the fuel and fire-
places. The beds, moreover, were disguised as elegant
sofas. Under each sofa were two deep drawers containing the bedding and other night-gear, all of which
were hidden by a fringed valance. At night the sofa
was converted in a few minutes into an excellent bed,
while in the morning, with equal rapidity, the bed was
transformed into an ornamental piece of furniture.
Pictet occupied one-half of the charming dwelling.
Perfect freedom was given and enjoyed, and the learned
Genevese always tried to arrange his day's work so that
he might, if possible, engage his friend on some subject
of research common to them both.

A portion of the motive force of a man of Rumford's temperament may be described as irritability. During the possession of physical vigour and sound health, this force is held firmly by the will and directed by intelligence and tact. But when health slackens and physical vigour subsides, irritability often becomes an energy wanting adequate control. Rumford's success in managing all manner of men in Bavaria illustrates his pliancy as much as his strength. But before he started the Royal Institution his health had given way, and 'temper,' it is to be feared, had got the upper hand. In point of intellect, moreover, he came then into contact with people of larger calibre and more varied accomplishments than he had previously met. He could hardly count upon the entire sympathy of Young and Davy, though I believe he remained on friendly terms with them to the end. They were gems of a different water, if I may use the term, from Rumford. The chief object of his fostering care was mechanical invention, as applied to the uses of life. The pleasures of both Young and Davy lay in another sphere. To them science was an end, not a means to an end. The getting at the mind of Nature, and the revealing of that mind in great theories, were the objects of their efforts, and formed the occupation of their lives. Had they been as enthusiastic as Rumford himself in Rumford's own direction, the three united would probably have daunted opposition, and for a somewhat longer time endeavoured to realise Rumford's dream. But differences arose between him and the other Managers. 'It is very clear to me,' writes Dr. Bence Jones to Dr. Ellis, 'that Count Rumford fell out with Mr. Bernard and with Sir John Hippesley. The fact was that Rumford's idea of workshops and kitchen, industrial school, mechanics' institution, model exhibition, social clubhouse, and scientific committees to do everything, was much too big and unworkable for a private body, and was fitted only for an absolute and wealthy Government.' In 1803 Dr. Bence Jones informs us that difficulties were gathering round the Institution, and it was even proposed to sell it off. Rumford had quitted London and gone to Paris. By Davy's aid, Mr. Bernard and Sir John Hippesley carried on the work, but in a fashion different from that contemplated by Rumford —that is to say, ' without workshops, or mechanics' institute, or kitchen, or model exhibition.' The place of these was taken by experimental and theoretical re-searches, which, instead of dealing with things already achieved, carried the mind into unexplored regions of Nature, forgetful, if not neglectful, whether the dis-coveries made in that region had or had not a bearing on the arts, comforts, or necessities of material life.

Rumford and his Institution had to bear the brunt of ridicule, and he felt it; but men of ready wit have not abstained from exercising it on societies of greater age and higher claims. Shafts of sarcasm without number have been launched at the Royal Society. It is perfectly natural for persons who have little taste for scientific inquiry and less knowledge of the methods of Nature, to feel amused, if not scandalised, by the ap-parently insignificant subjects which sometimes occupy the scientific mind. They are not aware that in science the most stupendous phenomena often find their suggestion and interpretation in the most minute— that the smallest laboratory fact is connected by in-dissoluble ties with the grandest operations of Nature. Thus, the iridescences of the common soap-bubble, subjected to scientific analysis, have emerged in the conclusion that stellar space is a *plenum* filled with a material substance, capable of transmitting motion

with a rapidity which would girdle the equatorial earth
eight times in a second ; while the tremors of this substance, in one form, constitute what we call light, and,
in all forms constitute what we call radiant heat. Not
seeing this connection between great and small ; not
discerning that as regards the illustration of physical
principles there is no great and no small, the wits, considering the small contemptible, permitted sarcasm to
flow. But these things have passed away, otherwise it
would not be superfluous to remind this audience, as
a case in point, that the splendour which in the form
of the electric light now falls upon our squares and
thoroughfares, has its germ and ancestry in a spark
so feeble as to be scarcely visible when first revealed
within the walls of this Institution.

It is with reluctance that I take the slightest exception to what my American friends have written
regarding Rumford and his achievements. But what
they have written induces me to assure them that the
scientific men of England are not prone to stinginess
in recognising the merits of their fellow-labourers in
other lands ; and had Rumford, instead of accom-
plishing none of his work in the land of his birth,
accomplished the whole of it there, his recognition
among us here would not be less hearty than it is now.
As things stand, national prejudice, if it existed, might
be expected to lean to Rumford's side. But no such
prejudice exists, and to write as if it did exist is a
mistake. In reference to myself, Dr. Ellis, gently but
still reproachfully, makes the following remark : —
' Professor Tyndall, in his work on " Heat," has but
moderately recognised the claims and merit of Rumford, when, after largely quoting from his essay, he adds,
" When the history of the dynamical theory of heat is

written, the man who, in opposition to the scientific belief of his time, could experiment, and reason upon experiment, as did Rumford in the investigation here referred to, cannot be lightly passed over." ' In my opinion, the most dignified and impressive way of dealing with labours like those of Rumford, is to show by simple quotations, well selected, what their merits are. This I did in the book referred to by Dr. Ellis, which was published at least eight years in advance of his. But the expression of my admiration for Rumford was not confined to the passage above-quoted, which is taken from the appendix to one of my lectures. In that lecture I drew attention to Rumford's labours in the following words :—' I have particular pleasure in directing the reader's attention to an abstract of Count Rumford's memoir on the generation of heat by friction, contained in the appendix to this lecture. Rumford in this memoir annihilates the material theory of heat. Nothing more powerful on the subject has since been written.'

But I must not go too far, nor suffer myself to dwell with one-sided exclusiveness upon the merits of Rumford. The theoretic conceptions with which he dealt were not his conceptions, but had been the property of science long prior to his day. This, I fear, was forgotten when the following claim for Rumford was made by a writer who has done excellent service in diffusing sound science among the people of the United States :[1]— ' He was the man who first took the question of the nature of heat out of the domain of metaphysics, where it had been speculated upon since the time of Aristotle, and placed it upon the true basis of physical experiment.' The writer of this passage could hardly, when

[1] The late Dr. Youmans.

he wrote it, have been acquainted with the experiments
and the reasonings of Boyle and Hooke, of Leibnitz
and Locke. As regards the nature of heat, these men
were quite as far removed from metaphysical subtleties
as Rumford himself. They regarded heat as 'a very
brisk agitation of the insensible parts of an object
which produces in us that sensation from whence we
denominate the object hot; so what in our sensation
is heat, in the object is nothing but motion.' Locke,
from whom I here quote, and who merely expresses
the ideas previously enunciated by Boyle and Hooke,
gives his reasons for holding this theoretic conception.
'This,' he says, 'appears by the way heat is produced,
for we see that the rubbing of a brass nail upon a
board will make it very hot; and the axle-trees of carts
and coaches are often hot, and sometimes to a degree
that it sets them on fire, by the rubbing of the naves
of the wheels upon them. On the other side, the utmost degree of cold is the cessation of that motion of
the insensible particles which to our touch is heat.'
The precision of this statement could not, within its
limits, be exceeded at the present day.

There is a curious resemblance, moreover, between
one of the experiments of Boyle, and the most celebrated experiment of Rumford. Boyle employed three
men, accustomed to the work, to hammer nimbly a
piece of iron. They made the metal so hot, that it
could not be safely touched. As in the case of Rumford, people were looking on at this experiment, and
Boyle's people, like those of Rumford, were struck
with wonder, to see the sulphur of gunpowder ignited
by heat produced without any fire. Hooke is equally
clear as regards the nature of heat, and like Rumford
himself, but more than a century before him, he compares the vibrations of heat with sonorous vibrations.

That Rumford went beyond these men is not to be denied. It could not be otherwise with a spirit so original and penetrating. But to speak of the space between him and Aristotle as if it were a scientific vacuum is surely a mistake.

While in Paris, Rumford made the acquaintance of Madame Lavoisier, a lady of wealth, spirit, and social distinction ; and, it is to be added, a lady of temper. Her illustrious husband had suffered under the guillotine on May 8, 1794; and inheriting his great name, together with a fortune of three million francs, she gathered round her, in her receptions, the most distinguished society of Paris. She and Rumford became friends, the friendship afterwards passing into what was thought to be genuine affection. The Elector of Bavaria took great interest in Rumford's projected marriage, and when that consummation came near, settled upon him an annuity of 4,000 florins. Before their marriage he was joined by Madame Lavoisier at Munich, whence they made a tour to Switzerland. In a letter to his daughter he thus describes his bride-elect: ' I made the acquaintance of this very amiable woman in Paris, who, I believe, would have no objection to having me for a husband, and who in all respects would be a proper match for me. She is a widow without children, never having had any ; is about my own age [she was four years younger than Rumford], enjoys good health, is very pleasant in society, has a handsome fortune at her own disposal, enjoys a most respectable reputation, keeps a good house, which is frequented by all the first philosophers and men of eminence in the science and literature of the age, or rather of Paris; and, what is more than all the rest, is goodness itself.' He goes on to describe her as

L 2

having been very handsome in her day, 'and even now at forty-six or forty-eight is not bad-looking.' He describes her as rather *embonpoint*, with a great deal of vivacity, and as writing incomparably well.

Before the marriage could take place he was obliged to obtain from America certificates of his birth, and of the death of his former wife. All preliminaries having been arranged, Count Rumford and Madame Lavoisier were married in Paris, on October 24, 1805. He describes the house in which they lived, Rue d'Anjou, No. 39, as a paradise. ' Removed from the noise and bustle of the street, facing full to the south, in the midst of a beautiful garden of more than two acres, well planted with trees and shrubbery. The entrance from the street is through an iron gate by a beautiful winding avenue well planted, and the porter's lodge is by the side of this gate ; a great bell to be rung in case of ceremonious visits.' Long after this event Rumford's daughter commented on it thus :— ' It seems there had been an acquaintance between these parties of four years before the marriage. It might be thought a long space of time for perfect acquaintance. But, " ah Providence ! thy ways are past finding out." '

In a letter written to his daughter two months after his marriage, he describes their style of living as really magnificent; his wife as exceedingly fond of company, in the midst of which she makes a splendid figure. She seldom went out, but kept open house to all the great and worthy. He describes their dinners and evening teas, which must have been trying to a man who longed for quiet. He could have borne the dinners, but the teas and their gossip annoyed him. Instead of living melodious days, his life gradually became a discord, and on January 15, 1806, he confides to his daughter, as a family secret, that he is ' not at all sure that two certain persons were not wholly mistaken in their marriage as to each other's character.' The *dénouement* hastened ; and on the first anniversary of his marriage he writes thus to his daughter :—' My dear child, — This being the first year's anniversary of my marriage, from what I wrote two months after it you will be curious to know how things stand at present. I am sorry to say that experience only serves to confirm me in the belief that in character and natural propensities Madame de Rumford and myself are totally unlike, and never ought to have thought of marrying. We are, besides, both too independent in our sentiments and habits of life to live peaceably together—she having been mistress all her days of her actions, and I, with no less liberty, leading for the most part the life of a bachelor. Very likely she is as much disaffected towards me as I am towards her. Little it matters with me, but I call her a female dragon—simply by that gentle name ! We have got to the pitch of my insisting on one thing and she on another.'

On the second anniversary of his marriage, matters were worse. The quarrels between him and Madame had become more violent and open, and having used the word quarrels to his daughter, he gives the following sample of them :—' I am almost afraid to tell you the story, my good child, lest in future you should not be good ; lest what I am about relating should set you a bad example, make you passionate, and so on. But I had been made very angry. A large party had been invited I neither liked nor approved of, and invited for the sole purpose of vexing me. Our house being in the centre of the garden, walled around, with iron gates, I put on my hat, walked down to the porter's

lodge, and gave him orders, on his peril, not to let
any one in. Besides, I took away the keys. Madame
went down, and when the company arrived, she talked
with them,—she on one side, they on the other, of the
high brick wall. After that she goes and pours boiling
water on some of my beautiful flowers.'

Six months later, the sounds of lamentation and
woe are continued. There was no alteration for the
better. He thought of separation, but the house and
garden in the Rue d'Anjou being a joint concern, legal
difficulties arose as to the division of it. ' I have suffered,' he says to his daughter, ' more than you can
imagine for the last four weeks; but my rights are
incontestable, and I am determined to maintain them.
I have the misfortune to be married to one of the
most imperious, tyrannical, unfeeling women that ever
existed, and whose perseverance in pursuing an object
is equal to her profound cunning and wickedness in
framing it.' He purposed taking a house at Auteuil.
It would be unfortunate if, notwithstanding all the
bounties of the King of Bavaria, he could not live
more independently than with this unfeeling, cunning,
tyrannical woman. ' Alas! little do we know people
at first sight! Do you preserve my letters ? You will
perceive that I have given very different accounts of
this woman, for *lady* I cannot call her.' He describes
his habitation as no longer the abode of peace. He
breakfasts alone in his apartment, while to his infinite
chagrin most of the visitors are his wife's determined
adherents. He is sometimes present at her tea-parties,
but finds little to amuse him. ' I have waited,' he
says (which we may doubt), ' with great, I may say
unexampled patience, for a return of reason and a
change of conduct, but I am firmly resolved not to be
driven from my ground, not even by disgust. A sepa-

ration,' he adds, ' is unavoidable, for it would be highly improper for me to continue with a person who has given me so many proofs of her implacable hatred and malice.'

The lease of the villa at Auteuil was purchased by Rumford in 1808. The separation between him and his wife took place 'amicably' on the 13th of June, 1809. Ever afterwards, however, anger rankled in his heart. He never mentions his wife but in terms of repugnance and condemnation. His release from her fills him with unnatural jubilation. On the fourth anniversary of his wedding-day he writes to his daughter :— ' I make choice of this day to write to you, in reality to testify joy, but joy that I am away from her.' On the fifth anniversary he writes thus :—' You will perceive that this is the anniversary of my marriage. I am happy to call it to mind that I may compare my present situation with the three and a half horrible years I was living with that tyrannical, avaricious, unfeeling woman.' The closing six months of his married life he describes as a purgatory sufficiently painful to do away with the sins of a thousand years. Rumford, in fact, writes with the bitterness of a defeated man. His wife retained her friends, while he, who a short time previously had been the observed of all observers, found himself practically isolated. This was a new and bitter experience, the thought of which, pressing on him continually, destroyed all magnanimity in his references to her.

From 1772 to 1800, Rumford's house at Auteuil had been the residence of the widow of a man highly celebrated in his day as a freethinker, but whom Lange describes as 'the vain and superficial Helvetius.' It is also the house in which, in the month of January 1870, the young journalist Victor Noir was shot dead by Prince

Pierre Bonaparte. Here, towards the end of 1811, the
Count was joined by his daughter. They found pleasure
in each other's company, but the affection between them
does not appear to have been intense. In his conversations with her the source of his bitterness appears. ' I
have not,' he says, ' deserved to have so many enemies ;
but it is all from coming into France, and forming this
horrible connection. I believe that woman was born to
be the torment of my life.' The house and gardens were
beautiful ; tufted woods, winding paths, grapes in abund-
ance, and fifty kinds of roses. Notwithstanding his
hostility to his wife, he permitted her to visit him on
apparently amicable terms. The daughter paints her
character as admirable, ascribing their differences to
individual independence arising from having been accustomed to rule in their respective ways : ' It was a fine
match, could they but have agreed.' One day in driving
out with her father, she remarked to him how odd it
was that he and his wife could not get on together,
when they seemed so friendly to each other, adding that
it struck her that Madame de Rumford could not be in
her right mind. He replied bitterly, ' Her mind is,
as it has ever been, to act differently from what she
appears.'

The statesman Guizot was one of Madame de
Rumford's most intimate friends, and his account of her
and her house is certainly calculated to modify the
account of both given by her husband. Rumford became
her guest at a time when he enjoyed in public 'a splendid
scientific popularity. His spirit was lofty, his conversation was full of interest, and his manners were marked
by gentle kindness. He made himself agreeable to
Madame Lavoisier. He accorded with her habits, her
tastes, one might almost say with her reminiscences.

. . . . She married him, happy to offer to a distinguished man a great fortune and a most agreeable existence.' Guizot, who writes thus, goes on to state that their characters and temperaments were incompatible. They had both grown to maturity accustomed to independence, which it is not always easy even for tender affection to stifle. The lady had stipulated, on her second marriage, that she should be permitted to retain the name of Lavoisier, calling herself Madame Lavoisier de Rumford. This proved disagreeable to the Count, but she was not to be moved from her determination to retain the name. ' I have,' she says, ' at the bottom of my heart a profound conviction that M. de Rumford will not disapprove of me for it, and that on taking time for reflection, he will permit me to continue to fulfil a duty which I regard as sacred.' Guizot adds that the hope proved deceptive, and that ' after some domestic agitations, which M. de Rumford, with more of tact, might have kept from becoming so notorious, a separation became necessary.' Guizot describes her dinners and receptions during the remaining twenty-seven years of her life as delightful. Cultivated intellects, piquant and serious conversation, excellent music, freedom of mind and tongue, without personal antagonism or political bias, ' license of thought and speech without any distrust or disquiet as to what Authority might judge or say—a privilege then more precious than any one to-day imagines, just as one who has breathed under an air-pump can best appreciate the delight of free respiration.' One cannot, however, forget the pouring of boiling water over the ' beautiful flowers.'

The ' Gentleman's Magazine ' for 1814 describes the seclusion in which Rumford's later days were spent.

After the death of the illustrious Lagrange, he saw but two or three friends, nor did he attend the meetings of the National Institute, of which he was a member. Cuvier was then its perpetual secretary, and for him Rumford entertained the highest esteem. He differed from Laplace on the question of 'surface-tension,' and dissent from a man then standing so high in the mathematical world was probably not without its penal consequences. Rumford always congratulated himself on having brought forward two such celebrated men as the Bavarian general Wieden, who was originally a lawyer or land steward, and Sir Humphry Davy. The German, French, Spanish, and Italian languages were as familiar to the Count as English. He played billiards against himself; he was fond of chess, which however made his feet like ice and his head like fire. The designs of his inventions were drawn by himself with great skill; but he had no knowledge of painting and sculpture, and but little feeling for them. He had no taste for poetry, but great taste for landscape-gardening. In later life his habits were most abstemious, and it is said that his strength was in this way so reduced as to render him unable to resist his last illness. After three days' suffering from nervous fever he succumbed on August 21, 1814, when he was on the eve of returning to England. He was buried in the small cemetery of Auteuil, which has since been disused as a place of burial. The grave, says Dr. Ellis, is marked by a horizontal stone—*une pierre tumulaire*—and by a perpendicular monument 6 feet high, 6 feet in breadth, and $3\frac{1}{2}$ feet in thickness. Both are of marble, and bear inscriptions as follows. That on the monument is :—

À la Mémoire
de
BENJAMIN THOMPSON,
Comte de Rumford,
né en 1753, à Concord[1] près Boston,
en Amérique,
mort le 21 Août, 1814, à Auteuil.
Physicien célèbre,
Philanthrope éclairé,
ses découvertes sur la lumière
et la chaleur
ont illustré son nom.
Ses travaux pour améliorer
le sort des pauvres
le feront toujours chéri
des amis de l'humanité.

The flat stone is thus inscribed :—

En Bavière
Lieutenant-Général,
Chef de l'Etat-Major Général,
Conseiller d'Etat,
Ministre de la Guerre.
En France
Membre de l'Institut,
Académie des Sciences.

RUMFORD'S SCIENTIFIC WORK.

As a factor in human affairs, Rumford ascribed to gunpowder a dominant importance. No other invention had exercised so great an influence. Hence the arduous labour he expended in determining its action. At Stoneland Lodge, the country seat of Lord George Germain, in the year 1778, his inquiries into the force and applications of gunpowder began. He directed his attention to the position of the vent, the weight and pressure of the charge, its bursting power, the quickness of combustion, the weight and velocity of the projectile, the effect of windage, and to many other matters of interest to the gunner. On all these questions he threw

[1] Ought to be Woburn.

important light. The velocity was determined in two ways: first, by the ballistic pendulum, invented by his predecessor and namesake, Benjamin Robins; and secondly, from the recoil of the gun itself. The ballistic pendulum is a heavy mass, so suspended as to be capable of free oscillation. Against it the bullet is projected, and from the weight of the bullet, the weight of the pendulum, and the arc, or distance, through which it is urged by the bullet, the velocity of the latter may be calculated.

To determine the recoil of the gun, he had it suspended by a bifilar arrangement, which permitted it to swing back when it was fired. Action and reaction being equal, the momentum of the gun was the momentum of the bullet on leaving the gun, and from the weight of the piece, and the arc of recoil, the velocity of the bullet was computed. The agreement between the results obtained by these two methods was in many cases remarkable. Until quite recently, Rumford's experiments on the force of gunpowder were considered to be the best extant. A mind so observant could not fail to notice the heating effects produced by the percussion of the bullet against its target, and by the jar of the gun at the moment of its discharge. By such facts he was naturally led to reflect on that connection between mechanical power and heat which he afterwards did so much to illustrate and develop.

The phenomena both of light and heat fascinated him; and we accordingly find him from time to time abandoning practical aims, and seeking for knowledge which had no apparent practical outcome. Thus we see him experimenting on the action of green vegetables and other matters upon light, or rather the action of light on the green leaves of plants. From this inquiry he turned to estimate the quantities of moisture taken up by different substances in humid air. Sheep's wool he found to be the most absorbent, while cotton wool and ravellings of fine linen were among the least. These experiments he regarded as of the highest importance, as they explained, to his mind, the salubrity of flannel when worn next the skin. Its healthfulness he ascribed to its power of taking up the moisture of the body, sensible and insensible, and dispersing it by evaporation in the air.

The propagation of heat in fluids was but imperfectly understood when Rumford took the subject up. In various parts of his writings he dwells on the importance of what he calls accidental observations, deeming them more fruitful than those which have sprung from the more recondite thoughts of the philosopher. But accidents, however numerous, if they fail to reach the proper soil are barren. Rumford ascribed to accident the investigations now referred to. He had been experimenting upon liquids, employing bulbs of copper with glass tubes attached to them. On one occasion, having filled his bulb and tube with spirits of wine, and heated the liquid, he placed it to cool in a window where the sun happened to shine upon it. Particles of dust had found their way into the spirit, and the sun, shining on these particles, made their motions vividly apparent. Along the axis of his tube the illuminated particles rose ; along its sides they fell, thus making manifest the currents within the liquid. The reason of this circulation is obvious enough. The glass tube in contact with the cold air had its temperature lowered. The glass drew heat from the liquid in contact with it, which thereby being rendered more dense, fell along the sides of the tube, while, to supply its place, the lighter liquid rose along the axis. The motion here described is exactly that of the great geyser of Iceland. The water falls along the sides of the geyser tube, and rises along the axis. In this way, then, heat is propagated through liquids. It is a case of bodily transport by currents, and not one of true conduction from molecule to molecule.

It immediately occurred to Rumford to hamper this motion of convection. He called to mind an observation he had made at Baiæ, where the water of the sea being cool to the touch, the sand a few inches below the water was intolerably hot. This he ascribed to the impediment offered by the sand to the upward diffusion of the heat. The length of time required by stewed apples to cool also occurred to him. He had frequently burnt his mouth by a spoonful of apple taken from the centre of a dish after the surface had become cool. He devised thermometers with a view of bringing his notions to an experimental test. With pure water he compared water slightly thickened with starch, water containing eiderdown, and stewed apples bruised into a pulp which consisted almost wholly of water. In all cases he found the propagation of heat impeded, and cooling retarded, by everything that prevented the formation of currents. As he pursued his inquiries, the idea became more and more fixed in his mind that convection is the *only* means by which heat is diffused in liquids. He denied them all power of true conduction, and though his experiments did not, and could not prove this, they did prove that in the propagation of heat through the liquids he examined, which were water, oil, and mercury, conduction played an extremely subordinate part.

Rumford changes from time to time the tone of the philosopher for that of the preacher. He seems filled with religious enthusiasm on contemplating what he holds to be the wisdom and benevolence displayed in the arrangement of the physical world. One fact in

particular excited this emotion. De Luc had pointed
out that when water is cooled, it shrinks in volume, until
it reaches a temperature of about 40° Fahr. At this point
it ceases to contract, and expands when cooled still further. The expansion we now know to be due to incipi-
ent crystallisation, or freezing, which, when it once sets
in, greatly, and suddenly, enhances the expansion. A consequence of this is that ice floats as a lighter body upon
water. This fact riveted the attention of Rumford, and
its obvious consequences filled him with the enthusiasm
to which I have referred. He was strong, but untrained,
and his language was not always such as a truly disciplined man of science would employ. ' Let me,' he
says, ' beg the attention of my reader, while I endeavour
to investigate this most interesting subject, and let me
at the same time bespeak his candour and indulgence.
I feel the danger to which a mortal exposes himself who
has the temerity to undertake to explain the designs of
Infinite Wisdom. The enterprise is adventurous, but it
cannot surely be improper.'

He ' explains ' accordingly ; and notwithstanding
his professed humility, does not hesitate to brand those
who fail to see with his eyes as ' degraded, and quite
callous to every ingenuous and noble sentiment.' He
indulges in excursions of the imagination to show the
misfortunes that would accrue if the arrangement of the
world had been different from what it is. ' Had not
Providence, in a manner which may be well considered
as miraculous,' stopped the contraction of water before
it reached its freezing point, and caused it to expand
afterwards, a single winter would freeze every fresh-water
lake within the polar circle to a vast depth, ' and it is
more than probable that the regions of eternal frost would
have spread on every side from the poles, and, advancing
towards the equator, would have extended its dreary

and solitary reign over a great part of what are now
the most fertile and most inhabited climates of the
world!' He expands this thesis in various directions,
the whole argument being based on the assumption that
'all bodies are condensed by cold, without limitation,
WATER ONLY EXCEPTED.' Repeated disappointments in
such matters have taught us caution. Legitimate
grounds for wonder exist everywhere around us; but
wonder must not be cultivated at the expense of truth.
Brought to the proper test, the assumption on which
Rumford built his striking teleological argument is
found to be a mere quicksand. The fact he adduces
as unique is not an exception to a universal law.
There are other substances, to which his reasoning
has not the remotest application, which, like water,
expand before and during crystallisation. The conditions necessary to the life of our planet must have existed before life appeared ; but whether those conditions
had prospective reference to life, or whether its im-
manent energy did not seize upon conditions which
grew into being without any reference to life, we do not
know ; and it would be mere arrogance at the present
day to dogmatise upon the subject.

In the controversy whether heat was a form of
matter or a form of motion, Rumford espoused the
latter view. Now those who supposed heat to be matter
naturally thought that it might be ponderable, and experiments favourable to this notion had been executed.
Operating with a balance of extreme delicacy, Rumford
took up this question, and treated it with great skill and
caution. His conclusion from his experiments was that,
if heat be a substance—a fluid *sui generis*—it must be
something so infinitely rare, even in its condensed state,
as to baffle all our attempts to discover its gravity.

But ' if the opinion which has been adopted by many of our ablest philosophers, that heat is an intestine vibratory motion of the constituent parts of bodies, should be well founded, it is clear that the weights of bodies can be in no wise affected by such motion.' The weight of a bell, he urges in another place, is not affected by its sonorous vibration.

Early in the year 1803, he being then in Munich, Rumford broke ground in the domain of radiant heat. He prepared bright metallic vessels, filled them with hot water, placed them in a large and quiet room, and observed the time required to cool them down a certain number of degrees. Covering some of his vessels with Irish linen and leaving others bare, he found, to his surprise, that the covered vessels were more rapidly chilled than the naked ones. Comparing in the same room a thick glass bottle, filled with hot water, with a tin bottle of the same shape and size, he found that the water in the glass vessel cooled twice as rapidly as that in the tin one. When, moreover, he coated his metallic vessel with glue, the cooling process was hastened, as it had been by the linen. Applying a second, a third, and a fourth coating of glue, he found the chilling promoted. Here, however, he came to a point where the addition of any further coatings produced a retardation of the chilling. Painting some of his vessels black and some white, he found the times of cooling to be practically the same for both—a result which he seems to have afterwards forgotten. From these and other experiments of the same kind he drew the just conclusion that a hot body does not lose its heat by the mere communication of it to the air, but that a large proportion of the heat escapes in *rays*, the escape being facilitated by the substances with which his vessels were coated.

The more rapid chilling of the glass bottle was due, in like manner, to the fact that glass possesses a greater radiative power than tin.

He next applies himself with energy, zeal, and tenacity to prove that there are frigorific rays which act in all respects like calorific rays, and which enjoy an individuality quite as assured as that of the latter. He pictures his frigorific rays as produced by vibrations of a special kind. In Pictet's celebrated experiment of conjugate mirrors, and in many other experiments, chilling by a cold body showed itself to be so exactly analogous to heating by a warm one, that Rumford never could shake from his mind the notion of rays of cold. The fall of the thermometer in one focus when a lump of ice was placed in the other, was in his view caused by a positive emission of cold rays from the ice, and not by its absorption of the heat radiated against it by the thermometer. These frigorific rays, he says, were suspected by Bacon. Their existence was actually established by the academicians of Florence, but these learned gentlemen were so ' blinded by their prejudices respecting the nature of heat, that they did not believe the report of their own eyes.'

Rumford indulges in various untenable speculations and erroneous notions regarding the part played by clothing, by the blackness of the negro's skin, and by the oiled surface of the Hottentot. We are, he contends, kept warm by our clothing, not so much by confining our heat as by keeping off the frigorific rays which tend to cool us. He reverts to the respective cases of a black and a white man, and describes an experiment which elucidates his views. He covered two of his vessels with goldbeater's skin, and painted one of them black with Indian ink, leaving the other of its natural white colour. Filling both vessels with hot water, he left them to cool in the air of a large, quiet room. The vessel covered with the black skin represented a negro, the other vessel a white man ; and the result was that while the black required only $23\frac{1}{2}$ minutes to cool, the white man required 28 minutes. The practical issue of the experiment is thus stated :—'All I will venture to say on the subject is, that were I called to inhabit a very hot country, nothing should prevent me from making the experiment of blackening my skin, or at least of wearing a black shirt, in the shade, and especially at night, in order to find out if, by those means, I could not contrive to make myself more comfortable.'

There was at times a headstrong element, if I may use the term, in Rumford's scientific reasoning. He here overlooks the fact that in a former experiment he found scarcely an appreciable difference between white and black as regards their powers of cooling. He also forgets the possible influence of a second coating, which his former experiments had revealed. As regards the negro and the white man, Rumford's first experiment illustrated the case more correctly than his subsequent ones. There are, moreover, transparent substances which, used as varnishes, would not have impaired the whiteness of the goldbeater's skin, but which would have hastened the cooling even more than the Indian ink.

Those who are acquainted with Sir John Leslie's experiments on radiant heat will not fail to notice that he and Rumford travelled over common ground. With a view of setting this matter right Rumford wrote a paper entitled ' Historical Review of Experiments on the Subject of Heat,' in which he shows that his experiments were not only talked about and executed before learned societies, but that they were in part published prior to the appearance of Leslie's celebrated work in 1804. Still, the style of that work furnishes, I think, internal evidence of its perfectly independent character, while the extent and variety of Leslie's labours render it practically impossible that they could have been derived from anything that Rumford had previously done. The two philosophers had no personal knowledge of each other, and the credit to be awarded, where they deal with the same subject, belongs, I think, equally to both.

Rumford's experimental work was far smaller in quantity than that of Leslie, but in regard to theory he must be conceded the highest place. In theory Leslie was inconsistent and confused, while Rumford, judged by the circumstances of his time, was in the main clear and correct. The part played by the luminiferous ether in the phenomena of light had been revived and enforced by the powerful experiments of Dr. Thomas Young. The undulatory hypothesis was therefore at hand, and Rumford made ample use of it. He has written a paper entitled ' Reflections on Heat,' in which he describes the views regarding its nature that were prevalent in his time. ' Some,' he says, ' regard it as a *substance*, others as a *vibratory motion* of the particles of matter of which a body is composed.' The heating of a body is, on the one hypothesis, due to the accumulation within it of *caloric*, while others hold the heating to be due to the acceleration of the vibratory motion. ' On the hypothesis of vibratory motion, a body which has become cold is thought to have lost nothing except motion ; on the other hypothesis, it is supposed to have lost some material substance.' The loss of motion Rumford clearly apprehends to be due to its communication to ' an eminently elastic fluid—an ether which fills all space throughout the universe.' The theoretic notions thus expressed are, in point of clearness and correctness, far in advance of those entertained by Leslie.

As already mentioned, the fact of water changing its density at a temperature of 40° Fahr. powerfully affected the mind of Rumford. On this subject he made many experiments; and one of the minor applications of the knowledge thus derived may be here noted. In company with his friend Professor Pictet, of Geneva, he paid a visit to the Mer de Glace, and discovered in the ice a pit ' perfectly cylindrical, about seven inches in diameter, and more than four feet deep, quite full of water.' He was informed by his guides that these pits are formed in summer, and gradually increase in depth during the warm weather. How can these pits deepen? Rumford answers thus:—The warm winds which in summer blow over the surface of the column of ice-cold water, communicate some small degree of heat to the fluid. The water at the surface being thus rendered specifically heavier, sinks to the bottom of the pit, to which the heat thus carried down is communicated, melting the ice and increasing the depth of the pit. We have here a small specimen of Rumford's penetration, but it is a very interesting one. The sun's invisible rays, however, are probably more influential than the action of the warm wind in producing the observed effect.

Various interesting experiments were made by Rumford on what is now known as ' surface-tension.' From his experiments he inferred that the surface of a liquid—of water, for example—is covered by a pellicle which can be caused to tremble throughout by touching it with the point of a needle. He proposed to the geometricians of Paris to determine the shape of a drop resting on a horizontal surface, and restrained solely by the resistance of a pellicle exerting a given force on its surface. This pellicle he considers to be due to the adhesion of the particles of liquids to each other, and he

makes various ingenious calculations to determine the
size of a particle of heavy matter—of gold, for instance
—which would rest suspended in water because of its
inability to force asunder the particles of the liquid.
The diameter of a sphere of gold which would behave
in this way he found to be $\frac{1}{283500}$ of an inch.

Even among scientific men, probably few are aware
that Rumford experimented on the diffusion of liquids;
a field of investigation in which Graham afterwards
rendered himself so eminent. Into a glass cylinder, $1\frac{2}{3}$
inch in diameter and 8 inches high, he poured a layer
of saturated aqueous solution of muriate of soda 3 inches
thick; over this he carefully poured a layer of distilled
water of the same thickness; he then let a drop of the
oil of cloves fall into the vessel. This oil, being heavier
than the pure water and lighter than the solution,
rested as a sphere at the common boundary of the two
liquids. A layer of olive oil four lines in thickness was
then poured over the water, in order to shut off the air.
The object of the experiment was to ascertain whether
one liquid remained permanently superposed upon the
other without any mixing. If this proved to be the
case the position of the drop of oil, would remain constant; but if the heavy mineral solution rose into
the water overhead, the drop of oil, which Rumford
called his 'little sentinel,' would warn him of the
event by rising in the liquid. After twenty-four
hours he entered the cellar in which the experiment
was made, and found that the little ball of oil had risen
three lines. For six days it continued to rise at the
rate of about three lines a day. He afterwards experimented with other solutions, the result being 'that
the mixture went on continually, but very slowly, between the various aqueous solutions employed and the
distilled water resting upon them.' Rumford's experi-

ments were probably prompted by his views on molecular physics. He would hardly have thought of the foregoing arrangement were not the intestine motions of the ultimate particles of bodies present to his mind. He was, moreover, quite aware of the importance of the result here established. The subject had often occupied his thoughts, and he had at different times made ' a considerable number of experiments with a view of throwing light into the profound darkness with which the subject is shrouded on every side.'

He devoted his attention to steam, considered as a vehicle for transporting heat ; he sought for the means of increasing the heat obtained from fuel ; he devised a new steam-boiler, in which we have a forecast of the tubular boiler of George Stephenson. After some preliminary experiments on wood and charcoal, he definitely took up the important investigation of the quantity of heat developed in combustion, and in the condensation of vapours. He described the new calorimeter employed in the inquiry. It was a kind of worm through which the heated air and products of combustion were led, and in which the heat was delivered up to cold water surrounding the worm.

He experimented upon white wax, spirit of wine, alcohol, sulphuric ether, naphtha, charcoal, wood, and inflammable gases. Whenever it was possible he aimed at quantitative results, and in the present instance he ' estimated the calorific power of a body by the number of parts by weight, of water, which one part by weight of the body would, on perfect combustion, raise one degree in temperature. Thus 1 lb. of charcoal, in combining with $2\frac{2}{3}$ lbs. of oxygen, to form carbonic acid, evolves heat sufficient to raise the temperature of about 8,000 lbs. of water 1° C. Similarly, 1 lb. of hydrogen, in combining with 8 lbs. of oxygen, to form water, generates an amount of heat sufficient to raise 34,000 lbs. of water 1° C. The calorific powers, therefore, of carbon and hydrogen are as 8 : 34. The refined researches of Favre and Silbermann entirely confirm these determinations of Rumford ' (Percy.) Following the experiments on combustion, we have others made to determine the quantity of heat set free by the condensation of various vapours, and the capacity of various liquids for heat. We have also an elaborate inquiry into the structure of wood, the specific gravity of its solid parts, the liquids and elastic fluids contained in it, the quantity of charcoal to be obtained from it, and the heat generated by the combustion of wood of different kinds.

But the main object of Rumford's life and the subject which chiefly interested him was the practical management of fire, and the economy of fuel. Eighty-seven pages of the second volume of his collected works are devoted to this subject. The whole of the third volume is devoted to it, while a large portion of the fourth and last volume is occupied with kindred questions. Some of those essays are rather tiresome to a reader of the present day, and Rumford had a suspicion that they might appear so to contemporary readers. ' I believe,' he says, ' that I am sometimes too prolix for the taste of the age; but it should be remembered that the subjects I have undertaken to investigate are by no means indifferent to me ; that I conceive them to be intimately connected with the comforts and enjoyments of mankind ; and that a habit of revolving them in my mind, and reflecting on their extensive usefulness, has awakened my enthusiasm, and rendered it quite impossible for me to treat them with cold indifference.'

For the most part, it is only when Rumford is self-conscious that this tedium appears. He wishes to excite his reader's interest, and sometimes adopts means to this end which defeat themselves. Such is the case when he dwells with reiteration on the refined and exquisite pleasure which he derives from being of service to humanity. Some also would deem him tedious, though I deem him courageous, when he deals with the details of his schemes. He leaves no stone unturned in his effort to render himself clear. He is in many cases simply writing out a specification, to be followed in all particulars. He gives directions as to the manner in which a slice of hasty pudding is to be eaten. A small pit is to be dug in the centre of the cake, a piece of butter placed in the pit, while the removed bit is to be placed on the butter to aid in melting it. You then begin at the circumference of your pudding, and eat all round, dipping each piece in the butter before conveying it to the mouth. Such details were sure to provoke sarcasm, and they did provoke it. But amid the verbosity we have incessant flashes of practical wisdom and examples of intellectual force. When he ceases to think of the exquisite delight of his philanthropic labours— ceases to think of himself—and permits his own personality to be effaced by his subject, we see Rumford at his best ; and his best was excellent. Suggestion follows suggestion, experiment succeeds experiment, until he has finally exhausted his subject, or is pulled up by inability to proceed further.

He tested quantitatively the relative intensities of various lights, constructing, while doing so, his well-known photometer. Placing two lights in front of a white screen, and at the same distance from it, and fixing an opaque rod between the lights and the screen, he obtained two shadows corresponding to the two lights. When the lights were equally intense, the shadows were equally dark, but when one of the lights was more powerful than the other, the shadow corresponding to that other was rendered pale, because the light from the most intense source fell upon it. Removing the more intense light farther from the screen, until a point was reached when the shadows appeared equal, Rumford obtained all the elements necessary for the computation of the relative intensities of the lights. He had only to apply the law of inverse squares, which makes a double distance correspond to a fourfold intensity, a treble distance to a ninefold intensity, and so on. In connection with these experiments he dwells repeatedly upon a defect which harasses the official gas-examiners of the present day, and that is, the fluctuations of the candles used as standards of measurement.

These photometric measurements are succeeded by a brief but beautiful essay on 'Coloured Shadows,' which, in connection with another short essay on the 'Harmony of Colours,' strikingly illustrates Rumford's penetration and experimental skill. He produced two shadows, one from daylight, the other from candle-light. The daylight shadow being shone upon by the candle, was, as might be expected, yellow, because the candle sheds a yellow light. But the other shadow, instead of being colourless, was 'the most beautiful blue that it was possible to imagine.' He states clearly that the colour of one shadow is real, while that of the other is imaginary. He finds it 'impossible to produce *two shadows* at the same time from the same body, the one answering to a beam of daylight, and the other to the light of a candle or lamp, without these shadows being coloured, the one yellow, and the other *blue*.' He obtained shadows from a light coloured by means of interposed glasses, and compared them with shadows obtained from uncoloured light. The shadows were always coloured when the lights differed from each other in whiteness, and the colours of the shadows were always such as, when added together, produced a pure white. The real colour, in fact, evoked, or ' called up,' or summoned an imaginary complementary colour. Goethe probably derived the expression ' geförderte Farben,' which occurs so often in the ' Farbenlehre,' from the terminology of Rumford.

But the experiments and discussion on which the fame of Rumford mainly rests are described in an essay of twenty pages—a vanishing quantity when compared with the sum-total of his published work. A cannon foundry had been built under his superintendence at Munich, where the heat developed during the boring of cannon powerfully attracted his attention. Upon this heat he made numerous tentative experiments, which are described in the essay. With the view of determining its exact quantity, he cut a cylinder from the muzzle end of a gun not yet bored, partially hollowed out this cylinder, and fitted into it a borer which resembled a blunt chisel in shape. The borer being strongly pressed against the bottom of the cylinder, it was caused to rotate by horse-power. He surrounded his cylinder with a wooden box, filling the box with water which embraced the entire cylinder. Soon after the starting of the rotation, the water felt warm to the hand. In an hour it had risen to 107° in temperature. In two hours and twenty minutes it had risen to 200°, while in two hours and thirty minutes it actually boiled.

' Rumford carefully estimated the quantity of heat possessed by each portion of his apparatus at the conclusion of his experiment, and adding all together, found a total sufficient to raise 26·58 lbs. of ice-cold water to its boiling-point, or through 180° Fahr. By careful calculation he found this heat equal to that given out by the combustion of 2303·8 grains ($=4\frac{8}{10}$ oz. troy) of wax. He then determined the " celerity " with which the heat was generated, summing up thus :— " From the results of these computations it appears that the quantity of heat produced equably, or in a continuous stream, if I may use the expression, by the friction of the blunt steel borer against the bottom of the hollow metallic cylinder, was greater than that produced in the combustion of nine wax candles, each three-quarters of an inch in diameter, all burning together with clear bright flames."

' "One horse," he continues, " would have been equal to the work performed, though two were actually employed. Heat may thus be produced merely by the strength of a horse, and in a case of necessity, this heat might be used in cooking victuals. But no circumstances could be imagined in which this method of procuring heat would be advantageous ; for more heat might be obtained by using the fodder necessary for the support of a horse as fuel." '

This is an extremely significant passage, intimating, as it does, that Rumford saw clearly that the force of animals was derived from the food, no creation of force taking place in the animal body.

' By meditating on the results of all these experiments, we are naturally,' he says, ' brought to the great question which has so often been the subject of speculation among philosophers, namely, What is heat—is there any such thing as an igneous fluid ? Is there anything that, with propriety, can be called caloric ? '

' We have seen that a very considerable quantity of heat may be excited by the friction of two metallic surfaces, and given off in a constant stream or flux in all directions, without interruption or intermission, and without any signs of diminution or exhaustion. In reasoning on this subject, we must not forget that most remarkable circumstance, that the source of the heat generated by friction in these experiments appeared evidently to be inexhaustible. It is hardly necessary to add that anything which any insulated body or system of bodies can continue to furnish without limitation cannot possibly be a material substance; and it appears to me to be extremely difficult, if not quite impossible, to form any distinct idea of anything capable of being excited and communicated in those experiments, except it be Motion.' [1]

[1] *Heat a Mode of Motion*, Lecture II.

1884.

LOUIS PASTEUR, HIS LIFE AND LABOURS.

(A Review.) [1]

IN the early part of the present year the French
original of this work was sent to me from Paris
by its author. It was accompanied by a letter from
M. Pasteur, expressing his desire to have the work
translated and published in England. Responding
to this desire, I placed the book in the hands of
Messrs. Longman, who, in the exercise of their own
judgment, decided on publication. The translation was
confided, at my suggestion, to Lady Claud Hamilton.

The translator's task was not always an easy one,
but it has, I think, been well executed. A few slight
abbreviations, for which I am responsible, have been
introduced, but in no case do they affect the sense. It
was, moreover, found difficult to render into suitable
English the title of the original :—' M. Pasteur, Histoire d'un Savant par un Ignorant.' A less piquant
and antithetical English title was therefore substituted
for the French one.

This filial tribute, for such it is, was written, under
the immediate supervision of M. Pasteur, by his
devoted and admiring son-in-law, M. Valery Radot. It
is the record of a life of extraordinary scientific ardour
and success, the picture of a mind on which facts fall
like germs upon a nutritive soil, and, like germs so

[1] Written as an introduction to the English translation.

favoured, undergo rapid increase and multiplication. One hardly knows which to admire most—the intuitive vision which discerns in advance the new issues to which existing data point, or the skill in device, the adaptation of means to ends, whereby the intuition is brought to the test and ordeal of experiment.

In the investigation of microscopic organisms—the ' infinitely little,' as Pouchet loved to call them—and their doings in this our world, M. Pasteur has found his true vocation. In this broad field it has been his good fortune to alight upon a crowd of connected problems of the highest public and scientific interest, ripe for solution, and requiring for their successful treatment the precise culture and capacities which he has brought to bear upon them. He may regret his abandonment of molecular physics ; he may look fondly back upon the hopes with which his researches on the tartrates and paratartrates inspired him ; he may think that great things awaited him had he continued to labour in this line. I do not doubt it. But this does not shake my conviction that he yielded to the natural affinities of his intellect, that he obeyed its truest impulses, and reaped its richest rewards, in pursuing the line that he has chosen, and in which his labours have rendered him one of the most conspicuous scientific figures of this age.

With regard to the earliest labours of M. Pasteur, a few remarks supplementary to those of M. Radot may be introduced here. The days when angels whispered into the hearkening human ear secrets which had no root in man's previous knowledge or experience are gone for ever. The only revelation—and surely it deserves the name—now open to the wise arises from ' intending the mind ' on acquired knowledge. When, therefore, M. Radot, following M. Pasteur, speaks with such emphasis about 'preconceived ideas,' he cannot mean ideas without antecedents. Preconceived ideas—if out of deference to M. Pasteur the term be admitted—are the vintage of garnered facts. We in England should rather call them inductions which, as M. Pasteur truly says, inspire the mind and shape its course in the subsequent work of deduction and verification.

At the time when M. Pasteur undertook his investigation of the diseases of silkworms, which led to such admirable results, he had never seen a silkworm; but so far from this being considered a disqualification, his friend M. Dumas regarded his freedom as a positive advantage. His first care was to make himself acquainted with what others had done. To their observations he added his own, and then, surveying all, came to the conclusion that the origin of the disease was to be sought, not in the worms, not in the eggs, but in the moths which laid the eggs. I am not sure that this conclusion is happily described as 'a preconceived idea.' Every whipster may have his preconceived ideas; but the divine power, so largely shared by M. Pasteur, of distilling from facts their essences—of extracting from them the principles from which they flow—is given only to a few.

With regard to the discovery of crystalline facets in the tartrates, dwelt upon by M. Radot, a brief reference to antecedent labours may be here allowed. It had been discovered by Arago, in 1811, and by Biot, in 1812 and 1818, that a plate of rock-crystal, cut perpendicular to the axis of the prism, and crossing a beam of plane polarised light, caused the plane of polarisation to rotate through an angle, dependent on the thickness of the plate and the refrangibility of the light. It had, moreover, been proved by Biot that there existed two species of rock-crystal, one of which turned the plane of polarisation to the right, and the other to the left. They were called respectively, right-handed and left-handed crystals. No external difference of crystalline form was at first noticed which could furnish a clue to this difference of action. But closer scrutiny revealed upon the crystals minute facets, which, in the one class, were ranged along a right-handed, and, in the other, along a left-handed spiral. The symmetry of the hexagonal prism, and of the two terminal pyramids of the crystal, was disturbed by the introduction of these spirally-arranged facets. They constituted the outward and visible sign of that inward and invisible molecular structure which produced the observed action, and difference of action, on polarised light.

When, therefore, the celebrated Mitscherlich brought forward his tartrates and paratartrates of ammonia and soda, and affirmed them to possess the same atoms, the same internal arrangement of atoms, and the same outward crystalline form, one of them, nevertheless, causing the plane of polarisation to rotate, while the other did not, Pasteur, remembering no doubt the observations just described, instituted a search for facets like those discovered in rock-crystal, which, without altering chemical constitution, destroyed crystalline identity. He first found such facets in the tartrates, while he subsequently proved the neutrality of the paratartrate to be due to the equal admixture of right-handed and left-handed crystals, one of which, when the paratartrate was dissolved, exactly neutralised the other.

Prior to Pasteur the left-handed tartrate was unknown. Its discovery, moreover, was supplemented by a series of beautiful researches on the compounds of right-handed and left-handed tartaric acid ; he having previously extracted from the two tartrates acids which, in regard to polarised light, behaved like themselves. Such was the worthy opening of M. Pasteur's scientific career, which has been dwelt upon so frequently and emphatically by M. Radot. The wonder however is, not that a searcher of such penetration as Pasteur should have discovered the facets of the tartrates, but that an investigator so powerful and experienced as Mitscherlich should have missed them.

The idea of *molecular* dissymmetry, introduced by Biot, was forced upon Biot's mind by the discovery of a number of liquids, and of some vapours, which possessed the rotatory power. Some, moreover, turned the plane of polarisation to the right, others to the left. Crystalline structure being here out of the question, the notion of dissymmetry, derived from the crystal, was transferred to the molecule. 'To produce any such phenomena, says Sir John Herschel, ' the individual molecule must be conceived as unsymmetrically constituted.' The illustrations employed by M. Pasteur to elucidate this subject, though well calculated to give a general idea of dissymmetry, will, I fear, render but little aid to the reader in his attempts to realise *molecular* dissymmetry. Should difficulty be encountered here at the threshold of this work, I would recommend the reader not to be daunted by it, or prevented by it from going further. He may comfort himself by the assurance that the conception of a dissymmetric molecule is not a very precise one, even in the mind of M. Pasteur.

One word more with regard to the parentage of preconceived ideas. M. Radot informs us that at Strasburg M. Pasteur invoked the aid of helices and magnets, with a view to rendering crystals dissymmetrical at the moment of their formation. There can, I think, be but little doubt that such experiments were suggested by the pregnant discovery of Faraday, published in 1845. By both helices and magnets Faraday caused the plane of polarisation in perfectly neutral liquids and solids to rotate. If the turning of the plane of polarisation be a demonstration of molecular dissymmetry, then, in the twinkling of an eye, Faraday was able to displace symmetry by dissymmetry, and to confer upon bodies, which in their ordinary state were inert and dead, this power of rotation which M. Pasteur considers to be the exclusive attribute of life.

The conclusion of M. Pasteur here referred to, which M. Radot justly describes as 'worthy of the most serious consideration,' is sure to arrest the attention of a large class of people, who, dreading ' materialism,' are ready to welcome any generalisation which differentiates the living world from the dead. M. Pasteur considers that his researches point to an irrefragable physical barrier between organic and inorganic Nature. Never, he says, have you been able to produce in the laboratory, by the ordinary processes of chemistry, a dissymmetric molecule—in other words, a substance which, in a state of solution, where molecular forces are paramount, has the power of causing a polarised beam to rotate. This power belongs exclusively to derivatives from the living world. Dissymmetric *forces*, different from those of the laboratory, are, in Pasteur's mind, the agents of vitality. They alone build up dissymmetric molecules which baffle the chemist when he attempts to reproduce them. Such molecules trace their ancestry to life alone. ' Pourrait-on indiquer une séparation plus profonde entre les produits de la nature vivante, et ceux de la nature minérale, que cette dissymétrie chez les uns, et son absence chez les autres ? ' It may be worth calling to mind that molecular dissymmetry is the *idea*, or *inference*, the observed rotation of the plane of polarisation by masses of sensible magnitude being the *fact*, on which the inference is based.

That the molecule, or unit brick, of an organism should be different from the molecule of a mineral is only to be expected, for otherwise the profound distinction between them would disappear. And that one of the differences between the two classes of molecules should be the possession, by the one, of this power of rotation, and its non-possession by the other, would be a fact, interesting no doubt, but not surprising. The critical point here has reference to the power and range of chemical processes, apart from the play of vitality. Beginning with the elements themselves, can they not be so combined as to produce organic compounds? Not to speak of the antecedent labours of Wöhler and others in Germany, it is well known that various French investigators, among whom are some of M. Pasteur's illustrious colleagues of the Academy, have succeeded in forming substances which were once universally regarded as capable of being elaborated by plants and animals alone. Even with regard to the rotation of the plane of polarisation, M. Jungfleisch, an extremely able pupil of the celebrated Berthelot, affirms that the barrier erected by M. Pasteur has been broken down; and though M. Pasteur questions this affirmation, it is at least hazardous, where so many supposed distinctions between organic and inorganic have been swept away, to erect a new one. For my part, I frankly confess my disbelief in its permanence.

Without waiting for new facts, those already in our possession tend, I think, to render the association which M. Pasteur seeks to establish between dissymmetry and life insecure. Quartz, as a crystal, exerts a very powerful twist on the plane of polarisation. Quartz dissolved exerts no power at all. The molecules of quartz, then, do not belong to the same category as the crystal of which they are the constituents; the former are symmetrical, the latter is dissymmetrical. This, in my opinion, is a very significant fact. By the act of crystallisation, and without the intervention of life, the forces of molecules, possessing planes of symmetry, are so compounded as to build up crystals which have no planes of symmetry. Thus, in passing from the symmetrical to the dissymmetrical, we are not compelled to interpolate new forces; the forces extant in mineral nature suffice. The reasoning which applies to the dissymmetric crystal applies to the dissymmetric molecule. The dissymmetry of the latter, however pronounced and complicated, arises from the composition of atomic forces which, when reduced to their most elementary action, are exerted along straight lines. In 1865 I ventured, in reference to this subject, to define the position which I am still inclined to maintain. ' It is the compounding, in the organic world, of forces belonging equally to the inorganic that constitutes the mystery and the miracle of vitality.' [1]

Add to these considerations the discovery of Faraday already adverted to. An electric current is not an organism, nor does a magnet possess life; still, by their action, Faraday, in his first essay, converted over one hundred and fifty symmetric and inert aqueous solutions into dissymmetric and active ones.[2]

---

[1] Art. ' Vitality,' *Fragments of Science*, 6th edit., vol. ii. p. 50.

[2] In Faraday's *induced* dissymmetry, the ray having once passed through the body under magnetic influence, has its rotation doubled instead of neutralised, as in the case of quartz, on being reflected back through the body. Marbach has discovered that chlorate of soda produces circular polarisation in all directions through the crystal, while in quartz it occurs only in the direction of the axis.

Theory, however, may change, and inference may fade away, but scientific experiment endures for ever. Such durability belongs, in the domain of molecular physics, to the experimental researches of M. Pasteur.

The weightiest events of life sometimes turn upon small hinges; and we now come to the incident which caused M. Pasteur to quit a line of research the abandonment of which he still regrets. A German manufacturer of chemicals had noticed that the impure commercial tartrate of lime, sullied with organic matters of various kinds, fermented when it was dissolved in water and exposed to summer heat. Thus prompted, Pasteur prepared some pure right-handed tartrate of ammonia, mixed with it albuminous matter, and found that the mixture fermented. His solution, limpid at first, became turbid, and the turbidity he found to be due to the multiplication of a microscopic organism, which found in the liquid its proper aliment. Pasteur recognised in this little organism a *living ferment*. This bold conclusion was doubtless strengthened, if not prompted, by the previous discovery of the yeast-plant—the alcoholic ferment—by Cagniard-Latour and Schwann.

Pasteur next permitted his little organism to take the carbon necessary for its growth from the pure paratartrate of ammonia. Owing to the opposition of its two classes of crystals, a solution of this salt, it will be remembered, does not turn the plane of polarised light either to the right or to the left. Soon after fermentation had set in, a rotation to the left was noticed, proving that the equilibrium previously existing between the two classes of crystals had ceased.

Marbach also discovered facets upon his crystals, resembling those of quartz.

The rotation reached a maximum, after which it was found that all the right-handed tartrate had disappeared from the liquid. The organism thus proved itself competent to select its own food. It found, as it were, one of the tartrates more digestible than the other, and appropriated it, to the neglect of the other. No difference of chemical constitution determined its choice; for the elements, and the proportions of the elements, in the two tartrates were identical. But the peculiarity of structure which enabled the substance to turn the plane of polarisation to the right, also rendered it a fit aliment for the organism. This most remarkable experiment was successfully made with the seeds of our common mould, *Penicillium glaucum.*

Here we find Pasteur unexpectedly landed amid the phenomena of fermentation. With true scientific instinct he closed with the conception that ferments are, in all cases, living things, and that the substances formerly regarded as ferments are, in reality, the food of the ferments. Touched by this wand, difficulties fell rapidly before him. He proved the ferment of lactic acid to be an organism of a certain kind. The ferment of butyric acid he proved to be an organism of a different kind. He was soon led to the fundamental conclusion that the capacity of an organism to act as a ferment depended on its power to live without air. The fermentation of beer was sufficient to suggest this idea. The yeast-plant, like many others, can live either with or without free air. It flourishes best in contact with free air, for it is then spared the labour of wresting from the malt the oxygen required for its sustenance. Supplied with free air, however, it practically ceases to be a ferment; while in the brewing-vat, where the work of fermentation is active, the budding *torula* is completely cut off by the sides of the vessel,

and by a deep layer of carbonic acid gas, from all contact with air. The butyric ferment not only lives
without air, but Pasteur showed that air·is fatal to it.
He finally divided microscopic organisms into two
great classes, which he named respectively *aérobies* and
anaérobies, the former requiring free oxygen to maintain life, the latter capable of living without free
oxygen, but able to wrest this element from its combinations with other elements. This destruction of
pre-existing compounds and formation of new ones,
through the increase and multiplication of the organism, constitute the process of fermentation.

Under this head are also rightly ranked the phenomena of putrefaction. As M. Radot well expresses
it, the fermentation of sugar may be described as the
putrefaction of sugar. In this particular field M. Pasteur, whose contributions to the subject are of the
highest value, was preceded by Schwann, a man of
great merit, of whom the world has heard too little.[1]
Schwann placed decoctions of meat in flasks, sterilised
the decoctions by boiling, and then supplied them with
calcined air, the power of which to support life he
showed to be unimpaired. Under these circumstances
putrefaction never set in. Hence the conclusion of
Schwann, that putrefaction was not due to the contact
of air, as affirmed by Gay-Lussac, but to something
suspended in the air which heat was able to destroy.
This something consists of living organisms, which
nourish themselves at the expense of the organic
substance, and cause its putrefaction.

The grasp of Pasteur on this class of subjects
was embracing. He studied acetic fermentation, and

[1] It was late in the day when the Royal Society made him a
foreign member.

found it to be the work of a minute fungus, the *My-coderma aceti*, which, requiring free oxygen for its nutrition, overspreads the surface of the fermenting liquid. By the alcoholic ferment the sugar of the grape-juice is transformed into carbonic acid gas and alcohol, the former exhaling, the latter remaining in the wine. By the *Mycoderma aceti* the wine is, in its turn, converted into vinegar. Of the experiments made in connection with this subject one deserves special mention. It is that in which Pasteur suppressed all albuminous matters, and carried on the fermentation with purely crystallisable substances. He studied the deterioration of vinegar, revealed its cause, and the means of preventing it. He defined the part played by the little eel-like organisms which sometimes swarm in vinegar-casks, and ended by introducing important ameliorations and improvements in the manufacture of vinegar. The discussion with Liebig and other minor discussions of a similar nature, which M. Radot has somewhat strongly emphasised, I will not here dwell upon.

It was impossible for an inquirer like Pasteur to evade the question—Whence come these minute organisms which are demonstrably capable of producing effects which constitute the bases of industries whereon whole populations depend for occupation and sustenance? He thus found himself face to face with the question of 'spontaneous generation,' to which the researches of Pouchet had just given fresh interest. Trained as Pasteur was in the experimental sciences, he had an immense advantage over Pouchet, whose culture was derived from the sciences of observation. One by one the statements and experiments of Pouchet were explained or overthrown, and the doctrine of spon-

taneous generation remained discredited until it was
revived with ardour, ability, and, for a time with success, by Dr. Bastian.

A remark of M. Radot's on page 103 needs quali-
fication. ' The great interest of Pasteur's method
consists,' he says, ' in its proving unanswerably that
the origin of life in infusions which have been heated
to the boiling point is solely due to the solid particles
suspended in the air.' This means that living germs
cannot exist *in the liquid* when once raised to a temperature of 212° Fahr. No doubt a great number of
organisms collapse at this temperature ; some, indeed,
as M. Pasteur has shown, are destroyed at a temperature of 90° below the boiling point. But this is by no
means universally the case. The spores of the hay-
bacillus, for example, have, in numerous instances, successfully resisted the boiling temperature for one, two,
three, four hours ; while in one instance *eight hours'*
continuous boiling failed to sterilise an infusion of
desiccated hay. The knowledge of this fact caused me
a little anxiety some years ago when a meeting was
projected between M. Pasteur and Dr. Bastian. For
though, in regard to the main question, I knew that
the upholder of spontaneous generation could not win,
on the particular issue touching the death temperature
he would probably have come off victor.

The manufacture and maladies of wine next occupied
Pasteur's attention. He had, in fact, got the key to
this whole series of problems, and he knew how to use
it. Each of the disorders of wine was traced to its
specific organism, which, acting as a ferment, produced
substances the reverse of agreeable to the palate. By
the simplest of devices, Pasteur at a stroke abolished
the causes of wine disease. Fortunately the foreign

organisms which, if unchecked, destroy the best red wines, are extremely sensitive to heat. A temperature of 50° Cent. (122° Fahr.) suffices to kill them. Bottled wines once raised to this temperature, for a single minute, are secured from subsequent deterioration. The wines suffer in no degree from exposure to this temperature. The manner in which Pasteur proved this, by invoking the judgment of the wine-tasters of Paris, is as amusing as it is interesting.

Moved by the entreaty of his master, the illustrious Dumas, Pasteur took up the investigation of the diseases of silkworms at a time when the silk-husbandry of France was in a state of ruin. In doing so he did not, as might appear, entirely forsake his former line of research. Previous investigators had got so far as to discover vibratory corpuscles in the blood of the diseased worms, and with such corpuscles Pasteur had already made himself intimately acquainted. He was therefore to some extent at home in this new investigation. The calamity was appalling, all the efforts made to stay the plague having proved futile. In June 1865 Pasteur betook himself to the scene of the epidemic, and at once commenced his observations. On the evening of his arrival he had already discovered the corpuscles, and shown them to others. Acquainted as he was with the work of living ferments, his mind was prepared to see in the corpuscles the cause of the epidemic. He followed them through all the phases of the insect's life— through the eggs, through the worm, through the chrysalis, through the moth. He proved that the germ of the malady might be present in the eggs and escape detection. In the worm also it might elude microscopic examination. But in the moth it reached a development so distinct as to render its recognition imme-

diate. From healthy moths healthy eggs were sure to
spring; from healthy eggs healthy worms; from healthy
worms fine cocoons: so that the problem of the restoration to France of its silk-husbandry reduced itself to
the separation of the healthy from the unhealthy moths,
the rejection of the latter, and the exclusive employment of the eggs of the former. M. Radot describes
how this is now done on the largest scale, with the
most satisfactory results.

The bearing of this investigation on the parasitic
theory of communicable diseases was thus illustrated:
Worms were infected by permitting them to feed for
a single meal on leaves over which corpusculous matter
had been spread; they were infected by inoculation,
and it was shown how they infected each other by the
wounds and scratches of their own claws. By the association of healthy with diseased worms, the infection
was communicated to the former. Infection at a distance was also produced by the wafting of the corpuscles through the air. The various modes in which
communicable diseases are diffused among human populations were illustrated by Pasteur's treatment of the
silkworms. 'It was no hypothetical infected medium
—no problematical pythogenic gas—that killed the
worms. It was a definite organism.' [1] The disease
thus far described is that called *pébrine*, which was the
principal scourge at the time. Another formidable
malady was also prevalent, called *flacherie*, the cause
of which, and the mode of dealing with it, were also
pointed out by Pasteur.

Overstrained by years of labour in this field, Pasteur
was smitten with paralysis in October 1868. But this
calamity did not prevent him from making a journey

[1] These words were uttered at a time when the pythogenic theory
was more in favour than it is now.

to Alais in January 1869, for the express purpose of combating the criticisms to which his labours had been subjected. Pasteur is combustible, and contradiction readily stirs him into flame. No scientific man now living has fought so many battles as he. To enable him to render his experiments decisive, the French Emperor placed a villa at his disposal near Trieste, where silkworm-culture had been carried on for some time at a loss. The success here is described as marvellous, the sale of cocoons giving to the villa a net profit of twenty-six millions of francs.[1] From the Imperial villa M. Pasteur addressed to me a letter, a portion of which I have already published. It may perhaps prove usefully suggestive to our Indian or Colonial authorities if I reproduce it here:—

'Permettez-moi de terminer ces quelques lignes que je dois dicter, vaincu que je suis par la maladie, en vous faisant observer que vous rendriez service aux Colonies de la Grande-Bretagne en répandant la connaissance de ce livre, et des principes que j'établis touchant la maladie des vers à soie. Beaucoup de ces colonies pourraient cultiver le mûrier avec succès, et, en jetant les yeux sur mon ouvrage, vous vous convaincrez aisément qu'il est facile aujourd'hui, non seulement d'éloigner la maladie régnante, mais en outre de donner aux récoltes de la soie une prospérité qu'elles n'ont jamais eue.'

The studies on wine prepare us for the 'studies on beer,' which followed the investigation of silkworm diseases. The sourness, putridity, and other maladies of beer, Pasteur traced to special 'ferments of disease,' of a totally different form, and therefore easily distin-

[1] The work on *Diseases of Silkworms* was dedicated to the Empress of the French.

guishable from the true *torula* or yeast-plant. Many
mysteries of our breweries were cleared up by this inquiry. Without knowing the cause, the brewer not
unfrequently incurred heavy losses through the use of
bad yeast. Five minutes' examination with the microscope would have revealed to him the cause of the
badness, and prevented him from using the yeast. He
would have seen the true *torula* overpowered by foreign
intruders. The microscope is, I believe, now everywhere
in use. At Burton-on-Trent its aid was very soon invoked. At the conclusion of his studies on beer M.
Pasteur came to London, where I had the pleasure of
conversing with him. Crippled by paralysis, bowed
down by the sufferings of France, and anxious about his
family at a troubled and an uncertain time, he appeared
low in health and depressed in spirits. His robust
appearance when he visited London, on the occasion of
the Edinburgh Anniversary, was in marked and pleasing
contrast with my memory of his aspect at the time to
which I have referred.

While these researches were going on, the Germ
Theory of infectious disease was noised abroad. The
researches of Pasteur were frequently referred to as
bearing upon the subject, though Pasteur himself kept
clear for a long time of this special field of inquiry.
He was not a physician, and he did not feel called upon
to trench upon the physician's domain. And now I
would beg of him to correct me if, at this point of the
Introduction, I should be betrayed into any statement
that is not strictly correct.

In 1876 the eminent microscopist, Professor Cohn,
of Breslau, was in London, and he then handed me
a number of his ' Beiträge,' containing a memoir
by Dr. Koch on Splenic Fever (*Milzbrand, Charbon,*

Malignant Pustule), which seemed to me to mark an
epoch in the history of this formidable disease. With
admirable patience, skill, and penetration Koch followed up the life-history of *bacillus anthracis*, the contagium of this fever. At the time here referred to he
was a young physician holding a small appointment in
the neighbourhood of Breslau, and it was easy to pre-
dict, and indeed I predicted at the time, that he would
soon find himself in a higher position. When I next
heard of him he was head of the Imperial Sanitary
Institute of Berlin. Koch's recent history is pretty
well known in England, while his appreciation by the
German Government is shown by the rewards and
honours lately conferred upon him.

Koch was not the discoverer of the parasite of
splenic fever. Davaine and Rayer, in 1850, had observed the little microscopic rods in the blood of
animals which had died of splenic fever. But they
were quite unconscious of the significance of their
observation, and for thirteen years, as M. Radot informs
us, strangely let the matter drop. In 1863 Davaine's
attention was again directed to the subject by the
researches of Pasteur, and he then pronounced the
parasite to be the cause of the fever. He was opposed
by some of his fellow-countrymen ; long discussions
followed, and a second period of thirteen years, ending
with the publication of Koch's paper, elapsed before
M. Pasteur took up the question. I always, indeed,
assumed that from the paper of the learned German
came the impulse towards a line of inquiry in which
M. Pasteur has achieved such splendid results. Things
presenting themselves thus to my mind, M. Radot
will, I trust, forgive me if I say that it was with very
great regret that I perused the disparaging references to
Dr. Koch which occur in the chapter on splenic fever.

After Koch's investigation, no doubt could be entertained of the parasitic origin of this disease. It completely cleared up the perplexity previously existing as to the two forms—the one fugitive, the other permanent—in which the contagium presented itself. I may here remark that it was on the conversion of the permanent hardy form into the fugitive and sensitive one, in the case of *bacillus subtilis* and other organisms, that the method of sterilising by ' discontinuous heating ' introduced by me in February 1877 was founded. The difference between an organism and its spores, in point of durability, had not escaped the penetration of Pasteur. This difference Koch showed to be of paramount importance in splenic fever. He moreover proved that while mice and guinea-pigs were infallibly killed by the parasite, birds were able to defy it.

And here we come upon what may be called a hand-specimen of the genius of Pasteur, which strikingly illustrates its quality. Why should birds enjoy the immunity established by the experiments of Koch ? Here is the answer. The temperature which prohibits the multiplication of *bacillus anthracis* in infusions is 44° Cent. (111° Fahr.). The temperature of the blood of birds is from 41° to 42°. It is therefore close to the prohibitory temperature. But then the blood-globules of a living fowl are sure to offer a certain resistance to any attempt to deprive them of their oxygen—a resistance not experienced in an infusion. May not this resistance, added to the high temperature of the fowl, suffice to place it beyond the power of the parasite ? Experiment alone could answer this question, and Pasteur made the experiment. By the application of cold water he lowered the temperature of a fowl to 37° or 38°. He inoculated the fowl, thus chilled, with the splenic fever parasite, and in twenty-four hours it was dead. The argument was clinched by inoculating a chilled fowl, permitting the fever to come to a head, and then removing the fowl, wrapped in cotton-wool, to a chamber with a temperature of 45°. The strength of the patient returned as the career of the parasite was brought to an end, and in a few hours health was restored. The sharpness of the reasoning here is only equalled by the conclusiveness of the experiment, which is full of suggestiveness as regards the treatment of fevers in man.

Pasteur had little difficulty in establishing the parasitic origin of fowl-cholera : indeed, the parasite had been observed by others before him. But, by his successive cultivations, he rendered the solution pure. His next step will remain for ever memorable in the history of medicine. I allude to what he calls ' virus attenuation.' And here it may be well to throw out a few remarks in advance. When a tree, or a bundle of wheat or barley straw, is burnt, a certain amount of mineral matter remains in the ashes—extremely small in comparison with the bulk of the tree or of the straw, but absolutely essential to its growth. In a soil lacking, or exhausted of, the necessary mineral constituents, the tree cannot live, the crop cannot grow. Now contagia are living things, which demand certain elements of life just as inexorably as trees, or wheat, or barley ; and it is not difficult to see that a crop of a given parasite may so far use up a constituent existing in small quantities in the body, but essential to the growth of the parasite, as to render the body unfit for the production of a second crop. The soil is exhausted, and, until the lost constituent is restored, the body is protected from any further attack of the same disorder. Some such explanation of non-recurrent diseases naturally presents itself

to a thorough believer in the germ theory, and such
was the solution which, in reply to a question, I ventured to offer nearly fifteen years ago to an eminent
London physician.[1] To exhaust a soil, however, a parasite less vigorous and destructive than the really virulent one may suffice ; and if, after having by means of
a feebler organism exhausted the soil without fatal
result, the most highly virulent parasite be introduced
into the system, it will prove powerless.[2]

The general problem, of which Jenner's discovery
was a particular case, has been grasped by Pasteur, in
a manner, and with results, which five short years ago
were simply unimaginable. How much ' accident '
had to do with shaping the course of his inquiries I
know not. A mind like his resembles a photographic
plate, which is ready to accept and develop luminous
impressions, sought and unsought. In the chapter on
fowl-cholera is described how Pasteur first obtained his
attenuated virus. By successive cultivations of the
parasite, he showed that after it had been a hundred
times reproduced, it continued to be as virulent as at
first. One necessary condition was, however, to be
observed. It was essential that the cultures should
rapidly succeed each other—that the organism, before
its transference to a fresh cultivating liquid, should not
be left long in contact with air. When exposed to air
for a considerable time the virus becomes so enfeebled
that when fowls are inoculated with it, though they
sicken for a time, they do not die. But this ' attenuated '
virus, which M. Radot justly calls ' benign,' constitutes
a sure protection against the virulent virus. It so
exhausts the soil that the really fatal contagium fails

[1] Sir Thomas Watson.
[2] Recent researches suggest other explanations.

to find there the elements necessary to its reproduction and multiplication.

Pasteur affirms that it is the oxygen of the air which, by lengthened contact, weakens the virus and converts it into a true vaccine. He has also weakened it by transmission through various animals. It was this form of attenuation that was brought into play in the case of Jenner.

The secret of attenuation had thus become an open one to Pasteur. He laid hold of the murderous virus of splenic fever, and succeeded in rendering it, not only harmless to life, but a sure protection against the virus in its most concentrated form. No man, in my opinion, can work at these subjects so rapidly as Pasteur without falling into errors of detail. But this may occur while his main position remains impregnable. Such a result, for example, as that obtained in presence of so many witnesses at Melun must remain an ever-memorable conquest of science. Having prepared his attenuated virus, and proved by laboratory experiments its efficacy as a protective vaccine, Pasteur accepted an invitation from the President of the Society of Agriculture at Melun to make a public experiment on what might be called an agricultural scale. This act of Pasteur's is, perhaps, the boldest thing recorded in this book. It naturally caused anxiety among his colleagues of the Academy, who feared that he had been rash in closing with the proposal of the President.

But the experiment was made. A flock of sheep was divided into two groups, the members of one group being all vaccinated with the attenuated virus, while those of the other group were left unvaccinated. A number of cows were also subjected to a precisely similar treatment. Fourteen days afterwards, all the sheep and all the cows, vaccinated and unvaccinated, were inoculated with a very virulent virus; and three days subsequently more than two hundred persons assembled to witness the result. The 'shout of admiration,' mentioned by M. Radot, was a natural outburst under the circumstances. Of twenty-five sheep which had not been protected by vaccination, twenty-one were already dead, and the remaining four were dying. The twenty-five vaccinated sheep, on the contrary, were ' in full health and gaiety.' In the unvaccinated cows intense fever was produced, while the prostration was so great that they were unable to eat. Tumours were also formed at the points of inoculation. In the vaccinated cows no tumours were formed; they exhibited no fever, nor even an elevation of temperature, while their power of feeding was unimpaired. No wonder that ' breeders of cattle overwhelmed Pasteur with applications for vaccine.' At the end of 1881 close upon 34,000 animals had been vaccinated, while the number rose in 1883 to nearly 500,000.

M. Pasteur is now [1884] exactly sixty-two years of age; but his energy is unabated. At the end of this volume we are informed that he has already taken up and examined with success, as far as his experiments have reached, the terrible and mysterious disease of rabies or hydrophobia. Those who hold all communicable diseases to be of parasitic origin, include, of course, rabies among the number of those produced and propagated by a living contagium. From his first contact with the disease Pasteur showed his accustomed penetration. If we see a man mad, we at once refer his madness to the state of his brain. It is somewhat singular that in the face of this fact the virus of a mad dog should be referred to the animal's saliva. The saliva is no doubt infected, but Pasteur soon proved the real seat and empire of the disorder to be the nervous system.

The parasite of rabies had not been securely isolated when M. Radot finished his task. But last May, at the instance of M. Pasteur, a commission was appointed by the Minister of Public Instruction in France, to examine and report upon the results which he had up to that time obtained. A preliminary report, issued to appease public impatience, reached me before I quitted Switzerland this year. It inspires the sure and certain hope that, as regards the attenuation of the rabic virus, and the rendering of an animal by inoculation proof against attack, the success of M. Pasteur is assured. The Commission, though hitherto extremely active, is far from the end of its labours; but the results obtained so far may be thus summed up :—

Of six dogs unprotected by vaccination, three succumbed to the bites of a dog in a furious state of madness.

Of eight unvaccinated dogs, six succumbed to the intravenous inoculation of rabic matter.

Of five unvaccinated dogs, all succumbed to inoculation, by trepanning, of the brain.

Finally, of three-and-twenty vaccinated dogs, not one was attacked with the disease subsequent to inoculation with the most potent virus.

Surely results such as those recorded in this book are calculated, not only to arouse public interest, but to excite public hope and wonder. Never before, during the long period of its history, did a day like the present dawn upon the science and art of medicine. Indeed, previous to the discoveries of recent times, medicine was not a science, but a collection of empirical rules,

dependent for their interpretation and application upon
the sagacity of the physician.   How does England stand
in relation to the great work now going on around
her ?   She is, and must be, behindhand.   Scientific
chauvinism is not beautiful in my eyes.   Still, one can
hardly see, without deprecation and protest, the English
investigator handicapped in so great a race by short-
sighted and mischievous legislation.

A great scientific theory has never been accepted
without opposition.   The theory of gravitation, the
theory of undulation, the theory of evolution, the
dynamical theory of heat—all had to push their way
through conflict to victory.   And so it has been with
the Germ Theory of communicable diseases.   Some
outlying members of the medical profession dispute it
still.   I am told they even dispute the communica-
bility of cholera.   Such must always be the course of
things, as long as men are endowed with different
degrees of insight.   Where the mind of genius discerns the distant truth, which it pursues, the mind
not so gifted often discerns nothing but the extra-
vagance, which it avoids.   Names, not yet forgotten,
could be given to illustrate these two classes of minds.
As representative of the first class, I would name a
man whom I have often named before, who fought,
in England, the battle of the germ theory with
persistent valour, but whose labours broke him down
before he saw the triumph which he *fore*saw completed.
Many of my medical friends will understand that I
allude here to the late Dr. William Budd, of Bristol.

The task expected of me is now accomplished, and
the reader is here presented with a record, in which
the verities of science are endowed with the interest of
romance.

1884.

THE RAINBOW AND ITS CONGENERS.[1]

THE oldest historic reference to the rainbow is known to all : ' I do set my bow in the cloud, and it shall be for a token of a covenant between me and the earth. . . . And the bow shall be in the cloud ; and I shall look upon it, that I may remember the everlasting covenant between God and every living creature of all flesh that is upon the earth.' To the sublime conceptions of the theologian succeeded the desire for exact knowledge characteristic of the man of science. Whtaever its ultimate cause might have been, the proximate cause of the rainbow was physical, and the aim of science was to account for the bow on physical principles. Progress towards this consummation was very slow. Slowly the ancients mastered the principles of reflection. Still more slowly were the laws of refraction dug from the quarries in which Nature had embedded them. I use this language, because the laws were incorporate in Nature before they were discovered by man. Until the time of Alhazan, an Arabian mathematician, who lived at the beginning of the twelfth century, the views entertained regarding refraction were utterly vague and incorrect. After Alhazan came Roger Bacon and Vitellio,[2] who

---

[1] A Friday evening discourse at the Royal Institution.

[2] Whewell (*History of the Inductive Sciences*, vol. i. p. 345) describes Vitellio as a Pole. His mother was a Pole ; but Poggendorff (*Handwörterbuch d. exacten Wissenschaften*) claims Vitellio himself

made and recorded many observations and measurements on the subject of refraction. To them succeeded Kepler, who, taking the results tabulated by his predecessors, applied his amazing industry to extract from them their meaning—that is to say, to discover the physical principles which lay at their root. In this attempt he was less successful than in his astronomical labours. In 1604 Kepler published his 'Supplement to Vitellio,' in which he virtually acknowledged his defeat by enunciating an approximate rule, instead of an all-satisfying natural law. The discovery of such a law, which constitutes one of the chief corner-stones of optical science, was made by Willebrord Snell, about 1621.[1]

A ray of light may, for our purposes, be presented to the mind as a luminous straight line. Let such a ray be supposed to fall vertically upon a perfectly calm water-surface. The incidence, as it is called, is then perpendicular, and the ray goes through the water without deviation to the right or left. In other words, the ray in the air and the ray in the water form one continuous straight line. But the least deviation from the perpendicular causes the ray to be broken, or 'refracted,' at the point of incidence. What, then, is the law of refraction discovered by Snell? It is this, that no matter how the angle of incidence, and with it the angle of refraction, may vary, the relative magnitude of two lines, dependent on these angles, and called their sines, remains, for the same medium, perfectly unchanged. Measure, in other words, for various angles, each of these two lines with a scale, and divide the length of the longer one by that of the shorter; then, however the lines individually vary in length, the quotient

as a German, born in Thüringen. 'Vitellio' is described as a corruption of Witelo.
[1] Born at Leyden, 1591; died 1626.

yielded by this division remains absolutely the same. It is, in fact, what is called ' the index of refraction ' of the medium.

Science is an organic growth, and accurate measurements give coherence to the scientific organism. Were it not for the antecedent discovery of the law of sines, founded as it was on exact measurements, the rainbow could not have been explained. Again and again, moreover, the angular distance of the rainbow from the sun had been determined and found constant. In this divine remembrancer there was no variableness. A line drawn from the sun to the rainbow, and another drawn from the rainbow to the observer's eye, always enclosed an angle of 41°. Whence this steadfastness of position—this inflexible adherence to a particular angle? Newton gave to De Dominis [1] the credit of the answer ; but we really owe it to the genius of Descartes. He followed with his mind's eye the rays of light impinging on a raindrop. He saw them in part reflected from the outside surface of the drop. He saw them refracted on entering the drop, reflected from its back, and again refracted on their emergence. Descartes was acquainted with the law of Snell, and taking up his pen he calculated, by means of that law, the whole course of the rays. He proved that the vast majority of them escaped from the drop as *divergent* rays, and, on this account, soon became so enfeebled as to produce no sensible effect upon the eye of an observer. At one particular angle, however—namely, the angle 41° aforesaid—they emerged in a practically *parallel sheaf*. In their union was strength, for it was this particular

[1] Archbishop of Spalatro and Primate of Dalmatia. Fled to England about 1616 ; became a Protestant, and was made Dean of Windsor. Returned to Italy, and resumed his Catholicism ; but was handed over to the Inquisition, and died in prison (Poggendorff's *Biographical Dictionary*).

sheaf which carried the light of the 'primary' rainbow to the eye.

There is a certain form of emotion called intellectual pleasure, which may be excited by poetry, literature, Nature, or art. But I doubt whether among the pleasures of the intellect there is any more pure and concentrated than that experienced by the scientific man when a difficulty which has challenged the human mind for ages melts before his eyes, and re-crystallises as an illustration of natural law. This pleasure was doubtless experienced by Descartes when he succeeded in placing upon its true physical basis the most splendid meteor of our atmosphere. Descartes showed, moreover, that the 'secondary bow' was produced when the rays of light underwent two reflections within the drop, and two refractions at the points of incidence and emergence.

It is said that Descartes behaved ungenerously to Snell—that, though acquainted with the unpublished papers of the learned Dutchman, he failed to acknowledge his indebtedness. On this I will not dwell, for I notice on the part of the public a tendency, at all events in some cases, to emphasise such shortcomings. The temporary weakness of a great man is often taken as a sample of his whole character. The spot upon the sun usurps the place of his 'surpassing glory.' This is not unfrequent, but it is nevertheless unfair.

Descartes proved that, according to the principles of refraction, a circular band of light must appear in the heavens exactly where the rainbow is seen. But how are the colours of the bow to be accounted for? Here his penetrative mind came to the very verge of the solution, but the limits of knowledge at the time barred his further progress. He connected the colours of the rainbow with those produced by a prism ; but then these

latter needed explanation just as much as the colours
of the bow itself. The solution, indeed, was not possible
until the composite nature of white light had been
demonstrated by Newton. Applying the law of Snell
to the different colours of the spectrum, Newton proved
that the primary bow must consist of a series of concentric circular bands, the largest of which is red and the
smallest violet; while in the secondary bow these colours
must be reversed. The main secret of the rainbow, if
I may use such language, was thus revealed.

I have said that each colour of the rainbow is
carried to the eye by a sheaf of approximately parallel
rays. But what determines this parallelism? Here
our real difficulties begin, but they are to be surmounted
by attention. Let us endeavour to follow the course
of the solar rays before and after they impinge upon a
spherical drop of water. Take, first of all, the ray that
passes through the centre of the drop. This particular
ray strikes the back of the drop as a perpendicular, its
reflected portion returning along its own course. Take
another ray close to this central one and parallel to it
—for the sun's rays when they reach the earth are
parallel. When this second ray enters the drop it is
refracted; on reaching the back of the drop it is there
reflected, being a second time refracted on its emergence
from the drop. Here the incident and the emergent
rays enclose a small angle with each other. Take, again,
a third ray a little further from the central one than
the last. The drop will act upon it as it acted upon
its neighbour, the incident and emergent rays enclosing
in this instance a larger angle than before. As we
retreat farther from the central ray the enlargement of
this angle continues up to a certain point, where it
reaches a maximum, after which farther retreat from
the central ray diminishes the angle. Now, a maximum

resembles the ridge of a hill, or a watershed, from which the land falls in a slope at each side. In the case before us the divergence of the rays when they quit the raindrop would be represented by the steepness of the slope. On the top of the watershed—that is to say, in the neighbourhood of our maximum—is a kind of summit-level, where the slope for some distance almost disappears. But the disappearance of the slope indicates, in the case of our raindrop, the absence of divergence. Hence we find that at our maximum, and close to it, there issues from the drop a sheaf of rays which are nearly, if not quite, parallel to each other. These are the so-called ' effective rays ' of the rainbow.[1]

Let me here point to a series of measurements which will illustrate the gradual augmentation of the deflection just referred to until it reaches its maximum, and its gradual diminution at the other side of the maximum. The measures correspond to a series of angles of incidence which augment by steps of ten degrees.

i			d	i			d
10°	.	.	10°	60°	.	.	42° 28'
20°	.	.	19° 36'	70°	.	.	39° 48'
30°	.	.	28° 20'	80°	.	.	31° 4'
40°	.	.	35° 36'	90°	.	.	15°
50°	.	.	40° 40'				

[1] There is, in fact, a bundle of rays near the maximum, which, when they enter the drop, are converged by refraction almost exactly to the same point at its back. If the convergence were *quite* exact, then the symmetry of the liquid sphere would cause the rays to quit the drop as they entered it—that is to say, perfectly parallel. But inasmuch as the convergence is not quite exact, the parallelism after emergence is only approximate. The emergent rays cut each other at extremely sharp angles, thus forming a ' caustic ' which has for its asymptote the ray of maximum deviation. In the secondary bow we have to deal with a minimum, instead of a maximum, the crossing of the incident and emergent rays producing the observed reversal of the colours. (See Engel and Shellbach's published diagrams of the rainbow.)

The figures in the column i express the angles of incidence, while under d we have in each case the accompanying deviation, or the angle enclosed by the incident
and emergent rays. It will be seen that as the angle i
increases, the deviation also increases up to 42° 28', after
which, although the angle of incidence goes on augmenting, the deviation becomes less. The maximum
42° 28' corresponds to an incidence of 60°, but in reality
at this point we have already passed, by a small quantity,
the exact maximum, which occurs between 58° and 59°.
Its amount is 42° 30'. This deviation corresponds to
the red band of the rainbow. In a precisely similar
manner the other colours rise to their maximum, and
fall on passing beyond it ; the maximum for the violet
band being 40° 30'. The entire width of the primary
rainbow is therefore 2°, part of this width being due to
the angular magnitude of the sun.

We have thus revealed to us the geometric construction of the rainbow. But though the step here
taken by Descartes and Newton was a great one, it left
the theory of the bow incomplete. Within the rainbow proper, in certain conditions of the atmosphere,
are seen a series of richly-coloured zones, which were
not explained by either Descartes or Newton. They are
said to have been first described by Mariotte,[1] and they
long challenged explanation. At this point our difficulties thicken, but, as before, they are to be overcome by
attention. It belongs to the very essence of a maximum, approached continuously on both sides, that on
the two sides of it pairs of equal value may be found.
The maximum density of water, for example, is 39°
Fahr. Its density, when 5° colder and when 5° warmer
than this maximum, is the same. So also with

[1] Prior of St. Martin-sous-Beaune, near Dijon ; member of the
French Academy of Sciences. Died in Paris, May 1684.

regard to the slopes of a watershed. A series of pairs
of points of the same elevation can be found upon the
two sides of the ridge; and, in the case of the rainbow,
on the two sides of the maximum deviation we have a
succession of pairs of rays having the same deflection.
Such rays travel along the same line, and add their
forces together after they quit the drop. But light,
thus reinforced by the coalescence of non-divergent
rays, ought to reach the eye. It does so; and were
light what it was once supposed to be—a flight of
minute particles sent by luminous bodies through space
—then these pairs of equally-deflected rays would
diffuse brightness over a large portion of the area within
the primary bow. But inasmuch as light consists of
waves, and not of particles, the principle of interference comes into play, in virtue of which waves alter-
nately reinforce and destroy each other. Were the
distance passed over by the two corresponding rays
within the drop the same, they would emerge as they
entered. But in no case are the distances the same.
The consequence is that when the rays emerge from
the drop they are in a condition either to support or
to destroy each other. By such alternate reinforce-
ment and destruction, which occur at different places
for different colours, the coloured zones are produced
within the primary bow. They are called 'supernumerary bows,' and are seen, not only within the
primary, but sometimes also outside the secondary bow.
The condition requisite for their production is, that
the drops which constitute the shower shall all be of
nearly the same size. When the drops are of different
sizes, we have a confused superposition of the different
colours, an approximation to white light being the consequence. This second step in the explanation of the
rainbow was taken by a man the quality of whose

genius resembled that of Descartes or Newton, and who eighty-two years ago was appointed Professor of Natural Philosophy in the Royal Institution of Great Britain. I refer, of course, to the illustrious Thomas Young.[1]

But our task is not, even now, complete. The finishing touch to the explanation of the rainbow was given by our eminent Astronomer Royal, Sir George Airy. Bringing the knowledge possessed by the founders of the undulatory theory, and that gained by subsequent workers, to bear upon the question, Sir George Airy showed that, though Young's general principles were unassailable, his calculations were sometimes wide of the mark. It was proved by Airy that the curve of maximum illumination in the rainbow does not quite coincide with the geometric curve of Descartes and Newton. He also extended our knowledge of the supernumerary bows, and corrected the positions which Young had assigned to them. Finally, Professor Miller, of Cambridge, and Dr. Galle, of Berlin, illustrated by careful measurements with the theodolite the agreement which exists between the theory of Airy and the facts of observation. Thus, from Descartes to Airy, the intellectual force expended in the elucidation of the rainbow, though broken up into distinct personalities, might be regarded as that of an individual artist, engaged throughout this time in lovingly contemplating, revising, and perfecting his work.

We have thus cleared the ground for the series of experiments which constitute the subject of this discourse. During our brief residence in the Alps this year, we were favoured with some weather of matchless

[1] Young's *Works*, edited by Peacock, vol. i. pp. 185, 293, 357.

perfection; but we had also our share of foggy and drizzly weather. On the night of the 22nd of September the atmosphere was especially dark and thick. At 9 P.M. I opened a door at the end of a passage and looked out into the gloom. Behind me hung a small lamp, by which the shadow of my body was cast upon the fog. Such a shadow I had often seen, but in the present case it was accompanied by an appearance which I had not previously noticed. Swept through the darkness round the shadow, and far beyond, not only its boundary, but also beyond that of the illuminated fog, was a pale, white, luminous circle, complete except at the point where it was cut through by the shadow. As I walked out into the fog, this curious halo went in advance of me. Had not my demerits been so well known to me, I might have accepted the phenomenon as an evidence of canonisation. Benvenuto Cellini saw something of the kind surrounding his shadow, and ascribed it forthwith to supernatural favour. I varied the position and intensity of the lamp, and found even a candle sufficient to render the luminous band visible. With two crossed laths I roughly measured the angle subtended by the radius of the circle, and found it to be practically the angle which had riveted the attention of Descartes—namely, 41°. This and other facts led me to suspect that the halo was a circular rainbow. A week subsequently, the air being in a similar misty condition, the luminous circle was well seen from another door, the lamp which produced it standing on a table behind me.

It is not, however, necessary to go to the Alps to witness this singular phenomenon. Amid the heather of Hind Head I have had erected a hut, to which I escape when my brain needs rest or my muscles lack vigour. The hut has two doors, one opening to the north and the other to the south, and in it we have been able to occupy ourselves pleasantly and profitably during the recent misty weather. Removing the shade from a small petroleum lamp, and placing the lamp behind me, as I stood in either doorway, the luminous circles surrounding my shadow on different nights were very remarkable. Sometimes they were best to the north, and sometimes the reverse, the difference depending for the most part on the direction of the wind. On Christmas night the atmosphere was particularly favourable. It was filled with true fog, through which, however, descended palpably an extremely fine rain. Both to the north and to the south of the hut the luminous circles were on this occasion specially bright and well defined. They were, as I have said, swept through the fog far beyond its illuminated area, and it was the darkness against which they were projected which enabled them to shed so much apparent light. The 'effective rays,' therefore, which entered the eye in this observation gave *direction*, but not distance, so that the circles appeared to come from a portion of the atmosphere which had nothing to do with their production. When the lamp was taken out into the fog, the illumination of the medium almost obliterated the halo. Once educated, the eye could trace it, but it was toned down almost to vanishing. There is some advantage, therefore, in possessing a hut, on a moor or on a mountain, having doors which limit the area of fog illuminated.

I have now to refer to another phenomenon which is but rarely seen, and which I had an opportunity of witnessing on Christmas Day. The mist and drizzle in the early morning had been very dense; a walk before breakfast caused the nap of my somewhat fluffy pilot-dress to be covered with minute water-globules, which, against the dark background underneath, suggested the bloom of a plum. As the day advanced, the south-eastern heaven became more luminous ; and the pale disk of the sun was at length seen struggling through drifting clouds. At ten o'clock the sun had become fairly victorious, the heather was adorned by pendent drops, while certain branching grasses, laden with liquid pearls, presented, in the sunlight, an appearance of exquisite beauty. Walking across the common to the Portsmouth road my wife and I, on reaching it, turned our faces sunwards. The smoke-like fog had vanished, but its disappearance was accompanied, or perhaps caused, by the coalescence of its minuter particles into little globules, visible where they caught the light at a proper angle, but not otherwise. They followed every eddy of the air, upwards, downwards, and from side to side. Their extreme mobility was well calculated to suggest a notion prevalent on the Continent, that the particles of a fog, instead of being full droplets, are really little bladders or vesicles. Clouds are supposed to owe their power of flotation to this cause. This vesicular theory never struck root in England ; nor has it, I apprehend, any foundation in fact.

As I stood in the midst of these eddying specks, so visible to the eye, yet so small and light as to be per-fectly impalpable to the skin both of hands and face, I remarked, 'These particles must surely yield a bow of some kind.' Turning my back to the sun, I stooped down so as to keep well within the layer of particles, which I supposed to be a shallow one, and, looking towards the ' Devil's Punch Bowl,' saw the anticipated phenomenon. A bow without colour spanned the Punch Bowl. Though white and pale it was well defined, and exhibited an aspect of weird grandeur. Once or twice I fancied a faint ruddiness could be discerned on its outer boundary. The stooping was not necessary, and as we walked along the new Portsmouth road, with the Punch Bowl to our left, the white arch marched along with us. At a certain point we ascended to the old Portsmouth road, whence, with a flat space of very dark heather in the foreground, we watched the bow. The sun had then become strong, and the sky above us blue, nothing which could in any proper sense be called rain existing at the time in the atmosphere. Suddenly my companion exclaimed, ' I see the whole circle meeting at my feet ! ' At the same moment the circle became visible to me also. It was the darkness of our immediate foreground that enabled us to see the lower half of the pale luminous band projected against it. We walked round Hind Head Common with the bow almost always in view. Its crown sometimes disappeared, showing that the minute globules which produced it did not extend to any great height in the atmosphere. In such cases two shining buttresses were left behind, which, had not the bow been previously seen, would have lacked all significance. In some of the combes, or valleys, where the floating particles had collected in greater numbers, the end of the bow plunging into the combe emitted a light of more than the usual brightness. During our walk the bow was broken and re-formed several times ; and, had it not been for our previous experience both in the Alps and at Hind Head, it might well have escaped attention. What this colourless white bow lost in intensity, as compared with the ordinary coloured bow, was more than atoned for by its weirdness and its novelty to both observers.

The white rainbow (*l'arc-en-ciel blanc*) was first described by the Spaniard Don Antonio de Ulloa, Lieutenant of the Company of Gentlemen Guards of the

Marine. By order of the King of Spain, Don Jorge
Juan and Ulloa made an expedition to South America,
an account of which is given in two amply-illustrated
quarto volumes to be found in the library of the Royal
Institution. The bow was observed from the summit
of the mountain Pambamarca, in Peru. The angle
subtended by its radius was 33° 30', which is considerably less than the angle subtended by the radius
of the ordinary bow. Between the phenomenon observed by us on Christmas Day, and that described by
Ulloa, there are some points of difference. In his case
fog of sufficient density existed to enable the shadows
of him and his six companions to be seen, each however only by the person whose body cast the shadow.
Around the head of each were observed those zones of
colour which characterise the ' spectre of the Brockeñ.'
In our case no shadows were to be seen, for there was
no fog-screen on which they could be cast. This implies also the absence of the zones of colour observed
by Ulloa.

The white rainbow has been explained in various
ways. A learned Frenchman, M. Bravais, who has
written much on the optical phenomena of the atmosphere, and who can claim the additional recommenda-
tion of being a distinguished mountaineer, has sought
to connect the bow with the vesicular theory to which
I have just referred. This theory, however, is more
than doubtful, and it is not necessary.[1] The genius of
Thomas Young throws light upon this subject, as upon
so many others. He showed that the whiteness of the
bow was a direct consequence of the smallness of the
drops which produce it. In fact, the wafted water-

---

[1] The vesicular theory was combated very ably in France by the
Abbé Raillard, who has also given an interesting analysis of the
rainbow at the end of his translation of my *Notes on Light*.

specks seen by us upon Hind Head[1] were the very kind needed for the production of the phenomenon. But the observations of Ulloa place his white bow distinctly *within* the arc that would be occupied by the ordinary rainbow—that is to say, in the region of supernumeraries—and by the action of the supernumeraries upon each other Ulloa's bow was accounted for by Thomas Young. The smaller the drops, the broader are the zones of the supernumerary bows, and Young proved by calculation that when the drops have a diameter of $\frac{1}{3000}$th or $\frac{1}{4000}$th of an inch, the bands overlap each other, and produce white light by their mixture. Unlike the geometric bow, the radius of the white bow varies within certain limits, which M. Bravais shows to be 33° 30′ and 41° 46′ respectively. In the latter case the white bow is the ordinary bow deprived of its colour by the superposition due to the smallness of the drops. In all the other cases it is produced by the action of the supernumeraries.

The physical investigator desires not only to observe natural phenomena but to re-create them—to bring them, that is, under the dominion of experiment. From observation we learn what Nature is willing to reveal. In experimenting we place her in the witnessbox, cross-examine her, and extract from her knowledge in excess of that which would, or could, be spontaneously given. Accordingly, on my return from Switzerland last October, I sought to reproduce in the laboratory the effects observed among the mountains. My first object, therefore, was to obtain artificially a mixture of fog and drizzle like that observed from the

[1] Had our refuge in the Alps been built on the southern side of the valley of the Rhone, so as to enable us to look with the sun behind us into the valley and across it, we should, I think, have frequently seen the white bow.

door of our Alpine cottage. A strong cylindrical copper boiler, sixteen inches high and twelve inches in
diameter, was nearly filled with water and heated by gas-
flames until steam of twenty pounds pressure was produced. A valve at the top of the boiler was then
opened, when the steam issued violently into the
atmosphere, carrying droplets of water mechanically
along with it, and condensing above to droplets of a
similar kind. A fair imitation of the Alpine atmosphere was thus produced. After a few tentative experiments, the luminous circle was brought into view,
and having once got hold of it, the next step was to
enhance its intensity. Oil-lamps, the lime-light, and
the naked electric light were tried in succession, the
source of rays being placed in one room, the boiler in
another, while the observer stood, with his back to the
light, between them. It is not, however, necessary to
dwell upon these first experiments, surpassed as they
were by the arrangements subsequently adopted. My
mode of proceeding was this. The electric light being
placed in a camera with a condensing-lens in front,
the position of the lens was so fixed as to produce a
beam sufficiently broad to clasp the whole of my head,
and leave an aureole of light around it. It being desirable to lessen as much as possible the foreign light
entering the eye, the beam was received upon a distant
black surface, and it was easy to move the head until
its shadow occupied the centre of the illuminated area.

To secure the best effect it was found necessary to
stand close to the boiler, so as to be immersed in the
fog and drizzle. The fog, however, was soon discovered
to be a mere nuisance. Instead of enhancing, it blurred
the effect, and I therefore sought to abolish it. Allowing the steam to issue for a few seconds from the
boiler, on closing the valve, the cloud rapidly melted

away, leaving behind it a host of minute liquid
spherules floating in the beam. An intensely-coloured
circular rainbow was instantly seen in the air in front
of the observer. The primary bow was duly attended
by its secondary, with the colours, as usual, reversed.
The opening of the valve for a single second caused
the bows to flash forth. Thus, twenty times in succession, puffs could be allowed to issue from the boiler,
every puff being followed by the appearance of this
splendid meteor. The bows produced by single puffs
are evanescent, because the little globules rapidly disappear. Greater permanence is secured when the valve
is left open for an · interval sufficient to discharge a
copious amount of drizzle into the air.[1]

Many other appliances for producing a fine rain
have been tried, but a reference to two of them will
suffice. The rose of a watering-pot naturally suggests
a means of producing a shower ; and on the principle
of the rose I had some spray-producers constructed.
In each case the outer surface was convex, the thin
convex metal plate being pierced by orifices too small
to be seen by the naked eye. Small as they are, fillets
of very sensible magnitude issue from the orifices,
but at some distance below the orifices the fillets shake
themselves asunder and form a fine rain. The small

---

[1] It is perhaps worth noting here, that when the camera and lens
are used, the beam which sends its 'effective rays' to the eye may
not be more than a foot in width, while the circular bow engendered
by these rays may be, to all appearance, fifteen or twenty feet in
diameter. In such a beam, indeed, the drops which produce the bow
must be very near the eye, for rays from the more distant drops
would not attain the required angle. The apparent distance of the
circular bow is often great in comparison with that of the originat-
ing drops. Both distance and diameter may be made to undergo
variations. In the rainbow we do not see a localised object, but
receive a luminous impression, which is often transferred to a portion of the field of view far removed from the bow's origin.

orifices are very liable to get clogged by the particles
suspended in London water. In experiments with the
rose, filtered water was therefore resorted to. A large
vessel was mounted on the roof of the Royal Institution, from the bottom of which descended vertically
a piece of compo-tubing, an inch in diameter and
about twenty feet long. By means of proper screw-
fittings, a single rose, or when it is desired to
increase the magnitude or density of the shower, a
group of two, three, or four roses, was attached to
the end of the compo-tube. From these, on the turning on of a cock, the rain fell. The circular bows
produced by such rain are far richer in colour than
those produced by the smaller globules of the condensed steam. To see the effect in all its beauty and
completeness, it is necessary to stand well within the
shower, not outside of it. A waterproof coat and cap
are therefore needed, to which a pair of goloshes may
be added with advantage. A person standing outside
the beam may see bits of both primary and secondary
bows in the places fixed by their respective angles ; but
the colours are washy and unimpressive. Within the
shower, with the shadow of the head occupying its
proper position on the screen, the brilliancy of the
effect is extraordinary. The primary clothes itself in
the richest tints, while the secondary, though less
vivid, shows its colours in surprising strength and
purity.

But the primary bow is accompanied by appearances
calculated to attract and rivet attention almost more
than the bow itself. I have already mentioned the
existence of effective rays over and above those which
go to form the geometric bow. They fall within the
primary, and, to use the words of Thomas Young,
' would exhibit a continued diffusion of fainter light,

but for the general law of interference which divides
the light into concentric rings.' One could almost
wish for the opportunity of showing Young how literally his words are fulfilled, and how beautifully his
theory is illustrated, by these artificial circular rainbows. For here the space within the primaries is
swept by concentric supernumerary bands, coloured
like the rainbow, and growing gradually narrower as
they retreat from the primary. These spurious bows,
as they are sometimes called,[1] which constitute one of
the most striking illustrations of the principle of interference, are separated from each other by zones of
darkness, where the light-waves on being added together destroy each other. I have counted as many
as eight of these beautiful bands, concentric with the
true primary. The supernumeraries are formed next
to the most refrangible colour of the bow, and therefore occur *within* the primary circle. But in the
secondary bow, the violet, or most refrangible colour,
is on the *outside*; and, following the violet of the
secondary, I have sometimes counted as many as five
spurious bows. Some notion may be formed of the
intensity of the primary, when the secondary is able to
produce effects of this description.

An extremely handy spray-producer is that employed to moisten the air in the Houses of Parliament.
A fillet of water, issuing under strong pressure from a
small orifice, impinges on a little disk placed at a
distance of about one-twentieth of an inch from the
orifice. On striking the disk, the water spreads later-
ally, and breaks up into an exceedingly fine spray. Here
also I have used the spray-producer both singly and
in groups, the latter arrangement being resorted to
when showers of special breadth and density were re-

---

[1] A term, I confess, not to my liking.

quired. In regard to primaries, secondaries, and super-numeraries, extremely brilliant effects have been obtained with this form of spray - producer. The quantity of water called upon being much less than that required by the rose, the fillet-and-disk instrument produces less flooding of the locality where the experiments are made. In this latter respect, the steam-boiler spray is particularly handy. A puff of two seconds' duration suffices to bring out the bows, the subsequent shower being so light as to render the use of waterproof clothing unnecessary. In other cases, the inconvenience of flooding may be avoided to a great extent by turning on the spray for a short time only, and then cutting off the supply of water. The vision of the bow being, however, proportionate to the duration of the shower, will, when the shower is brief, be evanescent. Hence, when quiet and continued contemplation of all the phenomena is desired, the observer must make up his mind to brave the rain.[1]

In one important particular the spray-producer last described commends itself to our attention. With it we can operate on substances more costly than water, and obtain rainbows from liquids of the most various refractive indices. To extend the field of experiment in this direction, the following arrangement has been devised : A strong cylindrical iron bottle, wholly or partly filled with the liquid to be experimented on, is tightly closed by a brass cap. Through the cap passes a metal tube, soldered air-tight where it crosses the cap, and ending near the bottom of the iron bottle. To the free end of this tube is attached the spray-producer. A second tube passes also through the cap, but ends above the surface of the liquid. This second tube,

[1] The rays which form the artificial bow emerge, as might be expected, polarised from the drops.

which is long and flexible, is connected with a larger iron bottle, containing compressed air. Hoisting the small bottle to a convenient height, the tap of the larger bottle is carefully opened, the air passes through the flexible tube to the smaller bottle, exerts its pressure upon the surface of the liquid therein contained, drives it up the other tube, and causes it to impinge with any required degree of force against the disk of the spray-producer. From this it falls in a fine rain. A great many liquids, including coloured ones,[1] have been tested by this arrangement, and very remarkable results have been obtained. I will confine myself here to a reference to two liquids, which commend themselves on account of their cheapness and of the brilliancy of their effects. Spirit of turpentine, forced from the iron bottle and caused to fall in a fine shower, produces a circular bow of extraordinary intensity and depth of colour. With paraffin oil or petroleum an equally brilliant effect is obtained.

Spectrum analysis, as generally understood, occupies itself with atomic or molecular action, but physical spectrum analysis may obviously be brought to bear upon our falling showers. A composite shower —that is to say, one produced by the mingled spray of two or more liquids—could, it seems plain, be analysed and made to declare its constituents by the production of the circular rainbows proper to the respective liquids. This was found to be the case. In the ordinary rainbow the narrowest colour-band is produced by its most refrangible light. In general terms, the greater the refraction, the smaller is the bow. Now, as spirit of turpentine and paraffin are both more refractive than water, it might be concluded that

---

[1] Rose-aniline, dissolved in alcohol, produces a splendid bow with specially broad supernumeraries.

in a mixed shower of water and paraffin, or water and turpentine, the smaller and more luminous circle of the latter would be seen within the larger circle of the former. The result was exactly in accordance with this anticipation. Beginning with water and producing its two bows, then allowing the turpentine to shower down and mingle with the water—within the large and beautifully-coloured water-wheel the more richly-coloured circle of the turpentine makes its appearance. Or beginning with turpentine and forming its concentrated iris ; on turning on the water-spray, though to the eye the shower seems absolutely homogeneous, its true character is instantly declared by the flashing out of the larger concentric aqueous bow. The water primary is accompanied by its secondary close at hand. Associated, moreover, with all the bows, primary and secondary, are the supernumeraries which belong to them ; and a more superb experimental illustration of optical principles it would be hardly possible to witness. It is not the less impressive because extracted from the simple combination of a beam of light and a shower of rain.

In the 'Philosophical Transactions' for 1835 the late Colonel Sykes gave a vivid description of a circular solar rainbow, observed by him in India during periods when fogs and mists were prevalent in the chasms of the Ghâts of the Deccan.

'It was during such periods that I had several opportunities of witnessing that singular phenomenon, the circular rainbow, which, from its rareness, is spoken of as a possible occurrence only. The stratum of fog from the Konkun on some occasions rose somewhat above the level of the top of a precipice forming the north-west scarp of the hill-fort of Hurreechundurghur, from 2,000 to 3,000 feet perpendicular, without coming

over upon the table-land. I was placed at the edge of
the precipice, just without the limits of the fog, and
with a cloudless sun at my back at a very low elevation. Under such a combination of favourable circumstances the circular rainbow appeared quite perfect,
of the most vivid colours, one-half above the level on
which I stood, the other half below it. Shadows in
distinct outline of myself, my horse, and people appeared in the centre of the circle as a picture, to which
the bow formed a resplendent frame. My attendants
were incredulous that the figures they saw under such extraordinary circumstances could be their own shadows,
and they tossed their arms and legs about, and put
their bodies into various postures, to be assured of the
fact by the corresponding movements of the objects
within the circle; and it was some little time ere the
superstitious feeling with which the spectacle was
viewed wore off. From our proximity to the fog, I
believe the diameter of the circle at no time exceeded
fifty or sixty feet. The brilliant circle was accompanied
by the usual outer bow in fainter colours.'

Mr. E. Colbourne Baber, an accomplished and in-
trepid traveller, has recently enriched the 'Transactions' of the Royal Geographical Society by a paper of
rare merit, in which his travels in Western China are
described. He made there the ascent of Mount O—
an eminence of great celebrity. Its height is about
11,000 feet above the sea, and it is flanked on one side
by a cliff 'a good deal more than a mile in height.'
From the edge of this cliff, which is guarded by posts
and chains, you look into an abyss, and if fortune, or
rather the mists, favour you, you see there a miracle,
which is thus described by Mr. Baber:—

'Naturally enough it is with some trepidation that
pilgrims approach this fearsome brink, but they are

drawn to it by the hope of beholding the mysterious apparition known as the " Fo-Kuang," or " Glory of Buddha," which floats in mid-air half-way down. So many eye-witnesses had told me of this wonder that I could not doubt ; but I gazed long and steadfastly into the gulf without success, and came away disappointed, but not incredulous. It was described to me as a circle of brilliant and many-coloured radiance, broken on the outside with quick flashes, and surrounding a central disc as bright as the sun, but more beautiful. Devout Buddhists assert that it is an emanation from the aureole of Buddha, and a visible sign of the holiness of Mount O.

' Impossible as it may be deemed, the phenomenon does really exist. I suppose no better evidence could be desired for the attestation of a Buddhist miracle than that of a Baptist missionary, unless indeed it be, as in this case, that of *two* Baptist missionaries. Two gentlemen of that persuasion have ascended the mountain since my visit, and have seen the Glory of Buddha several times. They relate that it resembles a golden sun-like disc, enclosed in a ring of prismatic colours more closely blended than in the rainbow. . . . The missionaries inform me that it was about three o'clock in the afternoon, near the middle of August, when they saw the meteor, and that it was only visible when the precipice was more or less clothed in mist. It appeared to lie on the surface of the mist, and was always in the direction of a line drawn from the sun through their heads, as is certified by the fact that the shadow of their heads was seen on the meteor. They could get their heads out of the way, so to speak, by stooping down, but are not sure if they could do so by stepping aside. Each spectator, however, could see the shadows of the bystanders as well as his own projected on to the

appearance. They did not observe any rays spreading from it. The central disc they think is a reflected image of the sun, and the enclosing ring is a rainbow. The ring was in thickness about one-fourth of the diameter of the disc, and distant from it by about the same extent ; but the recollection of one informant was that the ring touched the disc, without any intervening space. The shadow of a head, when thrown upon it, covered about one-eighth of the whole diameter of the meteor. The rainbow ring was not quite complete in its lower part, but they attribute this to the interposition of the edge of the precipice. They see no reason why the appearance should not be visible at night when the moon is brilliant and appositely placed. They profess themselves to have been a good deal surprised, but not startled, by the spectacle. They would consider it remarkable rather than astonishing, and are disposed to call it a very impressive phenomenon.'

It is to be regretted that Mr. Baber failed to see the ' Glory,' and that we in consequence miss his own description of it. There seems a slight inadvertence in the statement that the head could be got out of the way by stooping. The shadow of the head must have always occupied the centre of the ' Glory.'

Thus, starting from the first faint circle seen in the thick darkness at Alp Lusgen, we have steadily followed and developed our phenomenon, and ended by rendering the ' Glory of Buddha ' a captive of the laboratory. The result might be taken as typical of larger things.

[On Sept. 25, 1890, my friend M. Sarasin and myself witnessed at Alp Lusgen a very perfect example of the white bow. See page 329.]

ADDRESS DELIVERED AT THE BIRKBECK
INSTITUTION ON OCTOBER 22, 1884.

OUR lives are interwoven here below, frequently—
indeed most frequently—without our knowing it.
We are in great part moulded by unconscious interaction. Thus, without intending it, the present representative of the Birkbeck family in Yorkshire has helped
to shape my life. In 1856, or thereabouts, Mr. John
Birkbeck aided in founding on the slope of a Swiss
mountain the Æggischhorn Hotel. The success of this
experiment provoked in the neighbouring commune a
spirit of rivalry and imitation, and accordingly, upon a
bold bluff overlooking the great Aletsch glacier, was
subsequently planted the Bel Alp Hotel. To the Bel
Alp I went in my wanderings. Seeing it often I liked
it well, until at length the thought dawned upon me of
building a permanent nest there. Before doing so, however, I imitated the birds—chose, and was chosen by,
a mate who, like myself, loved the freedom of the
mountains, and we built our nest together.

From that nest I have come straight to the Birkbeck
Institution, so that the following chain of connection
stretches between Mr. John Birkbeck and me. Without him there would have been no Æggischhorn ; without the Æggischhorn there would have been no Bel
Alp ; without Bel Alp there would have been no Tyndall's
nest, and without that nest the person who now
addresses you would undoubtedly be a different man
from what he is. His bone would have been different

bone; his flesh different flesh; nay, the very grey matter of his brain, which is said to be concerned in the production of thought, would have been different from what it now is. The inference is obvious that, should this lecture prove a failure and a bore, or should any hitch occur to cause me to break down in the middle of it, you are bound in common fairness to lay upon the shoulders of Mr. John Birkbeck, who has tampered so seriously with my bodily and mental constituents, a good round share of the blame.

Thus I seek to shirk responsibility in regard to this lecture; and I dare say you would forgive me if I went a little further in this somewhat ignoble line. It is the fashion of the hour. Some of England's most conspicuous sons at the present day would seem to trace their moral pedigree to that mean old gardener who threw upon his wife the whole blame of eating the forbidden fruit. In reference to the present occasion, I wrote to Mr. Norris from the Alps, asking him to choose between a purely scientific lecture and an address based on the experiences of my own life. He chose the latter. I do not, however, ask you to blame Mr. Norris, but to blame me if a chapter from the personal history of a worker, instead of proving a stimulus and an aid, should seem to you flat, stale, and unprofitable.

Every operation of husbandry, every stroke of statesmanship, every movement of philanthropy, to be effectual and successful, must be executed at the proper time. If we sow in the autumn what ought to be sown in the spring, or if we sow in the spring what ought to be sown in the autumn, we can only reap disappointment. Every public movement is tested by the question, ' Does it live ? ' and this may be translated into the question, ' Does it grow ? ' For growth and multiplication constitute the evidence of life. Brought to this test the movement inaugurated by Dr. George Birkbeck returned a full and conclusive answer. It responded, at the proper time, to a national need and to a need of human nature. Not only in the various districts of London, but also in various towns throughout the country, and even beyond the bounds of England, institutions sprang up, founded on the model of the London Mechanics' Institution, which afterwards became the famous Birkbeck Institution, the anniversary of which we celebrate to-day.

Speaking of the opportune beneficence of Dr. Birkbeck's movement reminds me that in the days of my youth, personally and directly, I derived profit from that movement. In 1842, and thereabouts, it was my privilege to be a member of the Preston Mechanics' Institution—to attend its lectures and make use of its library. A learned and accomplished clergyman, named, if I remember aright, John Clay, chaplain of the House of Correction, lectured from time to time on mechanics. A fine earnest old man, named, I think, Moses Holden, lectured on astronomy, while other lecturers took up the subjects of general physics, chemistry, botany, and physiology. My recollection of it is dim, but the instruction then received entered, I doubt not, into the texture of my mind, and influenced me in after-life. One experiment made in these lectures I have never forgotten. Surgeon Corless, I think it was, who lectured on respiration, explaining among other things the changes produced by the passage of air through the lungs. What went in as free oxygen came out bound up in carbonic acid. To prove this he took a flask of lime-water and, by means of a glass tube dipped into it, forced his breath through the water. The carbonic acid from the lungs seized upon the dissolved lime, converting it into carbonate of lime, which, being practically insoluble, was precipitated. All this was predicted beforehand by the lecturer; but the delight with which I saw his prediction fulfilled, by the conversion of the limpid lime-water into a turbid mixture of chalk and water, remains with me as a memory to the present hour. The students of the Birkbeck Institution may therefore grant me the honour of ranking myself among them as a fellow-student of a former generation.

At the invitation of an officer of the Royal Engineers, who afterwards became one of my most esteemed and intimate friends, I quitted school in 1839, to join a division of the Ordnance Survey. The profession of a civil engineer having then great attractions for me, I joined the Survey, intending, if possible, to make myself master of all its operations as a first step towards becoming a civil engineer. Draughtsmen were the best paid, and I became a draughtsman. But I habitually made incursions into the domains of the calculator and computer, and thus learned all their art. In due time the desire to make myself master of field operations caused me to apply for permission to go to the field. The permission was granted by my excellent friend General George Wynne,[1] who then, as Lieutenant Wynne, observed and did all he could to promote my desire for improvement. Before returning to the office I had mastered all the mysteries of ordinary field work. But there remained a special kind of field work which had not been mastered—the taking of trigonometrical observations. By good fortune some work of this kind was required at a time when all the duly-recognised observers were absent. Under the tutelage of a clever master, named Conwill, I had acquired, before quitting

[1] Died at Cologne on June 27, 1890; and was buried there with military honours on June 30.

school, a sound knowledge of elementary geometry and trigonometry. Relying on this to carry me through, I volunteered to make the required observations. After some hesitation, and a little chaff, a theodolite was confided to me.

The instrument, you know, embraces an accurately-graduated horizontal circle for the measurement of horizontal angles, and a similarly graduated vertical circle for the taking of vertical angles. It is moreover furnished with a formidable array of clamp-screws, tangent-screws and verniers, sufficient to tax a novice to unravel them. My first care before applying the instrument was to understand its construction. This accomplished, I took the field with two assistants, who had to measure uphill and downhill along the sides of large triangles into which the whole country had been previously divided. At the same time angles of elevation had to be taken uphill and angles of depression downhill, and from these the true horizontal distance had to be calculated. The heights above the sea-level of the corners of the large triangles had been previously fixed with the utmost accuracy by a very powerful theodolite, and the measurements with my smaller instrument had to come pretty close to the accurate determination to save my work from rejection. Happily I succeeded, though there had been bets against me. The pay upon the Ordnance Survey was very small, but having ulterior objects in view, I considered the instruction received as some set-off to the smallness of the pay. It may prevent some of you young Birkbeckians from considering your fate specially hard, or from being daunted because from a very low level you have to climb a very steep hill, when I tell you that on quitting the Ordnance Survey in 1843, my salary was a little under twenty shillings a week. I have often wondered since at the amount of genuine happiness which a young fellow of regular habits, not caring for either pipe or mug, may extract even from pay like this.

Then came a pause, and after it the mad time of the railway mania, when I was able to turn to some account the knowledge gained upon the Ordnance Survey. In Staffordshire, Cheshire, Lancashire, Durham, and Yorkshire, more especially the last, I was in the thick of the fray. It was a time of terrible toil. The day's work in the field usually began and ended with the day's light, while frequently in the office, and more especially as the awful 30th of November drew near, there was little difference between day and night, every hour of the twenty-four being absorbed in the work of preparation. The 30th of November was the latest date at which plans and sections of projected lines could be deposited at the Board of Trade, failure in this particular often involving the loss of thousands of pounds. One of my last pieces of field work in those days was the taking of a line of levels from the town of Keighley to the village of Haworth in Yorkshire. On a certain day, under grave penalties, these levels had to be finished, and this particular day was one of agony to me. The atmosphere seemed filled with mocking demons, laughing at the vanity of my efforts to get the work done. My levelling-staves were snapped and my theodolite was overthrown by the storm. When things are at their worst a kind of anger often takes the place of fear. It was so in the present instance; I pushed doggedly on, and just at nightfall, when barely able to read the figures on my levelling-staff, I planted my last ' bench-mark ' on a tombstone in Haworth Churchyard. Close at hand was the vicarage of Mr. Brontë, where the genius was nursed which soon afterwards burst forth and astonished the world.

Among the legal giants of those days Austin and Talbot stood supreme. There was something grand as well as merciless in the power wielded by those men in entangling and ruining a hostile witness; and yet it often seemed to me that a clear-headed fellow, who had the coolness, honesty, and courage not to go beyond his knowledge, might have foiled both of them. Then we had the giants of the civil engineers—Stephenson, Brunel, Locke, Hawkshaw, and others. Judged by his power of fence, his promptness in calculation, and his general readiness of retort, George Bidder as a witness was unrivalled. I have seen him take the breath out of Talbot himself before a committee of the House of Lords. Strong men were broken down by the strain and labour of that arduous time. Many pushed through, and are still amongst us in robust vigour. But some collapsed, while others retired, with large fortunes it is true, but with intellects so shattered that, instead of taking their places in the front rank of English statesmen, as their abilities entitled them to do, they sought rest for their brains in the quiet lives of country gentlemen. In my own modest sphere, I well remember the refreshment occasionally derived from five minutes' sleep on a deal table, with Babbage and Callet's Logarithms under my head for a pillow.

It was a time of mad unrest—of downright monomania. In private residences and public halls, in London reception-rooms, in hotels and in the stables of hotels, among gipsies and costermongers, nothing was spoken of but the state of the share market, the prospects of projected lines, the good fortune of the ostler or pot-boy who, by a lucky stroke of business, had cleared ten thousand pounds. High and low, rich and

poor, joined in the reckless game. During my profes-
sional connection with railways I endured three weeks'
misery. It was not defeated ambition; it was not a
rejected love-suit; it was not the hardship endured in
either office or field, but it was the possession of certain
shares which I had purchased in one of the lines then
afloat. The share list of the day proved the winding-
sheet of my peace of mind. I was haunted by the
Stock Exchange. Then, as now, I loved the blue span
of heaven; but when I found myself regarding it
morning after morning, not with the fresh joy which,
in my days of innocence, it had brought me, but solely
with reference to its possible effect, through the harvest,
upon the share market, I became at length so savage
with myself, that nothing remained but to go down to
my brokers and put away the shares as an accursed
thing. Thus began and thus ended, without either
gain or loss, my railway gambling.

During this arduous period of my life my old tendencies, chief among which was the desire to grow
intellectually, did not forsake me; and, when railway
work slackened, I accepted in 1847 a post as master
in Queenwood College, Hampshire—an establishment
which is still conducted with success by a worthy
Principal. There I had the pleasure of meeting Dr.
Frankland, who had charge of the chemical laboratory.
Queenwood College had been the Harmony Hall of the
Socialists, which, under the auspices of the philan-
thropist, Robert Owen, was built to inaugurate the
Millennium. The letters 'C of M,' Commencement of
Millennium, were actually inserted in flint in the
brickwork of the house. Schemes like Harmony Hall
look admirable upon paper; but inasmuch as they are
formed with reference to an ideal humanity, they go to
pieces when brought into collision with the real one.

At Queenwood I learned, by practical experience, that
two factors go to the formation of a teacher. In
regard to knowledge he must, of course, be master of
his work. But knowledge is not all. There may be
knowledge without power—the ability to inform without the ability to stimulate. Both go together in the
true teacher. A power of character must underlie and
enforce the work of the intellect. There are men who
can so rouse and energise their pupils—so call forth
their strength and the pleasure of its exercise—as to
make the hardest work agreeable. Without this power
it is questionable whether the teacher can ever really
enjoy his vocation—with it I do not know a higher,
nobler, more blessed calling, than that of the man who,
scorning the 'cramming' so prevalent in our day, converts the knowledge he imparts into a lever, to lift,
exercise, and strengthen the growing minds committed
to his care.

At the time here referred to I had emerged from
some years of hard labour the fortunate possessor of
two or three hundred pounds. By selling my services
in the dearest market during the railway madness the
sum might, without dishonour, have been made a large
one ; but I respected ties which existed prior to the
time when offers became lavish and temptation strong.
I did not put my money in a napkin, but cherished the
design of spending it in study at a German university.
I had heard of German science, while Carlyle's references
to German philosophy and literature caused me to regard them as a kind of revelation from the gods.
Accordingly, in the autumn of 1848, Frankland and I
started for the land of universities, as Germany is often
called. They are sown broadcast over the country, and
can justly claim to be the source of an important portion
of Germany's present greatness. A portion, but not all.

The thews and sinews of German men were not given by German universities. The steady fortitude and valiant laboriousness which have fought against, and triumphed over, the gravest natural disadvantages are not the result of university culture. But the strength and endurance which belong to the German, as a gift of race, needed enlightenment to direct it; and this was given by the universities. Into these establishments was poured that sturdy power which in other fields had made the wastes of Nature fruitful, and the strong and earnest character had thus superposed upon it the informed and disciplined mind. It is the coalescence of these two factors that has made Germany great; it is the combination of these elements which must prevent England from becoming small. We may bless God for our able journalists, our orderly Parliament, and our free press; but we should bless Him still more for ' the hardy English root ' from which these good things have sprung. We need muscle as well as brains, character and resolution as well as expertness of intellect. Lacking the former, though possessing the latter, we have the bright foam of the wave without its rock-shaking momentum.

Our place of study was the town of Marburg in Hesse Cassel, and a very picturesque town Marburg is. It clambers pleasantly up the hillsides, and falls as pleasantly towards the Lahn. On a May day, when the orchards are in blossom, and the chestnuts clothed with their heavy foliage, Marburg is truly lovely. It has, moreover, a history. It was here that Saint Elizabeth shed her holy influence and dispensed her mercies. The noble double-spired church which bears her name, and contains her dust, stands here to commemorate her. On a high hilltop which dominates the town rises the fine old castle where, in the Rittersaal, Luther and

Zwingli held their famous conference on Consubstantiation and Transubstantiation. Here for a time lived William Tyndale, translator of the Bible into English, who was afterwards strangled and burnt in Vilvorden. Here Wolff expounded his philosophy, and here Denis Papin invented his digester, and is said to have invented a working steam-engine. The principal figure in the university at the time of our visit was Bunsen, who had made his name illustrious by chemical researches of unparalleled difficulty and importance, and by his successful application of chemical and physical principles to explain the volcanic phenomena of Iceland. It was he who first laid bare the secret of the geysers of Iceland and gave the true theory of their action. A very worthy old professor named Gerling kept the Observatory and lectured on physics. Professor Stegmann, an excellent teacher, lectured on mathematics, Ludwig and Fick were at the Anatomical Institute, Waitz lectured on philosophy and anthropology, Hessel expounded crystallography, while my accomplished friend Knoblauch arrived subsequently from Berlin. The university at the time numbered about three hundred students, and it suited my mood and means far better than one of the larger universities.

In the excellent biography of Dr. Birkbeck recently published by Dr. Godard, which to the writer of it was evidently a labour of love, the name of Birkbeck is referred to the little river of that name which rises in the 'Birkbeck fells' in Westmoreland. 'Beck' is stream in the North, and 'Birk' is birch, so that 'Birkbeck' means Birchstream. Turned into German there would be very little change. For here also Birch is *Birk*, while Beck is *Bach*. In Marburg I lived on the Ketzerbach, a street through the middle of which ran an open brook fringed with acacias. Before the

Reformation had gathered sufficient strength to put a stop to such things, a number of honest people, differing in belief from a number of equally honest people who possessed the will and power to murder them, were here burnt to death, their calcined bones being thrown into the brook. Hence the name Ketzerbach—Heretics' Brook—which survives to this hour. My lodging was a very homely one—two rooms at the top of the house, one a study, the other a bedroom. I was immediately visited by a personage who offered his services as master of the robes. Bearing as he did a good character, he was at once engaged. This Stiefelwichser, or boot-cleaner, whose name was Steinmetz, carried with him besides his brushes a little cane about two feet long, and his vocation was to enter the rooms of the student early in the morning, gather up his clothes and boots, retire to the landing, whence after a few minutes' vigorous beating and brushing, he returned with everything clean, neat, and presentable for the day.

My study was warmed by a large stove. At first I missed the gleam and sparkle from flame and ember, but soon became accustomed to the obscure heat. At six in the morning a small *milchbrod* and a cup of tea were brought to me. The dinner-hour was one, and for the first year or so I dined at a hotel. In those days living was cheap in Marburg. There was no railway to transport local produce to a distance, and this rendered it cheap at home. Our dinner consisted of several courses, roast and boiled, and finished up with sweets and dessert. The cost was a pound a month, or about eighteenpence per dinner. You must not suppose that I partook of all the courses. I usually limited myself to one of them, using even it in moderation, being already convinced that eating too much was quite as sinful, and almost as ruinous, as drinking too much. Watch and ward were therefore kept over the eating. By attending to such things I was able to work, without weariness, for sixteen hours a day.

With my Stiefelwichser I was soon at war. It was not a 'declared war.' It was not a 'war of reprisals.' It was not even a struggle for supremacy, but a modest contest on my part for mere equality. Preferring working in the early morning to working late at night, I thought five o'clock a fair hour at which to begin the day. But my Stiefelwichser chose to come at four. For a time I allowed him so to come, without changing my hour ; but shame soon began to take possession of me. I considered his case, and compared his aims and inducements with my own. For the services he rendered me I allowed him the usual pay—a few thalers for the Semester, or term. The thaler was three shillings. I asked myself what my aims and aspirations were worth if they were unable to furnish a motive power equal to that which this poor fellow extracted from his scanty wage. I tried to take refuge in a text of Scripture, and said to myself soothingly, ' The children of this world are always in their generation wiser than the children of light.' It was very comforting for the moment to think of poor Steinmetz as a child of this world, and of his employer as a child of light. But in those days there existed under the same skin two John Tyndalls, one of whom called the other a humbug, accompanying this descriptive noun by a moral kick which, in the matter of getting up, effectually converted into a child of this world the child of light. For a long time I was always in a condition to look Steinmetz in the face, and return his ' Guten Morgen ' when he arrived. We afterwards relaxed, and made our hour of meeting five ; and for the

last year or so, having climbed my roughest eminences,
and not feeling a continuance of the strain to be
necessary, I was content if found well submerged in my
tub before the clock of St. Elizabeth had finished ringing out six in the morning.

Early risers are sometimes described as insufferable
people. They are, it is said, self-righteous—filled with
the pharisaical ' Lord, I thank thee that I am not as
other men are ! ' It may be so, but we have now to
deal not with generalisations but with facts. My
going to Germany had been opposed by some of my
friends as quixotic, and my life there might perhaps
be not unfairly thus described. I did not work
for money; I was not even spurred by ' the last
infirmity of noble minds.' I had been reading Fichte,
and Emerson, and Carlyle, and had been infected by
the spirit of these great men. Let no one persuade
you that they were not great men. The Alpha and
Omega of their teaching was loyalty to duty. Higher
knowledge and greater strength were within reach of
the man who unflinchingly enacted his best insight.
It was a noble doctrine, though it may sometimes have
inspired exhausting disciplines and unrealisable hopes.
At all events it held me to my work, and in the long
cold mornings of the German winter, defended by a
Schlafrock lined with catskin, I usually felt a fresh-
ness and strength—a joy in mere living and working,
derived from perfect health—which was something
different from the malady of self-righteousness.

At Marburg I attended the lectures of many of the
eminent men above mentioned, concentrating my chief
attention, however, on mathematics, physics, and
chemistry. I should like to have an opportunity of
subjecting these lectures, especially those of Bunsen, to
a riper judgment than mine was at that time. I

learnt German by listening to Bunsen, and as my knowledge of the language increased the lectures grew more and more fascinating. But my interest was alive from the first, for Bunsen was a master of the language of experiment, thus reaching the mind through the eye as well as through the ear. The lectures were full of matter. Notes of them are still in my possession which prove to me how full they were, and how completely they were kept abreast of the most advanced knowledge of the day. This is a use and a sense of the word 'advanced' which may be safely commended to your sympathetic attention. In many directions it is easy enough to become advanced, but not in this one. Bunsen was a man of fine presence, tall, handsome, courteous, and without a trace of affectation or pedantry. He merged himself in his subject : his exposition was lucid, and his language pure ; he spoke with the clear Hanoverian accent which is so pleasant to English ears ; he was every inch a gentleman. After some experience of my own, I still look back on Bunsen as the nearest approach to my ideal of a university teacher. He sometimes seemed absent-minded, and, as he gazed through the window at the massive Elizabethen Kirche, appeared to be thinking of it rather than of his lecture. But there was no interruption, no halting or stammering to indicate that he had been for a single moment forgetful. He lectured every day in winter, and twice a day in summer, beginning his course on organic chemistry at seven in the morning. After the lectures, laboratory work continued till noon. During this time no smoking was allowed in the laboratory, but at noon liberty as regards the pipe began, and was continued through the day. Bunsen himself was an industrious smoker. Cigars of a special kind were then sold in Marburg, called 'Bunsen'sche Cigarren' ; they were

very cheap and very bad, but they were liked by my
illustrious friend, and were doubtless to him a source of
comfort. Dr. Debus, the late distinguished professor of
chemistry at the Royal Naval Cóllege, Greenwich, was
Bunsen's laboratory assistant at this time, and to him I
was indebted for some lessons in blowpipe chemistry.
Bunsen afterwards took me under his own charge, giving
me Icelandic trachytes to analyse, and other work.
Besides being a chemist, he was a profound physicist.
His celebrated ' Publicum ' on electro-chemistry, to
which we all looked forward as a treat of the highest
kind, was physical from beginning to end. He was the
intimate friend of W. Weber of Göttingen, and was well
acquainted with the labours of that great electrician.
Breaking ground in frictional electricity, he passed on
to the phenomena and theory of the Voltaic pile. He
was a great upholder of the famous Contact Theory,
which had many supporters in Germany at the time, one
of the foremost of these being the genial-minded Kohl-
rausch. This theory, as you are well aware, has undergone profound modifications. There are, no doubt,
eminent philosophers amongst us who would pronounce
the theory, in its first form, unthinkable, inasmuch as
it implied the creation of force out of nothing. But
the fact that some of the most celebrated scientific
men in the world, with the illustrious Volta himself as
their leader, accepted and saw nothing incongruous in
the theory, shows how ' unthinkability' depends upon
the state of our knowledge. The laws of Ohm were expounded with great completeness by Bunsen. Various
modes of electric measurement were illustrated ; the
electric light from the carbon battery, invented by
himself, was introduced, the electric telegraph was explained, Steinheil's researches in regard to the ' earth
circuit ' were developed ; and it was in these lectures

that I first heard an honouring and appreciative reference to *der Englische Bierbrauer, Joule.*

Stegmann, the professor of mathematics, was also a man of strong individuality. He lectured in a small room on the flat which he occupied. This was the usual arrangement; each professor had a lecture-room on his own floor, and the students in passing from lecture to lecture had sometimes to go from one end of Marburg to the other. The desks were of the most primitive description, and into them the inkhorns were securely fixed by means of spikes at the bottom. Besides attending his lectures I had private lessons from Professor Stegmann. He was what I have already described him to be, an excellent teacher. He lectured on analysis, on analytical geometry of two and three dimensions, on the differential and integral calculus, on the calculus of variations, and on theoretical mechanics. In mathematics he appeared to be entirely at home. I have sometimes seen him, after he had almost wholly covered his blackboard with equations, suddenly discover that he had somewhere made a mistake. When this occurred he would look perplexed, shuffle his chalk vaguely over the board, move his tongue to and fro between his lips, until he had hit upon the error. His face would then flush, and he would dash forward with redoubled speed and energy, clearing up every difficulty before the end of the lecture. It was he who gave me the subject of my dissertation when I took my degree. Its title in English was—' On a Screw Surface with Inclined Generatrix, and on the Conditions of Equilibrium on such Surfaces.' One evening, after he had given me this subject, I met him at a party and asked him a question, which I did not dream of as touching the solution of the problem. But he smiled and said, ' Yes, Herr Tyndall, but if

I tell you that I must tell you a great deal more.' I thought he meant to insinuate that I wished for illegitimate aid in the working out of my theme. I shrank together, and resolved that if I could not, without the slightest aid, accomplish the work from beginning to end, it should not be accomplished at all. Wandering among the pinewoods, and pondering the subject, I became more and more master of it; and when my dissertation was handed in to the Philosophical Faculty, it did not contain a thought that was not my own.

One of my experiences at Marburg may be worth noting. For a good while I devoted myself wholly to the acquisition of knowledge; heard lectures and worked in the laboratory abroad, and studied hard at home. When a boy at school I had read an article, probably by Addison, on the importance of order in the distribution of our time, and for the first year or so my time was ordered very stringently, specified hours being devoted to special subjects of study. But in process of time I began the attempt of adding to knowledge as well as acquiring it. My first little physical investigation was on a subject of extreme simplicity, but by no means devoid of scientific interest—' Phenomena of a Waterjet.' Among other things, I noticed that the musical sound of cascades and rippling streams, as well as the sonorous voice of the ocean, was mainly if not wholly due to the breaking of air-bladders entangled in the water. There is no rippling sound of water unaccompanied by bubbles of air. This inquiry was followed by others of a more complicated and difficult kind. Well, over and over again after work of this description had begun, I found myself infringing my programme of study. Discontent and self-reproach were the first result. But it was soon evident that a rigid ordering of time would now be out of place. You could not

call up at will the spirit of research. It was like
that other spirit which cometh when it listeth, and
greater wisdom was shown in following out at the time
a profitable line of thought, than in adhering to a
fixed lesson-plan. By degrees all discontent vanished,
and I became acclimatised to my new intellectual conditions. Continuing to work strenuously but happily
till the autumn of 1850, I then came to England.
But I soon returned to Germany, being this time
accompanied by my lifelong friend, Mr. Thomas Archer
Hirst, late Director of Studies in the Royal Naval
College.

To those Marburg days I look back with warm
affection, both in regard to Nature and to man. The
surrounding landscape with its various points of interest
and beauty is still present to my mind's eye: the
Dommelsberg, the Kirchspitze, Spielslust, Marbach,
Werda, and farther off Kirchain, with its neighbouring
spurt of basaltic rock. On this huge wart stands a
Catholic church, many Catholic crosses, and a village
containing a purely Catholic population. It might be
described as an oasis of Catholicism amid a howling
desert of Protestantism, for Protestantism was regnant
everywhere around. And then there were the various
places of refreshment dotted over the neighbourhood,
to which we resorted in little parties from time to time.
Close at hand was Ockershausen, where the students
used to enjoy their pancakes and sour milk, without a
thought that the sourness was produced by little growing microscopic rods—the lactic acid ferment. The
mention of this living ferment reminds me that during
my time at Marburg existed a delicacy which is now
eaten with precautions. On slices of black bread were
nicely spread layers of fresh butter, and on these again
thin slices of *rohe Schinken*—raw ham. The discovery

of *trichinæ* encysted in the muscles of the pig, which when eaten had the power of reproducing themselves in multitudes in human muscle, and of destroying life, has interfered with the luxury of *rohe Schinken*. During the *Semester* we had our excursions abroad and our social gatherings at home. *Kränzchen* means a small chaplet or crown of leaves ; but it also means a little circle or club, and we had our English *Kränzchen*, the members of which used to meet at each other's houses once a week, for the reading of Shakespeare and Tennyson.

Rumours of the great men of Berlin reached Marburg from time to time. Their names and labours were frequently mentioned in the lectures. Having previously learned that I should have the privilege of working there in the laboratory of Professor Magnus, to Berlin I went in the beginning of 1851. Magnus had made his name famous by physical researches of the highest importance. The finish and completeness of his experiments were characteristic. He was a wealthy man, and spared neither pains nor expense to render his apparatus not only effective but beautiful. His experiments on the deviation of projectiles may be noted as special illustrations here. But on everything he touched he sought to confer completeness. The last years of his life were, for the most part, occupied in a discussion with myself on one of the most difficult subjects of experimental physics—the interaction of radiant heat and matter in the gaseous state of aggregation. It was also my privilege to meet Dove, who was renowned in various ways as a physicist. He had won fame in optics, acoustics, and electricity, but his greatest labours were in the field of scientific meteorology The two Roses were there, Heinrich and Gustav, genial and admirable men, the one a great chemist, the other a great geologist. I met Mitscherlich, whose researches in crystallographic chemistry and physics had rendered his name illustrious. With Ehrenberg I had various conversations on microscopic organisms. I wanted at the time some amorphous carbonate of lime, and thought that Ehrenberg's microscopic chalk shells might serve my purpose ; but I was thrown back by learning that the shells, small as they were, were built of crystals smaller still. I made the acquaintance of Riess, the foremost exponent of frictional electricity, who more than once opposed to Faraday's radicalism his own conservatism as regards electric theory. Du Bois-Reymond was there at the time, full of power, both physical and mental. His fame had been everywhere noised abroad in connection with his researches on animal electricity. Du Bois-Reymond is now Perpetual Secretary to the Academy of Sciences in Berlin, and his discourses before that learned body show that his literary power takes rank with his power as an investigator. At the same time I met Clausius, known all over the world through his researches on the mechanical theory of heat, and whose first great paper on the subject I translated before quitting Marburg. Wiedemann was there, whose own researches have given him an enduring place in science ; and who has applied his vast powers of reading and of organisation to throwing into a convenient form the labours of all men and nations on voltaic electricity. Poggendorff, a very able experimenter, was also there. He is chiefly known in connection with the famous journal which so long bore his name. From all these eminent men I received every mark of kindness, and formed with some of them enduring friendships. Helmholtz was at this time in Königsberg. He had written his renowned essay on the Conservation of Energy, which I translated, and he

had just finished his experiments on the velocity of
nervous transmission—proving this velocity, which
had previously been regarded as instantaneous, or at all
events as equal to that of electricity, to be, in the
nerves of the frog, only 93 feet a second, or about one-
twelfth of the velocity of sound in air of the ordinary
temperature. In his own house I had the honour of an
interview with Humboldt. He rallied me on having
contracted the habit of smoking in Germany, his
knowledge on this head being derived from my little
paper on a water-jet, where the noise produced by the
rupture of a film between the wet lips of a smoker is
referred to. He gave me various messages to Faraday,
declaring his belief that he (Faraday) had referred the
annual and diurnal variation of the declination of the
magnetic needle to their true cause—the variation of
the magnetic condition of the oxygen of the atmosphere.
I was interested to learn from Humboldt himself that,
though so large a portion of his life had been spent in
France, he never published a French essay without
having it first revised by a Frenchman. In those days
I not unfrequently found it necessary to subject myself
to a process which I called depolarisation. My brain,
intent on its subjects, used to acquire a set resembling
the rigid polarity of a steel magnet. It lost the pliancy
needful for free conversation, and to recover this I used
to walk occasionally to Charlottenburg or elsewhere.
From my experiences at that time I derived the notion
that hard thinking and fleet talking do not run together.

In trying to carry out the desire of Mr. Norris I
have spoken as a worker to workers; believing that
though the word *I* has occurred so frequently in this
address, far from seeing in it a display of egotism, you
will accept it as a fragment of the life of a brother who

has felt the scars of the battle in which many of you are now engaged. Duty has been mentioned as my motive force. In Germany you hear this word much more frequently than the word 'glory.' The philosophers of Germany were men of the loftiest moral tone. In fact, they were preachers of religion as much as expounders of philosophy. Shall we say that from them the land took its moral colour? It would be to a great extent true to say so; but it should be added that the German philosophers were themselves products of the German soil, probably deriving the basis of their moral qualities from a period anterior to their philosophy. Let me tell you an illustrative anecdote. In the summer of 1871 I met at Pontresina two Prussian officers—a captain and a lieutenant—who had come there to recruit themselves after the hurts and sufferings of the war. We had many walks and many talks together. It was particularly pleasant to listen to the way in which they spoke of the kindness and the sympathy shown by the French peasantry towards the suffering German soldiers, whether wounded or broken down upon the march. I once asked them how the German troops behaved when going into battle. Did they cheer and encourage each other? The reply I received was this: 'Never in our experience has the cry, " Wir müssen siegen " [We must conquer], been heard from German soldiers; but in a hundred instances we have heard them resolutely exclaim, " Wir müssen unser Pflicht thun " [We must do our duty].' It was a sense of duty rather than love of glory that strengthened those men, and filled them with an invincible heroism. We in England have always liked the iron ring of the word 'duty.' It was Nelson's talisman at Trafalgar. It was the guiding-star of Wellington. When, on the death of Wellington, he wrote his immortal 'Ode,' our Laureate poured into the praise of Duty the full strength of his English heart :—

> Not once or twice in our rough island-story
> The path of duty was the way to glory :
> He that walks it, only thirsting
> For the right, and learns to deaden
> Love of self, before his journey closes,
> He shall find the stubborn thistle bursting
> Into glossy purples, which outredden
> All voluptuous garden roses.
> Not once or twice in our fair island-story
> The path of duty was the way to glory.

1886.

THOMAS YOUNG.[1]

EARLY LIFE AND STUDIES.

FOUR great names are indissolubly associated with
the establishment in which we are here assembled
—its founder, Benjamin Thompson, better known as
Count Rumford; its Chemical Professor, Humphry
Davy; its Professor of Natural Philosophy, Thomas
Young; and, finally, the man whom so many of us have
the privilege to remember, Michael Faraday. Of the
character and achievements of the third of the great
men here named, less seems to be publicly known than
ought to be known. Even a portion of this audience
may possibly have some addition made to its knowledge
by reference to the life of a man who served the Institution in the opening of the present century. I
therefore thought that such a brief account of him as
could be compressed into an hour might not be without
interest and instruction at the present time.

Thomas Young was born at Milverton, in Somersetshire, June 13, 1773. His parents were members
of the Society of Friends. Nearly seven years of
his childhood were spent with his maternal grandfather. He soon evinced a precocity which might have
been expected to run to seed and die rapidly out.
When he was two years old he was able to read with
considerable fluency, and before he had attained the

---

[1] My last lecture in the Royal Institution, delivered Jan. 22,
1886. Chief authority : Dean Peacock's *Life of Young*.

age of four years, he had read the Bible twice through. At the age of six he learnt by heart the whole of Goldsmith's 'Deserted Village.' His first formal teachers were not successful, and an aunt in those early days appears to have been more useful to him than anybody else. When not quite seven years of age, he was placed at what he calls a miserable boarding-school at Stapleton, near Bristol. But he soon became his own tutor, distancing in his studies those who were meant to teach him.

In March 1782 he was sent to the school of Mr. Thompson, at Compton, in Dorsetshire, of whose liberality and largeness of mind Young spoke afterwards with affectionate recognition. Here he worked at Greek and Latin, and read a great many books in both languages. He also studied mathematics and book-keeping. Of pregnant influence on his future life was the reading of Martin's 'Lectures on Natural Philosophy,' and Ryland's 'Introduction to the Newtonian Philosophy.' He read with particular delight the optical portions of Martin's work. An usher of the school, named Jeffrey, taught him how to make telescopes and to bind books. The early years of Young and Faraday thus inosculate, the one, however, pursuing bookbinding as an amusement, and the other as a profession. Young borrowed a quadrant from an intelligent saddler named Atkins, and with it determined the principal heights in his neighbourhood. He took to botany for a time, but was more and more drawn towards optics. He constructed a microscope. The disentangling of difficult problems was his delight. Seeing some fluxional symbols in Martin's work, he attacked the study of fluxions. Priestley on Air was read and understood, while the Italian language was mastered by the aid of one of his schoolfellows named Fox.

After leaving Compton, he devoted himself to the study of Hebrew. Mr. Toulmin, of whom Young speaks with affection, lent him grammars of the Hebrew, Chaldee, Syriac, and Samaritan languages, all of which he studied with diligence and delight. Mr. Toulmin also lent him the Lord's Prayer in more than a hundred languages, the examination of which gave him extraordinary pleasure. Through one of those accidents which enter so largely into the tissue of human life, Young found himself at Youngsbury, near Ware, in Hertfordshire. It was a strong testimony to his talent and character, that Mr. Barclay at this time accepted him as the preceptor of his grandson, Mr. Hudson Gurney, although Young was then little more than fourteen, and his pupil only a year and a half younger than himself. Thus began a lifelong friendship between him and Hudson Gurney. Young spent five years at Youngsbury, which he deemed the most profitable years of his life. He spent the winter months in London, visiting booksellers' shops and hearing occasional lectures. He kept a journal in Hertfordshire, the first entry of which informs us that he had written out specimens of the Bible in thirteen different languages. It is recorded of Young that, when requested by an acquaintance, who presumed somewhat upon his youthful appearance, to exhibit a specimen of his handwriting, he very delicately rebuked the inquiry by writing a sentence in his best style in fourteen different languages.

Although the catalogue of Young's books might give the impression that he was a great reader, his reading was limited; but whatever he read, he completely mastered. Fichte compared the reading of Reviews to the smoking of tobacco, affirming that the two occupations were equally pleasant, and equally profitable. Young, in this sense, was not a smoker. Whatever

study he began, he never abandoned; and it was, says
Dean Peacock, in his 'Life of Young,' to his steadily
keeping to the principle of doing nothing by halves,
that he was wont in after-life to attribute a great part
of his success as a scholar and a man of science.

Young's mother was the niece of Dr. Brocklesby,
and this eminent London physician appears to have
taken the greatest interest in the development of his
youthful relative. He nevertheless occasionally gave
Young a rap over the knuckles for what he called his
'prudery.' We all know the strenuous and honourable
opposition that has been always offered to negro slavery
by the Society of Friends. In carrying out their principles, they at one time totally abstained from sugar, lest
by using it they should countenance the West Indian
planters. Young here imitated the conduct of his sect,
which Dr. Brocklesby stigmatised as 'prudery.' 'My
late excellent friend Mr. Day,' says the Doctor, 'the
author of "Sandford and Merton," abhorred the base
traffic in human lives as much as you can do; and even
Mr. Granville Sharp, one of the earliest writers on the
subject, has not done half as much service as Mr. Day
in the above work. And yet Mr. Day devoured daily
as much sugar as I do. Reformation,' adds the Doctor,
'must take its rise elsewhere, if ever there is a general
mass of public virtue sufficient to resist such private
interests.'

Over and above his classical reading, from 1790 to
1792, Young read Simpson's ' Fluxions,' the ' Principia '
and ' Optics ' of Newton, and many of the works of other
famous authors, including Bacon, Linnæus, Boerhaave,
Lavoisier, Higgins, and Black. He confined himself
to works of the highest stamp. He mastered Corneille
and Racine, read Shakespeare, Milton, Blackstone, and
Burke. But he was, adds his biographer, ' contented

to rest in almost entire ignorance of the popular literature of the day.'

I must, however, hasten over the early years and acquirements of this extraordinary personality. During his youth he had none of the assistance which is usually within the reach of persons of position in England. All that I have here mentioned, and a vast deal more, he had acquired without having entered either a public school or a university. As a classic, he was, we are assured, both precise and profound. As a mathematician, he was many-sided, original, and powerful. Such an education, however, though well calculated to develop the strength of the individual, was not, in Peacock's opinion, the best calculated to place Young in sympathy with the mind of his age. ' He was, throughout life, destitute of that intellectual fellow-feeling (if the phrase may be used) which is so necessary to form a successful teacher or lecturer, or a luminous and successful writer.'

Young was intended for the medical profession, and his medical studies began in 1792. He came to London, and attended the lectures of Dr. Baily, Mr. Cruikshanks, and John Hunter. He made the acquaintance of Burke, Windham, Frederick North, Sir Joshua Reynolds, and Dr. Lawrence. By the advice of Burke he studied the philosophical works of Cicero. The bent of Young's moral character may be inferred from the quotations which he habitually entered in his commonplace book. Here is one of them :—' For my part,' says Cicero, ' I think the man who possessed that strength of mind, that constitutional tendency to temperance and virtue, which would lead him to avoid all enervating indulgences, and to complete the whole career of life in the midst of labours of the body and efforts of the mind ; whom neither tranquillity nor relaxation, nor the flattering attentions of his equals in age and station, nor public games nor banquets would delight; who would regard nothing in life as desirable which was not united with dignity and virtue ;—such a man I regard as being, in my judgment, furnished and adorned with some special gifts of the gods.'

His medical studies were pursued with the thoroughness which marked everything Young took in hand. He was an assiduous attendant at the best lectures. His delight in optics naturally drew him to investigate the anatomical structure of the eye. In regard to this structure it will be remembered that in front is the cornea, holding behind it the aqueous humour ; then comes the iris, surrounding the aperture called the pupil, at the back of which we have the crystalline lens. Behind this, again, is the vitreous humour, which constitutes the great mass of the eye. Thus, optically considered, the eye is a compound lens of great complexity and beauty. Behind the vitreous humour is spread the screen of the retina, woven of fine nerve-fibres. On this screen, when any object looked at is distinctly seen, a sharply-defined image of the object is formed. Definition of the image is necessary to the distinctness of the vision. Were the optical arrangements of the eye rigid, distinct vision would be possible only at one definite distance. But the eye can see distinctly at different distances. It has what the Germans call an *Accommodationsvermögen*—a power of adjustment—which liberates it from the thrall of rigidity. By what mechanical arrangement is the eye enabled to adjust itself both for near and distant objects ? Young replied, ' By the alteration of the curvature of the crystalline lens.' His memoir on this subject was considered so meritorious, that it was

printed in the 'Transactions of the Royal Society';
and in the year following, at the age of twenty-one, he
was elected a Fellow of the Society.

Young's memoir evoked sharp discussion, both as
regards the priority and the truth of the discovery. It
was claimed by John Hunter, while its accuracy was
denied by Hunter's brother-in-law, Sir Everard Home,
who, jointly with Mr. Ramsden, affirmed that the
adjustment of the eye depended on the changed
curvature of the cornea. A couched eye—that is to say,
an eye from which the crystalline lens had been removed—they affirmed to be capable of adjustment. In
the face of such authorities Young, with the candour
of a true man of science, abandoned the views he had
enunciated. But it was only for a time. He soon
resumed his inquiries, and proved to demonstration
that couched eyes had no trace of the power ascribed
to them. Before the time of Young, moreover, weighty
authorities leaned to the view that the adjustment of
the eye depended on the variation of the distance
between the cornea and the retina. When near objects
were viewed, it was thought that the axis was lengthened, the retina or screen being thereby thrown farther
back. In distant vision the reverse took place. But
Young proved beyond a doubt that no such variation in
the length of the axis of the eye occurs; and this has
been verified in our own day by Helmholtz. The change
in the curvature of the crystalline lens has been also
verified by the most exact experiments. When we pass,
for instance, from distant to near vision, the image of a
candle-flame reflected from the front surface of the lens
becomes smaller, proving the lens to be then more
sharply curved. When we pass from near to distant
vision, the image becomes larger, proving the curvature
of the lens to have diminished. The radius of curvature

of the lens under these circumstances has been shown
to vary from six to ten millimeters. Young's theory
of the adjustment of the eye has been therefore completely verified. But it is still a moot point as to what
the mechanism is by which the change of curvature is
produced. Young thought that it was effected by the
muscularity of the lens itself. The muscles, however,
would require nerves to excite them, and it would be
hardly possible, in the transparent humours of the eye,
for such nerves to escape detection. They, however,
have never been detected.

While passing through Bath in 1794 Young, at the
instance of Dr. Brocklesby, called upon the Duke of
Richmond. The impression made by Young at this
time may be gathered from a note addressed by the
Duke to the Doctor in these terms :—' But I must tell
you how pleased we all are with Mr. Young. I really
never saw a young man more pleasing and engaging.
He seems to have already acquired much knowledge in
most branches, and to be studious of obtaining more.
It comes out without affectation on all subjects he talks
upon. He is very cheerful and easy without assuming
anything ; and even on the peculiarity of his dress and
Quakerism, he talked so reasonably, that one cannot
wish him to alter himself in any one particular. In
short, I end as I began, by assuring you that the
Duchess and I are quite charmed with him.' The Duke,
then Master of the Ordnance, was a very competent
man. He was well acquainted with the instruments
used in the great Trigonometrical Survey under his
control. He offered to Young the post of private
secretary. Young's acceptance would have brought
within his reach both honour and emolument. But to
his credit be it recorded he refused the post, because

its acceptance would have rendered necessary the abandonment of his costume as a member of the Society of
Friends. Soon afterwards he paid a visit to a celebrated cattle-breeder near Ashbourne, and describes with
vivid interest what Mr. Bickwell had accomplished by
the process of artificial selection. Facts like these,
presented afterwards to the pondering mind of Darwin,
caused the great naturalist to pass from artificial to
natural selection. Young visited Darwin's grandfather,
and criticised his 'Zoonomia.' The inspection of Dr.
Darwin's cameos, minerals, and plants, gave him great
delight, the supreme pleasure being derived from the
cameos. Dr. Darwin stated that he had borrowed much
of the imagery of his poetry from the graceful expression and vigorous conception which these cameos
breathe. His opinion of his visitor was pithily expressed in a letter of introduction to a friend in Edinburgh. 'He unites the scholar with the philosopher,
and the cultivation of modern arts with the simplicity
of ancient manners.'

Young went to Edinburgh to continue his studies in
medicine. His reputation had gone before him, and he
was welcomed in the best society of the northern capital.
He met Bostock, Bancroft, Turner, Gibbs, Gregory,
Duncan, Black, and Munroe. He dwells specially upon
the lectures of John Bell, whose demonstrations in
anatomy appeared to him to be of first-rate excellence.

There is nothing that I have met in Dean Peacock's
'Life of Young' to denote that he was fervently religious. The Ciceronian 'virtue,' rather than religious
emotion, seemed to belong to his character. The hold
which mere habit long exercised over him, and which
loyalty to his creed had caused him to maintain at a
period of temptation, became more and more relaxed.
He gradually gave up the formal practices of Quakerism

in regard to dress and other matters. He took lessons in dancing, and appeared to delight in that graceful art. I remember the late Mr. Babbage telling me that once, upon a London stage, by the untimely raising of a drop-scene, Young was revealed in the attitude of a dancer. He assiduously attended the theatre. So, it may be remarked, did the profoundly religious Faraday. On leaving Edinburgh he paid a farewell visit to his friend Cruikshanks, who took him aside, and after much preamble, 'told me,' says Young, 'that he had heard that I had been at the play, and hoped that I should be able to contradict it. I told him that I had been several times, and that I thought it right to go. I know you are determined to discourage my dancing and singing, and I am determined to pay no regard whatever to what you say.'

After completing his studies at Edinburgh, Young went to the Highlands. The houses in which he was received show the consideration in which he was held. He visited the chief seats of learning, and the principal libraries, as a matter of course; but he had also occasion to enjoy and admire 'the good sense, frankness, cordiality of manners, personal beauty, and accomplishments' of the Scottish aristocracy. So greatly was he delighted with his visit to Gordon Castle, that before quitting it he wrote thus : 'I could almost have wished to break or dislocate a limb by chance, that I might be detained against my will. I do not recollect that I have ever passed my time more agreeably, or with a party whom I thought more congenial to my own disposition.' He visited Staffa, but took more pleasure in Pennant's plates and descriptions than in Fingal's Cave, or the scenery of the island. From the Duchess of Gordon he carried a letter of introduction to the Duke of Argyll, and spent some time at Inverary. In riding out he was given his choice to proceed leisurely with the Duke, or to ride with the ladies and be galloped over. His reply was, that of all things he liked being galloped over, and he made his choice accordingly. He compares the two daughters of the Duke to Venus and Minerva, both being goddesses. He visited the Cumberland lakes. But here it may be said, once for all, that Young was somewhat stunted in his taste for natural scenery. He was a man of the town, fond of social intercourse, and of intellectual collision. He could not understand the possibility of any man choosing to live in the country if the chance of living in London was open to him. At Liverpool he dined with Roscoe, proceeding afterwards to Coalbrookdale and its ironworks. As previously at Carron, he was greatly impressed by the glare of the furnaces. Mr. W. Reynolds, who appeared to interest himself in physical experiments on a large scale, told him that he had the intention of making a flute 150 feet long and $2\frac{1}{2}$ feet in diameter, to be blown by a steam-engine and played upon by barrels. From Young's letters it is evident that he then saw the value and necessity of what we now call technical education.

In October 1795, he became a student in the University of Göttingen. He gives an account of his diurnal occupations, embracing attendance at lectures on history, on materia medica, on acute diseases, and on natural history. He is careful to note that he had also lessons twice a week from Blessmen, the academical dancing-master, and the same number of lessons on the clavichord from Forkel. Young's pursuit of 'personal accomplishments' is considered by his biographer to have been excessive. At Göttingen he attended, on Sundays, tea dances or supper dances. The mothers of handsome daughters appear to have been wary of the students, having reason ' to fear a traitor in every young man.' He made at Göttingen the acquaintance of many famous professors—of Heyne, Lichtenberg, Blumenbach, and others. He records a joke practised on the professor of geology which had serious consequences. The students were rather bored by the professor's compelling them to go with him to collect ' petrifactions ; ' and the young rogues, says Young, ' in revenge, spent a whole winter in counterfeiting specimens and buried them in a hill which the good man meant to explore, and imposed them upon him as the most wonderful *lusus naturæ*.' Peacock adds the remark that the unhappy victim of this ' roguery ' died of mortification when the imposition was made known to him.

Before taking his degree, it is customary for the student in German Universities to hand in a dissertation written by himself. This is circulated among the Professors and is followed by a public disputation. On July 16, Young did battle in the Auditorium, the subject chosen for discussion being the human voice. He acquitted himself creditably, was complimented by those present, and received his degree as doctor of physic, surgery, and midwifery. In the thesis chosen for discussion, Young broke ground in those studies on sound which, for intrinsic merit, and suggesting as they did his subsequent studies on light, will remain for ever famous in the history of science. During a pause in the lectures he visited the Hartz Mountains, making himself acquainted with the scene of Goethe's *Walpurgisnacht* on the summit of the Brocken. Wedgwood and Leslie accompanied him on this tour. The curious fossils dug up by the young men in the Unicorn's Cave at Schwarzfeld excited curiosity and wonder, but no-

thing more. Their significance at that time had not
been revealed. Hearing Kant so much spoken of in
Germany, Young naturally attacked the 'Critique of
Pure Reason,' but his other studies prevented him from
devoting much time to the Critical Philosophy. To
the portion of it which he read he attached no high
value. He admitted Kant's penetration, but dwelt
upon his confusion of ideas. The language of the
' Critique ' he thought unpardonably obscure.

He visited Brunswick, where, clothed in the proper
costume, he was presented at Court. After the reception came a supper, about twenty ladies sitting on one
side of a table, and twenty gentlemen on the other.
He endeavoured to converse with his neighbour, but
found him either sulky or stupid. The dowager duchess,
whom he likened to a spectre, made her appearance
and began to converse pleasantly. When told that
Young had studied at Göttingen, and that he was a
doctor of medicine, she asked him whether he could
feel a pulse, and whether the English or the Germans
had the best pulses. Young replied that he had felt
but one pulse in Germany—the pulse of a young lady,
and that it was a very good pulse. Göttingen was then
the foremost school of horsemanship in Europe. Young
was passionately fond of this exercise, and there were
no feats of horsemanship, however daring or difficult,
which he did not attempt or accomplish. His muscular power had been always remarkable, and he could
clear a five-barred gate without touching it. He was
better known among the students for his vaulting on a
wooden horse than for writing Greek. At a Court
masquerade he appeared in the character of harlequin,
which gave him an excellent opportunity of exhibiting
his personal activity. Notwithstanding all this, he did
not quite like his life in Göttingen. The professors of

the University were worked too hard to leave much
time for the receptions and social gatherings in which
Young delighted. So he quitted Göttingen on August
28, ' with as little regret as a man can leave any place
where he has resided nine months.'

From Göttingen he walked to Cassel, and thence by
Gotha, Erfurt, Weimar, and Jena, to Leipzig. He saw
everything which to him was worth seeing, and as he
carried letters of introduction from the most eminent
men of the age, he was welcomed everywhere. Most
of the professors were absent on their holiday, but at
Weimar he conversed with Herder, who, though well
versed in the English poets, cared nothing, it was said,
about rhyme. At Jena he found Bütmer, who, at the
age of eighty-three, was about to begin the publication
of a general dictionary of all existing languages. He
visited Dresden, the Saxon Switzerland, and the mines
of Freiberg. Here he made the acquaintance of the
celebrated Werner. From Freiberg he went to Berlin,
where he dined twice with the English Ambassador,
Lord Elgin, and once with Dr. Brown, a Welsh physician in great favour with the King. Over the monotonous sandy flat that lies between the two cities he
journeyed from Berlin to Hamburg. Detained here for
a time by adverse winds, he was treated with great
hospitality.

One word in conclusion regarding the German
schools of learning. Germany is now united and strong;
her sons are learned, and her prowess is proved. But
the units from which her blended vigour has sprung
ought not to be forgotten. These were the little
principalities and powers of which she was formerly
composed. Each of them asserted its individuality
and independence by the establishment of a local University, and all over Germany, in consequence, such

institutions are sown broadcast. In these nurseries of mind and body, not only Bismarck and Von Moltke, but numbers of the rank and file of the German Army, found nutriment and discipline; so that although, as long as her principalities remained separate, Germany as a whole was weak, the individual action of those small States educated German men so as to make them what we now find them to be.

Two epochs of Young's career as a medical student have been now referred to—his residence in Edinburgh, and his residence at Göttingen. Immediately after his return to England he became a fellow-commoner of Emanuel College, Cambridge. When the master of the college introduced him to those who were to be his tutors he jocularly said, ' I have brought you a pupil qualified to read lectures to his tutors.' On one occasion, in the Combination Room, Dr. Parr made some dogmatic observation on a point of scholarship. ' Bentley, sir,' said Young promptly and firmly, ' was of a different opinion.' ' A smart young man that,' said Parr when Young quitted the room. His lack of humour and want of knowledge of popular literature sometimes made him a butt at the dinner-table, but he bore the banter with perfect good humour. The materials for Young's life at Cambridge are very scanty; but there is one brisk and energetic letter, published by Dean Peacock, written by a man who was by no means partial to Young. ' Young,' he said, ' was beforehand with the world in perceiving the defects of English mathematicians. He looked down upon the science, and would not cultivate the acquaintance of any of our philosophers. He seemed never to have heard the names of the poets and literary characters of the last century, and hardly ever spoke of English literature.' According to Peacock's correspondent, there was about Young no pretence or assumption of superiority. ' He spoke upon the most difficult subjects as if he took it for granted that all understood the matter as well as himself. But he never spoke in praise of any of the writers of the day, and could not be persuaded to discuss their merits. He would speak of knowledge in itself—of what was known or what might be known— but never of himself or of any one else as having discovered anything, or as likely to do so. His language was correct, his utterance rapid, but his words were not those in familiar use, and he was therefore worse calculated than any man I ever knew for the communication of knowledge.' This writer heard Young lecture at the Royal Institution, but thought that nothing could show less judgment than the method he adopted. ' It was difficult to say how he employed himself at Cambridge. He read little ; [1] there were no books piled on his floor, no papers scattered on his table. His room had all the appearance of belonging to an idle man. He seldom gave an opinion, and never volunteered one ; never laid down the law like other learned doctors, or uttered sayings to be remembered. He did not think abstractedly. A philosophical fact, a difficult calculation, an ingenious instrument, or a new invention, would engage his attention ; but he never spoke of morals, or metaphysics, or religion. Of the last, I never heard him say a word. Nothing in favour of any sect, or in opposition to any doctrine.'

The impression made upon Young by Cambridge was, from first to last, entirely favourable. In those days, six years' study was indispensable before the degree of Bachelor of Medicine could be taken. Young

---

[1] Critics and commentators must be great readers ; but *creators* in science and philosophy do not always belong to this category.

graduated in 1803, when he was thirty years of age, and five years more had to elapse before he could take the degree of M.D. Meanwhile he had begun the practice of medicine. Dr. Brocklesby died in 1797, on the night of a day when he had entertained his relative and some other friends at dinner. During dinner he seemed perfectly well, but he expired a few minutes after he went to bed. He left Young his house and furniture in Norfolk Street, Park Lane, his library, his prints, a collection of pictures chiefly selected by Sir Joshua Reynolds, and about 10,000*l*. in money.

THE WAVE THEORY.

On January 16, 1800, Young communicated to the Royal Society a memoir entitled 'Outlines and Experiments respecting Sound and Light.' In this paper he treated of the 'interference' of sound, and his researches on this subject led him on to the discovery of the interference of light—'which has proved,' says Sir John Herschel, 'the key to all the more abstruse and puzzling properties of light, and which would alone have sufficed to place its author in the highest rank of scientific immortality, even were his other almost innumerable claims to such a distinction disregarded.' Newton considered the sensation of light to be aroused by the impinging of particles, inconceivably minute, against the retina. Huyghens, on the contrary, supposed the sensation of light to be aroused by the impact of minute waves. Young favoured the theory of undulation, and by his researches on sound he was specially equipped for its thorough examination. Before he formally attacked the subject he gave, in a paper dealing with other matters, his reasons for espousing the wave theory. The velocity of light, for instance, in the same medium is constant. All refractions are

attended with partial reflection. The dispersion of
light is no more incompatible with this than with any
other theory. Reflection and refraction are equally
explicable on both suppositions. Huyghens indeed
had proved this, and much more. Inflection may be
better explained by the wave theory than by its rival.
The colours of thin plates, which are perfectly unin-
telligible on the common hypothesis, admit of complete
explanation by the wave theory. In dealing with the
colours of thin films, of which the soap-bubble offers
a familiar example, Young first proved his mastery
over the undulatory theory. In the pursuit of this
great task he was able to apply to Newton's Theory
of Fits the Theory of Waves, and to determine the
lengths of the undulations corresponding to the different colours of the spectrum.

We now approach a phase of Young's career which
more specially concerns us. The Royal Institution,
as already stated, was founded by Count Rumford, supported by many of the foremost men in England. The
King was its patron, the Earl of Winchilsea its first
president, while Lord Morton, Lord Egremont, and Sir
Joseph Banks were its vice-presidents. On January 13,
1800, the Royal Seal was attached to the charter of the
Royal Institution. Dr. Thomas Garnett was appointed
Professor of Natural Philosophy and Chemistry. During
his previous residence in Bavaria, Rumford had ruled
with beneficent but despotic sway, and the habit of
mind thus engendered may have made itself felt in
his behaviour to Dr. Garnett. At all events, they did
not get on well together. On February 16, 1801, Davy
was appointed Assistant Lecturer in Chemistry, Director of the Chemical Laboratory, and Assistant Editor
of the Journals of the Institution. The post of Pro-

fessor of Natural Philosophy was offered to Young, and
he accepted it. The salary was to be 300*l.* a year. On
August 3, 1801, the following resolution was passed : —
' Resolved, that the Managers approve of the measures
taken by Count Rumford, and that the appointment of
Dr. Young be confirmed.' Young, it is said, was not
successful as a lecturer in the Institution, and this
Dr. Peacock ascribes to his early education, which gave
him no opportunity of entering into the intellectual
habits of other men. More probably the defect was
due to a mental constitution, not plastic, like that of
Davy or Faraday, in regard to exposition. Young now
fairly fronted the undulatory theory of light. Before
you is some of the apparatus he employed. I hold in
my hand an ancient tract upon this subject by the
illustrious Huyghens. It was picked up on a bookstall, and presented to me some years ago by Professor
Dewar. In this tract Huyghens deals with refraction
and reflection, giving a complete explanation of both ;
and here, also, he enunciated a principle which now
bears his name, and which forms one of the foundation-
stones of the undulatory theory.

The most formidable obstacle encountered by Young,
and one which he never entirely surmounted, was an
objection raised by Newton to the assumption of a fluid
medium as the vehicle of light. Looking at the waves
of water impinging on an isolated rock, Newton observed that the rock did not intercept the wave motion.
The waves, on the contrary, bent round the rock, and
set in motion the water at the back of it. Basing
himself on this and similar observations, he says, ' Are
not all hypotheses erroneous in which light is supposed
to consist of a pression or motion propagated through
a fluid medium ? If it consisted in pression or motion,
it would bend into the shadow.' He instances the case

of the sound of a bell being heard behind a hill which conceals the bell; of the turning of corners by sound; and then, with conclusive force, he points to the case of a planet coming between a fixed star and the eye, when the star is completely blotted out by the interposition of the opaque body. This, Newton urged, could not possibly occur if light were propagated by waves through a fluid medium, for such waves would infallibly stir the fluid behind the planet, and thus obliterate the shadow.

Young was firmly persuaded of the truth of the undulatory theory. The number of riddles that he had solved by means of it, the number of secrets he had unlocked, the number of difficulties he had crushed, rendered him steadfast in his belief; still, he never fairly got over this objection of Newton's. It was finally set aside by one of the most illustrious men that ever adorned the history of science. A young French officer of engineers, Augustin Fresnel, first really grappled with the difficulty and overthrew it. The principle of Huyghens, to which I have already referred, is, that every particle, in every wave, acts as if it alone were a centre of wave motion. When you throw a stone into the Serpentine, circular waves or ripples are formed, which follow each other in succession, retreating farther and farther from the point of disturbance.[1] Fix your attention on one of these circular waves. The *form* of the wave moves forward,

[1] 'I prove it thus, take heed now
By experience, for if that thou
Threw in water now a stone
Well wost thou it will make anone,
A little roundell as a cercle,
Peraventure as broad as a couercle, .
And right anone thou shalt see wele,
That whele cercle wil cause another whele,
And that a third and so forth brother,
Every cercle causing other.'—CHAUCER'S *House of Fame.*

but the motion of its individual particles, at any moment, is simply a vibration up and down. Now each oscillating particle of every moving wave, if left to itself, would produce a series of waves, not so high, but in other respects exactly similar to those produced by the stone. The coalescence of all these small waves produces another wave of exactly the same kind as that which started them. The principle that every particle of a wave acts independently of all other particles, while the waves produced by all the particles afterwards combine, is, as I have said, the great principle of Huyghens. Taken in conjunction with the interference of light, first established by Thomas Young, which proved that when waves coalesce or combine, they may either support each other or neutralise each other, the neutralisation being either total or partial, according as the opposition of the combining waves is complete or incomplete—taking, I say, the principle of interference in conjunction with that of Huyghens, Fresnel proved that although light does diverge behind an opaque body, as Newton supposed that it would diverge, these divergent waves completely efface each other, producing the shadow due to the tranquillity of the medium which propagates the light.

By reference to the waves of water, Young illustrates, in the most lucid manner, the interference of the waves of light. He pictures two series of waves generated at two points near each other in a lake, and reaching a channel issuing from the lake. If the waves arrive at the same moment, neither series will destroy the other. If the elevations of both series coincide, they will, by their joint action, produce in the channel a series with higher elevations. But if the elevations of one series correspond to the depressions of the other, the ridges will exactly fill the furrows, smooth water in the channel being the result. 'At least,' says Young, 'I can discover no alternative, either from theory or from experiment. Now,' he continues, gathering confidence as he reasons, 'I maintain that similar effects take place whenever two portions of light are thus mixed, and this I call the general law of the Interference of Light.'

The physical meaning of all the terms applied to light was soon fixed. *Intensity* depended upon the *amplitudes* of the waves. *Colour* depended on the *lengths* of the waves. Two series of waves coalesced and helped each other when one was any number of complete undulations, or, in other words, any *even* number of half-undulations, behind the other. Two series of waves extinguished each other when the one series was any *odd* number of semi-undulations behind the other. But inasmuch as white light is made up of innumerable waves of different lengths, such waves cannot all interfere at the same time. Some interfere totally, and destroy each other; some partially; while some add themselves together and enhance the effect. Thus, by interference, a portion only of the white light is destroyed, and the remaining portion is, as a general rule, coloured. Indeed most glowing and brilliant effects of coloration are thus produced. Young applied the theory successfully to explain the colours of striated surfaces which, in the hands of Mr. Rutherfurd and others, have been made to produce such splendid effects. The iridescences on the polished surfaces of mother-of-pearl are due to the striæ produced by the edges of the shell-layers, which are of infinitesimal thickness; the fine lines drawn by Coventry, Wollaston, and Barton upon glass also showed these colours. Barton afterwards succeeded in transferring the lines to steel and brass. Most of you are acquainted with the iridescence of Barton's buttons. A descendant of M.r. Barton has, I believe, succeeded in reproducing the instrument wherewith his grandfather obtained his brilliant effects.

But the greatest triumph of Young in this field was the explanation of the beautiful phenomenon known as Newton's rings. The colours of thin plates were profusely illustrated by the experiments of Hooke and Boyle, but Newton longed for more than illustrations. He desired quantitative measurement. The colour of the film was known to depend upon its thickness. Can this thickness be measured ? Here the unparalleled penetration of Newton came into play. He took a lens, consisting of a slice of a sphere of a diameter so large that a portion of the curved surface of the lens approximated to a plane surface. Upon this slightly convex surface he placed a plate of glass the surface of which was accurately plane. Squeezing them together, and allowing light to fall upon them, he observed those beautiful iris-circles with which his name will be for ever identified. The iris-colours were obtained when he employed white light. When monochromatic light was used he had simply successive circles of light and darkness. Here then, from the central point where the two glasses touched each other, Newton obtained a film of air which gradually increased in thickness as he retreated from the point of contact. Whence this wonderful recurrence of light and darkness ? The very constitution of light itself must be involved in the answer. His desire was now to ascertain the thickness of the film of air corresponding to the respective rings. Knowing the curvature of his lens, this was a matter of easy calculation. He measured the diameter of the fifth ring of the series. This might be accurately done with a pair of fine compasses, for the diameter was over the fifth of an inch in length. But it was the interval

between the glasses corresponding to this distance that
Newton required to know, and this he found by calculation to be $\frac{1}{37000}$th of an inch. This, be it remembered, is the distance corresponding to the fifth ring.
The interval corresponding to the first ring would
be only a fifth of this, or, in other words, about
$\frac{1}{180000}$th of an inch. Such are the magnitudes with
which we have to deal before the question 'What is
Light?' can be scientifically answered.

Newton's explanation of the rings which he was the
first to discover, though artificial in the highest degree,
is marked by his profound sagacity. He was hampered
by the notion of the 'corporeity' of light. He could
not get over the objection raised by himself as to the
impossibility of shadows in a fluid medium. He held
therefore that light was due to the darting forth of
minute particles in straight lines; and he threw out the
idea that colour might be due to the difference of 'big-
ness' in the particles. He endowed these particles with
what he called *fits* of easy transmission and reflection.
The dark rings, in his immortal experiment, were produced where the light-particles were in their trans-
missive 'fit.' They went through both surfaces of the
film of air, and were not thrown back to the eye. The
bright rings occurred where the light-particles were in
their reflective fit, and where, on reaching the second
surface of the film, they were thrown back to the eye.
The cardinal point here is, that Newton regarded the
recurrence of light and darkness as due to an action
confined to the *second surface* of the film. And here
it was that Young came into irreconcilable collision
with him, proving to demonstration that the dark rings
occurred where the portions of light reflected from *both
sides* of the film extinguished each other by interference, while the bright rings occurred where the light

reflected from the two surfaces coalesced to enhance the intensity.

Young next applied the wave theory to account for the diffraction or inflection of light—that is to say, the effects produced by its bending round the edges of bodies. When a cone of rays, issuing from a very minute point, impinges on an opaque body, so as to embrace it wholly, the shadow of the body, if received upon a screen, exhibits fringes of colour. They follow so closely the contour of the opaque body, that Sir John Herschel compared them to the lines along the sea-coast in a map. If a very thin slip of card, or a hair, be placed within such a cone, it is noticed that besides the fringes outside the shadow, bands of colour occur within it ; the central, or brightest, band being always *white* when white light is employed. It is a singular and somewhat startling fact, that by the interposition of an opaque body, say a small circle of tinfoil, the point on which we should expect the centre of the shadow to fall is, by the joint action of diffraction and interference, illuminated to precisely the same extent as it is when the opaque circle is withdrawn.[1] In reference to the interior fringes Young made the observation, which is of primary importance, that if you intercept the light passing by one of the edges of the strip of card or of the hair, the fringes disappear. It requires the inflection of the waves round *both* edges of the object, and their consequent interference, to produce these fringes.

Young's attempt to explain the phenomena of diffraction was a distinct advance on the extremely artificial hypothesis of Newton. Still his attempt was not so successful as his explanation of the colours of striated

[1] A similar diffraction has been proved by Lord Rayleigh to occur in the case of sound.

surfaces and of thin, thick, and mixed plates. Here the young officer of engineers to whom I have already referred—Fresnel—entered the field. He presented in 1815, to the French Institute, a memoir on Diffraction which marks an epoch in the history of the wave theory. It is usual, when such a paper is presented, to refer it to a ' Commission,' who consider it and report upon its merits. The *Commissionnaires* in this instance were Arago and Prony.

Arago had read the memoirs of Young in the ' Philosophical Transactions,' but had not understood their full significance. The study of Fresnel's memoir caused the full truth to flash upon him that his young countryman had been anticipated thirteen years previously by Dr. Young. Fresnel had re-discovered the principle of interference independently, and, with profound insight and unrivalled experimental skill, had applied it to the phenomena of diffraction. It was no light thing for Fresnel to find himself as regards the principle of interference suddenly shorn of his glory. He, however, bore the shock with resignation. He might have readily made claims which would have found favour with his countrymen and with the world at large. But he did nothing of the kind. The history of science indeed furnishes no brighter example of honourable fairness than that exhibited throughout his too short life by the illustrious young Frenchman. Once assured that he had been anticipated—whatever might have been the extent of his own labours, however independently he might have arrived at his results—he unreservedly withdrew all claim to the discovery. There is, I repeat, no finer example of scientific honour than that manifested by Augustin Fresnel.

Fresnel was a powerful mathematician, and well versed in the best mathematical methods of his day.

With enormous labour he calculated the positions where the phenomena of interference must display themselves in a definite way. He was, moreover, a most refined experimentalist, and having made his calculations, he devised instrumental means of the most exquisite delicacy with the view of verifying his results. In this way he swept the field of diffraction practically clear of difficulty, solving its problems where even Young had failed.

Truly, these were minds possessing gifts not purchasable with money! And round about the central labours of each, minor achievements of genius are to be found, which would be a fortune to less opulent men. I hardly know a more striking example of Young's penetration than his account of the spurious or supernumerary bows observed within the true primary rainbow. These interior bows are produced by interference. It is not difficult, by artificial means, to form them in great number and beauty. This is a subject on which I worked assiduously a couple of years ago.[1] And often, when looking at these wondrous interference circles, the words of Young seemed to me like the words of prophecy. The bows were the physical transcript of what he stated must occur; a transcript, moreover, which, compared with his words, was far more complete and impressive than any ever exhibited by the rainbow in Nature. Take another instance. The beautiful rings of colour observed when a point of light is looked at through the seeds of lycopodium shaken over a piece of glass, or shaken as a cloud in the air, are known to be produced by minute particles all of the same size. The iridescence of clouds seen sometimes in great splendour in the Isle of Wight, but more frequently in the Alps, is due to this equality in the size of the

[1] See art. ' The Rainbow and its Congeners,' in this volume.

cloud-particles. Now, the smaller the particles, the wider are the coloured rings, and Young devised an instrument, called the *Eriometer*, which enabled him from the measurement of the rings to infer the size of the particles.[1] Again, Ritter had discovered the ultraviolet rays of the spectrum, while Wollaston had noticed the darkening effect produced by these rays when permitted to fall on paper, or leather, which had been dipped in a solution of muriate of silver. Employing these invisible rays to produce invisible Newton's rings, Young projected an image of the rings upon the chemically-prepared paper. He thus obtained a distinct photographic image of the rings. This was one of the earliest experiments wherein a true photographic picture was produced. Young had little notion at the time of the vast expansions which the art of photography was subsequently to undergo.

But Young was not permitted to pursue his great researches in peace. The ' Edinburgh Review ' had at that time among its chief contributors a young man of vast energy of brain and vast power of sarcasm, without the commensurate sense of responsibility which might have checked and guided his powers. His intellect was not for a moment to be measured with that of Young ; but as a writer appealing to a large class of the public, he was, at that time, an athlete without a rival. He afterwards became Lord Chancellor of England. Young, it may be admitted, had given him some annoyance, but his retaliation, if such it were, was out of all proportion to Young's offence. Besides, whatever his per-

[1] In disorders of the eyes such particles sometimes escape into the humours, and produce vivid colours. The enlargement of the circles, which generally excites terror, is a good sign, as it indicates the increasing smallness of the particles by absorption.

sonal feelings were, it was not Young that he assailed
so much as those sublime natural truths of which Young
at the time was the foremost exponent. Through the
undulatory theory he attacked Young without scruple
or remorse. He sneered at his position in the Royal
Institution, and tried hard to have his papers excluded
from the 'Philosophical Transactions.' 'Has the Royal
Society,' he says, 'degraded its publications into bulletins
of new and fashionable theories for the ladies of the
Royal Institution ? Let the Professor continue to amuse
his audience with an endless variety of such harmless
trifles, but in the name of science let them not find
admittance into that venerable repository which contains the works of Newton and Boyle and Cavendish
and Maskelyne and Herschel.' The profound, complicated, and novel researches on which Young was then
engaged rendered an occasional change of view necessary. How does the reviewer interpret this praise-
worthy loyalty to truth ? 'It is difficult,' he says, ' to
deal with an author filled with a medium of so fickle
and vibratory a nature. Were we to take the trouble
of refuting him, he might tell us, "*my opinion is
changed, and I have abandoned that hypothesis.
But here is another for you.*" We demand if the
world of science which Newton once illuminated is to
be as changeable in its modes as the world of fashion,
which is directed by the nod of a silly woman or a
pampered fop? . . . . . We have a right to demand
that the hypothesis shall be so consistent with itself as
not to require perpetual mending and patching ; that
the child we stoop to play with shall be tolerably
healthy, and not of the puny and sickly nature of Dr.
Young's productions, which have scarcely *stamina* to
subsist until the fruitful parent has furnished us with a
new litter, to make way for which he knocks on the

head, or more barbarously exposes, the first.' The reviewer taunts Young with claiming the inheritance of Newton's queries, ' vainly imagining that he fulfils this destination by ringing changes on these hypotheses, arguing from them as if they were experiments or demonstrations, twisting them into a partial coincidence with the clumsy imaginations of his own brain, and pompously parading what Newton left as hints in a series of propositions with all the affectation of system.'

To Brougham's coarse invective Young replied in a masterly and exhaustive letter. A single copy, and one only, was sold by its publisher. There were at that time in the ranks of science no minds competent to understand the controversy. The poison worked without an antidote, and, for thirteen years, Young and his researches on light had no place in public thought. His discoveries remained absolutely unnoticed until their re-discovery by Fresnel lifted the pall which for so long a time had been thrown over this splendid genius.

Young lectured for two years at the Royal Institution, and he afterwards threw the lectures into a permanent form in a quarto volume of 750 pages, with 40 plates, and nearly 600 figures and maps. He also produced at the same time a second volume of the same magnitude, embracing his optical and other memoirs, and a most elaborate classed catalogue of works and papers, accompanied by notes, extracts, and calcula- tions. For this colossal work Young was to receive 1,000*l.* His publisher however became bankrupt, and he never touched the money. His lectures constitute a monument of Young's power almost equal to that of his original memoirs. They are replete with profound reflections and suggestions. In his eighth lecture, on ' Collision,' the term *energy*, now in such constant use, was first introduced and defined. By it he was able to avoid, and enable us to avoid, the confusion which had crept into scientific literature by the incautious employment of the word *force*. Further, the theory now known as the Young-Helmholtz theory, which refers all the sensations of colour to three primary sensations—red, green, and violet—was clearly enunciated by Young in his thirty-seventh lecture, on 'Physical Optics.' His views of the nature of heat were original and correct. He regarded the generation of heat by friction as an unanswerable confutation of the whole doctrine of material caloric. He gave appropriate illustrations of the manner in which he supposed the molecules of bodies to be shaken asunder by heat. 'All these analogies,' he says, 'are certainly favourable to the opinion of the vibratory nature of heat, which has been sufficiently sanctioned by the authority of the greatest philosophers of past times and by the most sober reasoners of the present.' In anticipation of Dr. Wells, Young had observed and recorded the fact, that a cloud passing over a clear sky sometimes causes the almost instantaneous rise of a thermometer placed upon the ground. The cloud he assumed acted as a vesture which threw back the heat of the earth. Radiant heat and light are here placed in the same category. William Herschel had already shown their kinship, by proving that the most powerful rays of the sun were entirely non-luminous. Subsequent to this, the polarisation of heat, by Principal James Forbes, rendered yeoman service in the propagation of the true faith.

Young's essay on the 'Cohesion of Fluids' is to be ranked amongst the most important and difficult of his labours. It embraced his views and treatment of the subject of capillary attraction. But as this topic is to be treated here next week by a spirit kindred to that of

Young himself,[1] I may be excused for saying nothing more about it. The essay drew Young into a controversy with the illustrious Laplace, in which the Englishman exhibited that scimitar-like sharpness of pen which more than once had drawn him into controversy.

Young resigned his post at the Royal Institution, believing that devotion to work alien to his profession would be sure to injure his prospects as a physician. In the summer of 1802 he visited Paris, and at one of the meetings of the Academy was introduced to the First Consul. In March 1803, he became M.B. of Cambridge—six years after entering the University—while five years more had to elapse before he was able to take the degree of M.D. In June 1804, he married Miss Eliza Maxwell, the daughter of J. P. Maxwell, Esq., of Trippendence, near Farnborough, in Kent.

As regards medical practice, Young, to be a popular physician, was probably too cool and cautious in the examination of his data, and trusted too little to the lancet and the calomel invoked in the vigorous practice of his time. After a somewhat strenuous contest he was appointed Physician to St. George's Hospital. The appointment was a strong proof of the esteem in which he was held. His lectures, however, were not so well attended as those of his colleagues, for he lacked the warmth and pliancy which usually commend a lecturer to young men. Young's medical works, embodying the results of great labour and research, were received with high consideration and esteem.

By the force of his sarcasm and the glamour of his rhetoric, Brougham had succeeded in inflicting a serious, if not an irreparable, wound on the science of his

[1] Sir William Thomson.

country. After Young's crushing reply, which produced
no effect whatever upon the public, the author of that
reply was practically forgottten as a factor in the ad-
vancement of physical optics. But science has always
before her the stimulus of natural problems demanding
solution, and after a temporary lull the desire to know
more of the nature of light grew in force. New stars
arose in France, while the strenuous industry and experimental discoveries of Brewster did much to hold us
in equipoise with the Continent. In Paris, Laplace,
Malus, Biot, and Arago were all actively engaged. The
first three proceeded strictly on Newtonian lines, and
by the memoir of Laplace on Double Refraction all
antagonism to the theory of emission was considered to
be for ever overthrown. In the 'Quarterly Review,'
Young criticised this memoir with sagacity and power,
and his criticism remains valid to the present time.
In accordance with the principles of the wave theory
Huyghens had given a solution of the problem of double
refraction in Iceland spar. The solution was opposed
to that of Laplace. Dr. Wollaston, a man of the
highest scientific culture and the most delicate experimental skill, subjected the theory of Huyghens to the
severest metrical tests, and his results proved entirely
favourable to that theory. Wollaston, however, lacked
the boldness which would have made him a commander
in those days of scientific strife. He saw opposed to
him the names of Newton and Laplace, and in the
face of such authority he shrank from closing with the
conclusions to which his own experiments so distinctly
pointed.

We now come to a critical point in the fortunes of
the wave theory. I need not again refer to the difference between the motion of a wave and the motions of
the particles which constitute a wave. A wave of

sound, for instance, passing through the air of this
room would have a velocity of about 1,100 feet a second,
while the particles which constitute the wave, and pro-
pagate it at any moment, may only move through inconceivably small spaces to and fro. Now, in the case
of sound, this to-and-fro motion occurs *in the direction*
in which the sound is propagated, and a little reflection
will make it clear that no matter how a ray of sound,
if we may use the term, is received upon a reflecting
surface, it will be reflected equally all round as long
as the angle inclosed between the reflecting surface and
the ray remains unchanged. In other words, the sound-
ray has no *sides* and no preferences as regards reflection. Now Malus discovered that in certain conditions
a beam of light shows such preferences. When caused
to impinge upon a plane glass mirror, placed in a certain position, it may be wholly reflected ; whereas when
the mirror is placed in the rectangular position it may
not be reflected at all.

Up to the hour when this discovery was made by
Malus light had been supposed to be propagated through
ether, exactly as sound is propagated through air. In
other words, the direction in which the particles of
ether were supposed to vibrate to and fro coincided
with that of the ray of light. Those who had previously held the undulatory theory were utterly stag-
gered by this new revelation, and their perplexity was
shared by Young. He was for a time unable to conceive of a medium capable of propagating the impulses
of light in a way different from the propagation of the
impulses of sound. To ascribe to the light-medium
qualities which would enable it to differ in its mechanical action from the sound-medium was an idea too
bold—I might indeed say too repugnant—to the scientific mind to be seriously entertained. Yet, deeply

pondering the question, Young was at length forced to the conclusion that the vibrations concerned in the propagation of light were executed *at right angles* to the direction of the ray. By this assumption of transverse vibrations, which removed all difficulty, Young also removed the ether from the class of aeriform bodies, and endowed it with the properties of a semi-solid.

Fresnel's memoir on Diffraction, upon which, as already stated, Arago had reported, initiated a lasting friendship between the two illustrious Frenchmen. They subsequently worked together. Fresnel, the more adventurous and powerful spirit of the two, came independently to the same conclusion that Young had previously enunciated. But so daring did the idea of transverse vibrations appear to Arago—so inconsistent with every mechanical quality which he could venture to assign to the ether—that he refused to allow his name to appear in conjunction with that of Fresnel on the title-page of the memoir in which this heretical doctrine was broached. Still, the heresy has held its ground, and the theory of transverse vibrations, as applied to the luminiferous ether, is now universally entertained.

Fresnel died in the fortieth year of his age.

Allow me to wind up this section of our labours by reference to a German estimate of Young's genius. ' His mind,' says Helmholtz, ' was one of the most profound that the world has ever produced ; but he had the misfortune of being too much in advance of his age. He excited the wonder of his contemporaries, who, however, were unable to rise to the heights at which his daring intellect was accustomed to soar. His most important ideas lay, therefore, buried and forgotten in the folios of the Royal Society, until a new generation gradually and painfully made the same discoveries, proving the truth of his assertions and the exactness of his demonstrations.'

Note on Energy.

The passage in which Young introduces and defines the term *energy* is so remarkable that I venture to reproduce it here.

'The term energy may be applied, with great propriety, to the product of the mass or weight of a body into the square of the number expressing its velocity. Thus, if a weight of one ounce moves with a velocity of a foot in a second, we may call its energy 1 ; if a second body of two ounces have a velocity of three feet in a second, its energy will be twice the square of three, or 18. This product has been denominated the living or ascending force, since the height of the body's vertical ascent is in proportion to it; and some have considered it as the true measure of the quantity of motion ; but although this opinion has been very universally rejected, yet the force thus estimated well deserves a distinct denomination. After the considerations and demonstrations which have been premised on the subject of forces, there can be no reasonable doubt with respect to the true measure of motion ; nor can there be much hesitation in allowing at once that since the same force, continued for a double time, is known to produce a double velocity, a double force must also produce a double velocity in the same time. Notwithstanding the simplicity of this view of the subject, Leibnitz, Smeaton, and many others, have chosen to estimate the force of a moving body by the product of its mass into the square of its velocity ; and though we cannot admit that this estimation of force is just, yet it may be allowed that many of the sensible effects of motion, and even the advantage of any mechanical power, however it may be employed, are usually proportional to this product, or to the weight of the moving body, multiplied by the height from which it must have fallen in order to acquire the given velocity. Thus, a bullet moving with a double velocity will penetrate to a quadruple depth in clay or tallow; a ball of equal size, but of one-fourth of the weight, moving with a double velocity, will penetrate to an equal depth; and, with a smaller quantity of motion, will make an equal excavation in a shorter time. This appears at first sight somewhat paradoxical; but on the other hand we are to consider the resistance of the clay or tallow as a uniformly retarding force, and it will be obvious that the motion, which it can destroy in a short time, must be less than that which requires a longer time for its destruction. Thus also when the resistance opposed by any body to a force tending to break it is to be overcome, the space through which it may be bent before it breaks being given, as well as the force exerted at every point of that space, the power of any body to break it is proportional to the energy of its motion, or to its weight multiplied by the square of its velocity.'

[*The foregoing Essay was prepared with the view of giving the members of the Royal Institution some notion of a man regarding whom many of them knew but little. I tried at the same time to draw up a brief account of Young's labours on the Hieroglyphics of Egypt. The subject lay far apart from my usual studies, and this fact, coupled with my anxiety to avoid offence in dealing with the relationship of Young and Champollion, threw upon me an amount of work to which my health at the time was unequal. Though not included in the Address delivered to the members, this account was published in the 'Proceedings of the Royal Institution.' Despite its inadequacy to give any just notion of the magnitude of Young's labours in this particular field, the record of his achievements will be rendered more complete by its introduction here.*]

HIEROGLYPHICAL RESEARCHES.

Young's capacity and acquirements in regard to languages have been already glanced at. As a classical scholar his reputation was very high. His Greek calligraphy was held to vie in elegance with that of Porson. A man so rounded in his culture could hardly be said to have an intellectual bent; but if he had one, the examination and elucidation of ancient manuscripts must have fallen in with it. It is quite possible, however, that, had he not been disheartened by the apparent success of Brougham, he would have clung more steadfastly to physical science. However this may be, we now find him in a new field. In October 1752 the first rolls of the papyri of Herculaneum, wearing the aspect of blackened roots, were discovered in what appeared to be the library of a palace near Portici. They had been covered to a depth of 120 feet with the mixed ashes, sand, and lava of Vesuvius. The inscriptions were for the most part written in Greek, but some of them were in Latin. The leaves were carbonised and hard, being glued together by heat to an almost homogeneous mass. Learned Italians had devoted great labour and ingenuity to the separating of the leaves and the deciphering of the inscriptions. To the credit of the Prince of Wales, afterwards George IV., let it be recorded that he manifested from the first an enlightened, liberal, and truly practical interest in these researches. He wrote to the Neapolitan Government, offering to defray all the expenses of unrolling and deciphering the papyri; and he sent out Mr. Hayter, a classical scholar of repute, to act as co-director with Rossini in the superintendence of the work. Mr. Hayter appears to have been unequal to the task committed to him. His trans-

lations were defective ; the gaps were serious and
numerous ; and he finally abandoned the manuscripts
when he fled from Naples, with the royal family, on
the French invasion in 1806. Some of the rolls, which
had been presented to the Prince of Wales, were committed to the care of the Royal Society, and placed by
the Society in the hands of Dr. Young. He spent many
months in devising and applying means for the opening
of the leaves ; and though only partially successful in
this respect,[1] he was able to correct many important
errors, and to fill many serious gaps in the work of his
predecessors.

The 'Quarterly Review' was established in 1809,
and Young was intimate with its leading contributors.
One of these, George Ellis, 'a man of ardent affections,'
had resented, almost as personal to himself, the attacks
on Young in the 'Edinburgh Review,' and Young's pen
was soon invoked to enrich and adorn the pages of its
rival. A great work, the 'Herculanensia,' on the ancient
condition of Herculaneum and its neighbourhood, had
been published. The review of this work was committed to Young, and his article upon it, embodying
his own views and researches, was published in 1810.
'The appearance of the article,' says Peacock, 'equally
remarkable for its critical acuteness and vigorous writing, at once placed its author, in the estimation of the
public, in the first class of the scholars of the age.'
Gifford, the editor of the 'Quarterly,' described the
article as 'certainly beyond all praise.' Ellis, at the
same time, wrote thus to Young :—'It is a consolation
to know that Brougham, who took advantage of the
growing circulation of the 'Edinburgh Review' to dis-

---

[1] Davy afterwards tried his hand upon the rolls, with imperfect
success.

seminate his vile abuse of you ; and Jeffrey, who permitted him to do so, should be condemned to hear your
praises upon all sides.' The tide had clearly turned in
Young's favour, even prior to his final and triumphant
vindication by Fresnel. From this time forward inscriptions of all kinds were sent to Young for discussion
or interpretation. They were found in numbers among
his papers after his death.[1]

It was a mind thus endowed and disciplined that
now turned to the formidable but fascinating task of
deciphering the hieroglyphics of Egypt. An adum-
bration of his researches, which must, under the circumstances, be weak and faint, I will endeavour to bring
before you.

The famous Rosetta stone was discovered by the
French in Egypt in 1799. It bore three inscriptions :

[1] 'In the Appendix,' says Young's biographer, ' to Captain
Light's Travels in Egypt, Nubia, Palestine, and Cyprus, he furnished
translations and restorations of several Greek inscriptions ; and when
Barrow gave an account in the *Quarterly Review* of recent researches
in Egypt, more especially those of Caviglia on the Great Sphinx, it
was from Young that he obtained the restoration of the inscription
on the second digit of the great paw.' In the third volume of Young's
Works, this inscription, taken from the nineteenth volume of the
Quarterly Review, is given, with translations into modern Greek,
Latin, and English. The last-mentioned runs thus :—

' Thy form stupendous here the gods have placed,
   Sparing each spot of harvest-bearing land ;
And with this mighty work of art have graced
   A rocky isle, encumber'd once with sand ;
   And near the Pyramids have bid thee stand :
Not that fierce Sphinx that Thebes erewhile laid waste,
   But great Latona's servant mild and bland ;
Watching that prince beloved who fills the throne
Of Egypt's plains, and calls the Nile his own.
That heavenly monarch [who his foes defies],
Like Vulcan powerful [and like Pallas wise].'

the first, Hieroglyphical or sacred; the second, Enchorial [1]—a name given by Young to the common language employed by the Egyptians in the time of the
Ptolemies ; and the third, Greek. At the end are given
the following directions :—' What is here decreed shall
be engraved on a block of hard stone, in sacred, in
native, and in Greek characters, and placed in each
temple of the first and second and third gods.' All
three inscriptions were more or less mutilated and
effaced when the stone was discovered. Porson and
Heyne had, however, succeeded in almost completely
restoring the Greek one. It had been a custom with
Young to pay an annual visit to Worthing, and to
pursue there for a portion of the year his practice as
a physician. The Society of Antiquaries had caused
copies of the three inscriptions of the Rosetta stone to
be made and published. In the summer of 1814
Young took all of them to Worthing, where he subjected them to a severe comparative examination.

Baron Sylvestre de Sacy, an eminent Orientalist,
had discovered in the native Egyptian certain groups
of characters answering to proper names, while Akerblad, a profound Coptic scholar, had not only added to
the number, but attempted to establish an alphabet
answering to the native Egyptian inscription. Young
took up the researches of these distinguished men as
far as they could be relied on. Assuming all three inscriptions to express the same decree, one of them being
in a language known to scholars, it was inferred by
Young that a strict comparison of line with line, word
with word, and character with character, would lead
him by the sure method of science from the known to

[1] Called in the Greek ' ENCHORIA GRAMMATA,' or letters of the
country. Young deprecates the introduction, afterwards, by Champollion, of the term ' DEMOTIC,' or popular.

the unknown. He rapidly passed his predecessors. De
Sacy had determined three proper names in the Egyptian; Akerblad nine others, and five or six Coptic
words; while Young, soon after, detected the rudiments
of fifty or sixty Coptic words, which, however, formed
but a very small fraction of the whole inscription.

Here an unexpected stumbling-block was encountered. The effort of Akerblad to reduce the whole
Enchorial inscription to Coptic had failed,[1] and it soon
became evident to Young that every such attempt must
of necessity fail. His conviction and its grounds are
first mentioned in a letter to Mr. Gurney, written in
August 1814. 'I doubt,' he writes, 'if it will be ever
possible to reduce much more of it to Coptic, especially
*as I have fully ascertained that some of the characters
are hieroglyphics.*' As bearing upon the *derivative*
origin of the Enchorial inscription, the discovery here
announced is obviously of the highest importance.
Young continues : 'I have, however, made out the sense
of the whole sufficiently for my purpose, and, by means
of variations from the Greek, I have been able to effect
a comparison with the hieroglyphics which it would have
been impossible to do satisfactorily without this intermediate step.' In a letter to the Archduke John of
Austria, dated August 2, 1816, Young announced that
he had 'now fully demonstrated the hieroglyphical
origin' of the Enchorial inscription.[2]

[1] 'Notwithstanding this failure, his name,' says Peacock, 'should
ever be held in honour as one of the founders of our knowledge of
Egyptian literature, to the investigation of which he brought no
small amount of patient labour and philological learning.'

[2] 'The same discovery,' says the editor of the third volume of
Young's Works, 'was announced by M. Champollion, as his own, in
his memoir, *De l'Ecriture Hiératique des Anciens Égypients*, published at Grenoble in 1821. . . . . This memoir contained several
plates in which the hieroglyphic and hieratic characters are com-

'I had thought it necessary,' says Young, in an essay written to clear the air on this and various other points some years afterwards, ' to make myself in some measure familiar with the remains of the old Egyptian language as they are preserved in the Coptic and Thebaic versions of the Scriptures ; and I had hoped, with the assistance of this knowledge, to be able to find an alphabet which would enable me to read the Enchorial inscription, at least into a kindred dialect. But in the progress of the investigation I had gradually been compelled to abandon this expectation, and to admit the conviction that no such alphabet would ever be discovered, because it had never been in existence. I was led to this conclusion, not only by the untractable nature of the inscription itself—which might have depended on my own want of information and address—but still more decidedly by the manifest occurrence of a multitude of characters which were obviously imperfect imitations of the more intelligible pictures that were observable among the distinct hieroglyphics of the first inscription, such as a priest, a statue, a mattock, or plough, which were evidently, in their primitive state, delineations of the objects intended to be denoted by them, and which were, as evidently, introduced among the Enchorial characters.'

Young, as we have seen, had begun his labours on

pared, on the same plan as Dr. Young's specimens in the *Encyclopædia Britannica*, published in 1819. He sent a copy of them to Dr. Young, but withheld the letterpress. Dr. Young accordingly remained for several years under the impression that this work had been published at a much earlier period. Writing to Sir William Gell in 1827 in reference to this point, Young remarks : ' I never knew till now how much later his publication was, for he gave it to me without the text.' The publication was Champollion's *Comparative Table of Hieroglyphics*, ' containing,' says Young, ' what I had published in 1816,' five years earlier.

the Rosetta stone in May 1814. In the month of
August he was able to announce to Mr. Gurney his
discovery that some of the Enchorial characters were
hieroglyphics. Prior to Young, no human being had
dreamt of the transfer of the characters of the first
inscription to the second. The first was pictural and
symbolic; the second, to all appearance, a purely alphabetical running-hand. It had always been regarded as
such. By means of the funeral papyri Young still
further established the relationship between the first
and second inscriptions. In 1816 he obtained from
Mr.William Hamilton a loan of the noble work entitled
' Description de l'Egypte,' in which were carefully published several of the papyrus manuscripts. Many of the
inscriptions dealt with the same text, and by comparing
them one with another Young was able to trace the
gradual departure from the original hieroglyphic characters. Probably with a view to more rapid writing, these
had passed through various phases of degradation, until
they reached the stage corresponding to the Enchorial
inscription of the Rosetta stone, 'which,' says Young,
' resembled in its general appearance the most un-
picturesque of these manuscripts.' Long before the
time of Young, learned men had tried their hands on
the Rosetta characters, but no relationship like that
here indicated had ever been discerned.

Pre-eminent among the Egyptologists of that time
was the celebrated Champollion, librarian at Grenoble.
In his very first reference to Champollion, Dean
Peacock speaks thus of the illustrious Frenchman:—
' He had made the history, the topography, and anti-
quities of Egypt, as well as the Coptic language and its
kindred dialects, the study of his life, and he started
therefore upon this inquiry with advantages that prob-

ably no other person possessed ; and no one who is
acquainted with his later writings can call in doubt his
extraordinary sagacity in bringing to bear upon every
subject connected with it, not merely the most apposite, but also the most remote, and sometimes the most
unexpected, illustrations.' Thus equipped, however,
Champollion made next to no progress before the advent of Young. 'With the exception,' says Peacock,
' of the identification of a few additional Coptic words,
very ingeniously elicited from the Egyptian text, he
made no important advance on what had already been
done by Akerblad. Like him, also, he abandoned the
task of identifying the hieroglyphical inscription, or
portions of it, with those corresponding to them in the
Egyptian or Greek text, as altogether hopeless, in consequence of the very extensive mutilations which it
had undergone.'

Young, however, had determined about ninety or
one hundred characters of the mutilated hieroglyphic
inscription (the funeral papyri enabled him afterwards
to more than double the number), and these sufficed
to prove, ' first, that many simple objects were represented by their actual delineations ; secondly, that
many other objects, represented graphically, were used
in a figurative sense only, while a great number of the
symbols, in frequent use, could be considered as the
pictures of no existing objects whatever ; thirdly, that
a dual was denoted by a repetition of the character,
but that three characters of the same kind following
each other implied an indefinite plurality, more compendiously represented by three lines or bars attached
to a single character ; fourthly, that definite numbers
were expressed by dashes for units, and arches, either
round or square, for tens ; fifthly, that all hieroglyphic
inscriptions were read from front to rear, as the ob-

jects naturally follow each other; sixthly, that proper
names were included by the oval ring, or border, or
cartouche ;[1] and seventhly, that the name of Ptolemy
alone existed on this pillar, having only been completely
identified by help of the analysis of the Enchorial inscription. And,' adds Young, ' as far as I have ever
heard or read, *not one* of these particulars had ever
been established and placed on record by *any other
person*, dead or alive.'

No man was a better judge of intellectual labour
than Dean Peacock. The whole of Young's writings,
preparatory and otherwise, were before him when he
wrote; and he states emphatically, that it is impossible to estimate, either the vast extent to which Dr.
Young had carried his hieroglyphical investigations, or
the progress which he had made in them, without an
inspection of these manuscripts. In reference to an
article entitled ' Egypt,' written by Young in 1818,
and published in the ' Encyclopædia Britannica ' for
1819, a writer in the ' Edinburgh Review' for 1826
delivers the following weighty opinion: ' We do
not hesitate to pronounce this article the greatest
effort of scholarship and ingenuity of which modern
literature can boast.' Even to an outsider it offers
proof of astonishing learning and research. Still,
Peacock assures us that this publication of 1819 could
hardly be considered more than a popular and superficial sketch of the vast mass of materials on which it
was founded.

Young was limited to what Peacock here calls ' a
popular and superficial sketch ' by the fact that the

---

[1] Young's editor adds here : ' The discovery was long afterwards
made by Champollion that the cartouches were confined to the
names of royal personages.'

article in the 'Encyclopædia Britannica' was written
for ordinary readers, rather than for critics or learned
men. In this article, however, we are allowed a
glimpse of Young's mode of collating and comparing
the different inscriptions. He looks at the Enchorial
inscription, and notices certain recurrent groups of
characters ; he looks at the Greek inscription, and
finds there words with the same, or approximately the
same, periods of recurrence. Thus, 'a small group of
characters occurring very often, in almost every line,
might be either some termination, or some very common particle ; it must therefore be reserved till it is
found in some decisive situation, after some other words
have been identified, and it will then easily be shown
to mean *and*. The next remarkable collection of
characters is repeated twenty-nine or thirty times in
the Enchorial inscription ; and we find nothing that
occurs so often in the Greek, except the word *king*. . . .
A fourth assemblage of characters is found fourteen
times in the Enchorial inscription, agreeing sufficiently
well in frequency with the name of *Ptolemy*. . . .
By a similar comparison, the name of Egypt is identified. . . . Having thus,' says Young, 'obtained a
sufficient number of common points of subdivision, we
next proceed to write the Greek text over the Enchorial, in such a manner that the passages ascertained
may all coincide as nearly as possible ; and it is obvious that the intermediate parts of each inscription
will then stand very near to the corresponding passages
of the other. . . . By pursuing the comparison of
the inscriptions thus arranged, we ultimately discover
the signification of the greater part of the individual
Enchorial words.'

Having thus compared the Greek text with the
Enchorial, Young next proceeded to compare the En-

chorial with the hieroglyphical. About half the lines
of the latter were obliterated, and the rest were considerably defaced. Towards the ends, however, both
inscriptions were fairly well preserved; and these were
the portions subjected to the scrutiny of Young.
Making allowance for the differences of space occupied
by the two inscriptions, and measuring from the final
words of the inscription proportional distances, determined by the Enchorial characters for *God*, *King*,
Priest and *Shrine*, the meaning of which had been
well established, Young sought at the places indicated
by these measurements for the corresponding hieroglyphics. He soon found that *Shrine* and *Priest* were
denoted by pictures of the things themselves. The
other terms, *God* and *King*, were still more easily
ascertained, from their situation near the name of
Ptolemy. Having thus fixed his points of orientation,
Young placed them side by side, and subjected the
characters lying between them to a searching comparison. He offers in his article of 1819 the last line
of the sacred characters, with the corresponding parts
of the other inscriptions, as a 'fair specimen of the
result that has been attained from these operations.'

Up to the time of which we now speak, although
profoundly learned men had attempted to decipher the
funeral papyri of Egypt, if we omit the labours of
Young, very little progress had been made in this
direction; while in regard to the decipherment of the
hieroglyphics *nothing* had been done.

A vast extension of our knowledge of Egyptian
writing is to be ascribed to the following accident.
An Italian named Cassati had brought to Paris several
manuscripts from Upper Egypt. One was written exactly in the Enchorial character of the second inscrip-

tion on the Rosetta stone. It was a deed of sale, and
on the back of the manuscript was an endorsement
in Greek. When in Paris, Young had received from
Champollion a tracing of the Enchorial deed, but not
of the Greek endorsement. About the same time,
Mr. Grey, an English traveller, brought to England a
number of manuscripts, which he placed in the hands
of Dr. Young. One of them was written entirely in
Greek, and Young immediately perceived that it was
a perfect copy of the Enchorial deed of sale. He
wrote immediately to Champollion, informing him of
the fact, and begging him to send a copy of the Greek
endorsement. Champollion did not comply with this
request, but his countryman, Raoul Rochette, courteously
and promptly responded to Young's application, and sent
him a correct copy of the whole Cassati manuscript.

The possession of the Greek translation was of
course an immense help to Young in his efforts to
decipher the Enchorial deed, on which he was at this
very time engaged. 'I could not,' he says, 'but conclude that a most extraordinary chance had brought
into my possession a document which was not very
likely, in the first place, ever to have existed, still less
to have been preserved uninjured for my information
through a period of near two thousand years. But
that this very extraordinary translation should have
been brought safely to Europe, to England, and to me,
at the very moment when it was most of all desirable
to me to possess it, as the illustration of an original
which I was then studying, but without any reasonable
hope of being able to fully comprehend it,—this combination would in other times have been considered as
affording evidence of my having become an Egyptian
sorcerer.'

Grey's manuscript related, not to the sale of a house

or field, but to portions of the collections and offer-
ings made from time to time for the benefit of a
certain number of mummies. The persons of whom
the mummies were the remains were described at length
in bad Greek; but though bad, a comparison between
it and the Enchorial writing gave important information regarding the orthography of ancient Egypt. Mr.
Grey's collection contained three other similar deeds,
all written in the Enchorial character of the Rosetta
stone, and endorsed with the Greek registry. The
dates of these documents closely corresponded with that
of the Cassati manuscript, which was 146 years before
Christ. They refer to the sale of land, the boundaries
of which were very clearly defined.[1] In those days, as
we know, the Egyptians were the best land-surveyors
in the world. The comparison of these documents
formed, as might be expected, an epoch in the history
of Egyptian literature.

[1] And the persons concerned were equally well defined. In this
respect the Egyptians might vie with the writers of Continental
passports. The following is a translation of the famous papyrus of
Anastasy, recording a deed of sale :—' There was sold by Pamonthes,
aged about forty-five, of middle size, dark complexion, and handsome figure, bald, round faced, and straight nosed ; by Snachomneus,
aged about twenty, of middle size, sallow complexion, likewise
round faced and straight nosed ; and by Semmuthis Persineï, aged
about twenty-two, of middle size, sallow complexion, round faced,
flat nosed, and of quiet demeanour ; and by Tathlyt Persineï, aged
about thirty, of middle size, sallow complexion, round face, and
straight nose with their principal Pamonthes a party in the sale ;
the four being of the children of Petepsais of the leather-cutters
of the Memnonia ; out of the piece of level ground which belongs
to them in the southern part of the Memnonia, 8,000 cubits of
open field. . . . It was bought by Nechutes the less, the son of
Asos, aged about forty, of middle size, sallow complexion, cheerful
countenance, long face, and straight nose, with a scar upon the
middle of his forehead, for 601 pieces of brass, the sellers standing
as brokers, and as securities for the validity of the sale. It was
accepted by Nechutes the purchaser.'

We now approach a period of stormy discussion regarding the claims of different discoverers. And as the tempest raged chiefly round Young and Champollion, it is desirable to fix with precision, if that be possible, the position of the learned Frenchman before he came into contact with Young. This a work published by Champollion at Grenoble in 1821 enables us to do. After speaking of the notions previously entertained regarding the hieroglyphical and epistolographic characters of the Egyptians, and of the opinion, universally diffused, that the Egyptian manuscripts, like those of to-day, are alphabetical, the author states his case thus:—
‘Une longue étude, et surtout une comparaison attentive des textes hiéroglyphiques avec ceux de la seconde espèce, regardés comme alphabétiques, nous ont conduit à une conclusion contraire.

Il résulte, en effet, de nos rapprochements :—

1° Que l’écriture des manuscrits égyptiens de la seconde espèce (l'hiératique) n’est point alphabétique ;

2° Que ce second système n’est qu’une simple modification du système hiéroglyphique, et n’en diffère uniquement que par la forme des signes ;

3° Que cette seconde espèce d’écriture est l’hiératique des auteurs grecs, et doit être regardée comme une tachygraphie hiéroglyphique ;

4° Enfin, que les caractères hiératiques sont DES SIGNES DE CHOSES, ET NON DES SIGNES DE SONS.’

There is no mention here of the name of Young, though he had, many years previously, made known to the world, as the result of his own researches, the first, second, and third of these propositions. Immediately after the publication of this work in 1821, Champollion became acquainted with the ‘ popular and superficial sketch ’—in reality, the transcendently able article of

Young—published in the ' Encyclopædia Britannica '
for 1819.

Peacock's analysis of what next occurred is not
agreeable reading. Champollion's memoir of 1821
was rapidly suppressed, and soon became so scarce that
it has been passed over by almost every author who
has written on the subject. In the following year
Champollion addressed a letter to M. Dacier, in which,
to use the language of Peacock, we suddenly find him
pushed forward into the inmost recesses of the sanc-
tuary, reached by Young five years before. The plates,
moreover, of the suppressed memoir were circulated,
without dates and without letterpress. A copy of these
plates was given by Champollion to Young, who was
left in entire ignorance of the date of publication.
' The suppression of a work,' writes Peacock, in strong
reproof, ' expressing opinions which its author has subsequently found reason to abandon, may sometimes be
excused, but rarely altogether *justified; but under no
circumstances can such a justification be pleaded
when the suppression is either designed or calculated
to compromise the claims of other persons with reference to our own.* The memoir in question very
clearly showed that, so late as the year 1821, Champollion had made no real progress in removing the
mysterious veil which had so long enveloped the ancient
literature of Egypt. The article " Egypt," written by
Young, had meantime confessedly come under his
observation. He saw the errors of his views and
suppressed them, without giving due credit to the
man who had first struck into the true path.' In
reference to an account given by Champollion of
the labours of Young, Peacock remarks. ' It would
be difficult to point out in the history of literature
a more flagrant example of the disingenuous sup-

pression of the real facts bearing upon an important discovery.'

And yet the Dean of Ely is by no means stingy in his praise of Champollion. It would be unjust, he says, to refuse to Champollion the honour due to his rare skill and sagacity, not merely in the application of a principle already known, but in its rapid extension to a multitude of other cases, so as not merely to point out its character and use, but also to determine the principal elements of a phonetic alphabet. His long-continued studies, Peacock remarks, had fitted him more than any other living man, Young himself hardly excepted, to deal with this subject, 'and the rapidity of his progress, when once fully started on his career of discovery, was worthy of the highest admiration.' Peacock, moreover, describes his work as ever memorable in the history of hieroglyphical research, not only from the vast range of knowledge which it displays, but from the clear and lucid order in which it is arranged. 'It was,' he continues, 'singularly unfortunate that one who possessed so much of his own should have been so much wanting in a proper sense of justice to those who had preceded him in these investigations as materially to lessen his claims to the respect and reverence which would otherwise have been most willingly conceded to him.'

With regard to the lack of literary candour, thus so strongly commented on, it is of interest to note the views concerning Champollion held by one of his own countrymen. Soon after the researches of Young had begun, an extremely interesting correspondence was established between him and De Sacy. As early as October 1814 Young was able to submit to his correspondent a 'conjectural translation' into Latin of the Egyptian Rosetta inscription. He subsequently sent him an English translation, the receipt of which is acknowledged by De Sacy in a letter dated Paris, July 20, 1815. The opening paragraph of this letter contains an allusion of considerable historic importance :—' Outre la traduction latine de l'inscription égyptienne, que vous m'avez communiquée, j'ai reçu postérieurement une autre traduction anglaise imprimée, que je n'ai pas en ce moment sous les yeux, *l'ayant prêtée à M. Champollion sur la demande que son frère m'en a faite d'après une lettre qu'il m'a dit avoir reçue de vous.*' In view of the statement of Champollion in a *Précis* of his researches published in 1824, that he had arrived at results similar to those obtained by Dr. Young without having any knowledge of Young's opinions, the foregoing extract is significant. De Sacy goes on to recognise formally the progress which had been made by Young at the date of the foregoing letter. He asks some questions regarding Young's method, which in certain cases appeared to him enigmatical. The requisite explanations were promptly given by Young. In a labour of the kind here under consideration, that force of genius which we vaguely term *intuition* must come conspicuously into play ; and it is not always easy for him to whom the exercise of this force is habitual, to make plain to others the nature and results of its action.

De Sacy embodies in the letter above quoted some personal remarks which, were it not that their omission would involve a virtual injustice to Young, one would willingly pass over. ' Si j'ai un conseil à vous donner,' writes the Baron, ' c'est de ne pas trop communiquer vos découvertes à M. Champollion. Il se pourrait faire qu'il prétendât ensuite à la priorité. Il cherche en plusieurs endroits de son ouvrage à faire croire qu'il a découvert beaucoup des mots de l'inscription égyptienne de Rossette. J'ai bien peur que ce ne soit là que du charlatanisme : j'ajoute même que j'ai de fortes raisons de le penser.' The work of Champollion here referred to was entitled ' L'Égypte sous les Pharaons, ou recherches sur la géographie, la religion, la langue, les écritures et l'histoire de l'Éygpte avant l'invasion de Cambyses.' Two volumes of the work were published in 1814, but it was never completed.

In a letter written towards the end of 1815, Young passes the following judgment upon this book in regard to its relation to the Rosetta inscriptions :—

' I have only spent literally five minutes in looking over Champollion, turning, by means of the index, to the parts where he has quoted the inscription of Rosetta. He follows Akerblad blindly, with scarcely any acknowledgment. But he certainly has picked out the sense of a few passages in the inscription by means of Akerblad's investigations ; although in four or five Coptic words which he pretends to have found in it, he is wrong in all but one, and that is a very short and a very obvious one.'

Our neighbours, the French, have been always fond, perhaps rightly fond, of national glory, not only in military matters, but also in science and literature. They rallied round Champollion. Even De Sacy, who had previously warned Young against him, eventually joined in the general pæan. Arago also, who, in regard to the optical discoveries of Young had behaved so honourably, delivered an Eloge of Young, founded, according to Peacock, on the most imperfect and narrow views of the case. In fact, patriotism came into play where cosmopolitanism ought to have been supreme. Arago seeks to make out that Young stands in the same relation to Champollion as Hooke, in regard

to the doctrine of interferences, stands to Young. This
is certainly a bold comparison. Arago himself gives
his reasons for entering the controversy, and they are
these:—' Que l'interprétation des hiéroglyphes égyptiens
est l'une des plus belles découvertes de notre siècle ; que
Young a lui-même mêlé mon nom aux discussions dont
elle a été l'objet ; qu'examiner enfin, si la France peut
prétendre à ce nouveau titre de gloire, c'est agrandir la
mission que je remplis en ce moment, c'est faire acte
de bon citoyen. Je sais d'avance tout ce qu'on trouvera
d'étroit dans ces sentimens ; je n'ignore pas que le
cosmopolitisme a son beau côté ; mais en vérité, de
quel nom ne pourrais-je pas le stigmatiser si, lorsque
toutes les nations voisines énumèrent avec bonheur les
découvertes de leurs enfans, il m'était interdit de cher-
cher dans cette enceinte même, parmi des confrères dont
je ne me permettrai pas de blesser la modesté, la preuve
que la France n'est pas dégénérée, qu'elle aussi apporte
chaque année son glorieux contingent dans le vaste
dépôt des connaissances humaines ? '

The Copley medal of the Royal Society of London
was awarded to Arago in 1825, and on November 30 of
that year, on handing over the medal to the gentleman
deputed to receive it, Sir Humphry Davy, then President of the Society, used the following words :—' Fortunately science, like that Nature to which it belongs,
is neither limited by time nor by space. It belongs to
the world, and is of no country and no age.' I do not
hesitate to say that I prefer the sentiment of Davy to
that of Arago.

Still, even in France, Young did not lack defenders.
M. de Paravey, Inspecteur de l'Ecole Royale Polytech-
nique, for example, speaking of himself in the third
person, makes the following remarks in a letter dated
February 1835, six years after Young's death :—' Il y

admira la science avec laquelle M. le Docteur Young
avait rétabli la chronologie des Rois d'Egypte, ne com-
mençant leur série qu'à la XVIII dynastie de Manethon,
en regardant les séries antérieures comme inadmissi-
bles ; résultats auxquels des travaux tout différents
avaient également conduit M. de Paravey : et, en outre,
il jugea, et il juge encore, que le premier il entrait
d'une manière plausible et sûre dans l'interprétation des
hiéroglyphes, *fournissant ainsi à M. Champollion le
jeune une clef sans laquelle ce dernier n'aurait jamais
pu arriver aux résultats importants et curieux que
depuis il a obtenus.*'

In the same sense, and almost in the same words,
writes Sir Gardner Wilkinson, an ardent admirer of
Champollion, and his chivalrous defender against the
assaults made upon him after his death. After speaking
of him as the kindler into a flame of the spark obtained
by Young, he continues thus :—' Had Champollion been
disposed to give more credit to the value and originality of Dr. Young's researches, and to admit that the
real discovery of the *key* to the hieroglyphics, which in
his dexterous hand proved so useful in unlocking those
treasures, was the result of his [Young's] labours, he
would unquestionably have increased his own reputation without making any sacrifice.'

Peacock speaks with wondering admiration of the
modesty and forbearance which Young invariably
showed in regard to Champollion. He complained a
little, but he threw no doubt or insinuation upon the
Frenchman's honour. He confined himself exclusively
to his own published writings, and made no reference to
the loads of labour which lay upon his shelves unpublished. Peacock complains, and justly complains, of
the unfairness of comparing the Champollion of 1824

with the Young of 1816. Young was the initiatory
genius. He gave Champollion the key, which he used
subsequently with that masterly skill and sagacity
which have rendered his name illustrious. But Peacock emphatically affirms that Champollion passed
over Young's special researches in connection with the
papyri of Grey and Cassati; that whatever principle
of discovery had been perceived and established or made
known, was appropriated without acknowledgment, the
dates which would have proved the unquestionable
priority of Dr. Young being carefully suppressed.

The Dean of Ely obviously felt very sore in regard
to the treatment of Dr. Young. ' It is not our object,'
he says, ' to underrate the merits of the great contributions which were made by Champollion to our knowledge of hieroglyphical literature, but to protest against
the persevering injustice with which he treated the
labours of Dr. Young; and we feel more especially
called upon to do so in consequence of finding that an
author like Bunsen, occupying so high a position
among men of letters, should have supported with the
weight of his authority some of the grossest of his mis-
representations.' Peacock acknowledges his own obligations to the valuable labours of his friend Mr. Leitch;
but he also claims to have pursued an independent
course by consulting the unpublished documents in his
possession, which were unknown even to himself until
he was compelled to study them in connection with the
publications which had been founded upon them. ' It
was only,' he says, 'after this perusal that I became
fully aware how imperfectly the published writings of
Dr. Young represented either the extent or the character of his researches; or the real progress he had made
in the discovery of phonetic hieroglyphics many years before Champollion had made his appearance in the field.'

Within the walls of the village church of Farnborough, Kent, in the family vault of his wife, lie the remains of Thomas Young. On the church wall is a white marble slab with an inscription, the original of which, composed by Mr. Hudson Gurney, stands under Chantrey's medallion of Young in Westminster Abbey. It runs thus :—

SACRED TO THE MEMORY OF

THOMAS YOUNG, M.D.,

FELLOW AND FOREIGN SECRETARY OF THE ROYAL SOCIETY,
MEMBER OF THE NATIONAL INSTITUTE OF FRANCE;
A MAN ALIKE EMINENT
IN ALMOST EVERY DEPARTMENT OF HUMAN LEARNING.
PATIENT OF UNINTERMITTED LABOUR,
ENDOWED WITH THE FACULTY OF INTUITIVE PERCEPTION,
WHO, BRINGING AN EQUAL MASTERY
TO THE MOST ABSTRUSE INVESTIGATIONS
OF LETTERS AND OF SCIENCE,
FIRST ESTABLISHED THE UNDULATORY THEORY OF LIGHT,
AND FIRST PENETRATED THE OBSCURITY
WHICH HAD VEILED FOR AGES

THE HIEROGLYPHICS OF EGYPT.

ENDEARED TO HIS FRIENDS BY HIS DOMESTIC VIRTUES,
HONOURED BY THE WORLD FOR HIS UNRIVALLED ACQUIREMENTS,
HE DIED IN THE HOPES OF THE RESURRECTION OF THE JUST.

BORN AT MILVERTON, IN SOMERSETSHIRE, JUNE 13TH, 1773;
DIED IN PARK SQUARE, LONDON, MAY 10TH, 1829
IN THE 56TH YEAR OF HIS AGE.

1887.

LIFE IN THE ALPS.[1]

LEAVING England in July, and returning in October,
I spend three months of every year among the
Swiss mountains. Various and striking are the aspects
of nature witnessed during these long visits : Sunshine
from unclouded skies, dense fog, mountain mist, furious
rain and hail, snow so deep that, were not my wife and
I thorough children of the hills, and well acquainted
with their ways, we should sometimes fear imprisonment
in our highland home. We have also our due share of
thunder-storms—the peals sometimes rolling at safe
distances, but sometimes breaking so close to us that
we can hear the hiss of the rocks which precedes the
deafening crash. A long roll of echoes follows, undulat-
ing in loudness, and finally dying away amid the rocky
halls of the mountains. This is the state of things so
vividly described by Lord Byron :—

> From peak to peak the rattling crags among
> Leaps the live thunder.

As regards thunder-storms, however, we are far
better off than our neighbours in Northern Italy, whose
hills, acting as lightning-conductors, partially drain the
clouds of their electricity before we receive the shots of
their 'red artillery.' We can see from our perch the

[1] Written for *The Youth's Companion*, Boston, Mass. With
additions.

wonderful thrilling of these Italian thunder-storms, beyond the great mountain range at the farther side of
the valley of the Rhone.

At night it is one of the grandest of spectacles.
Flash rapidly follows flash, while at times the light
bursts simultaneously from different parts of the
heavens, every cloud and mountain-top appearing then
' white-listed through the gloom.' At night the eye is
far more sensitive than it is by day, the more vivid
lightning thrills being then quite dazzling. Meanwhile, no sound is heard; and an observer might be
disposed to conclude that the lightning was without
thunder—*Blitz ohne Donner*, as the Germans say.
Among the southern mountains, however, where the
flashes occur, there is one, called the Monte Generoso,
on which stands a hotel in telegraphic communication
with the lower world. Thither I have telegraphed on
various occasions; and invariably, when the lightning
was thrilling silently in the manner just described, I
have been informed that a terrific thunderstorm was
raging over Northern Italy.[1] From our position here
the peals were too far away to be heard.

The region where we dwell was chosen by Mrs.
Tyndall and myself on account of its surpassing beauty
and grandeur. I first made its acquaintance twenty-
nine years ago, having previously become familiar with
Mont Blanc and its glaciers, and with other glaciers,
both in Switzerland and the Tyrol. It is in the Roman
Catholic canton of Valais, which, notwithstanding the
success of the Reformation in adjoining cantons, has, up
to the present day, maintained its ancient religion.

---

[1] These messages once caused the worthy proprietor of the
Monte Generoso Hotel to remark that he knew Englishmen to be interested in the weather; but that a gentleman at the Bel Alp
seemed positively crazy on the subject.

Here we live on the friendliest terms with both the priests and the people.

Switzerland is made up of a number of cantons, which are subdivided into communes, each possessing its own president and council, and making its own local laws. The communal laws are, however, subject to the revision of the cantonal government. We live, for instance, in the commune of Naters. The sale of the land on which our châlet stands was first agreed to by the vote of the assembled burghers of the commune ; but their vote had to be afterwards ratified by the ' high government ' of Sion, the chief town of the canton. Naters, the name of the commune, is also the name of its principal village.

I had the honour this year of being unanimously elected an honorary burgher of the commune. This confers upon me certain rights and privileges not previously enjoyed. I can pasture cows upon the alps—a name given by the inhabitants, not to the snow-capped mountains, but to the grassy slopes stretching far below the snows. I am also entitled to a certain allowance of fuel from the pine-woods. Finally, I can build a châlet on the communal ground. Prizing, however, the goodwill of the people more than these material advantages, if I at all avail myself of my newly acquired rights and privileges, it will be to a very moderate extent.

The well-known Bel Alp Hotel stands some two or three hundred feet below our cottage. The name ' Bel ' is derived from a little hamlet of huts planted in the midst of grassy pastures, or alps, about half an hour distant from the hotel. The ancient name of the alp on which we have built our nest is Lusgen Alp, and this is the name that we have given to our cottage. I have called it a châlet, but it is by no means one of the picturesque wooden edifices to which this term is usually applied. It has to bear, at times, the pressure of a mighty mass of snow. The walls are therefore built of stone, and are very thick.

I could give you many illustrations of the breakages produced by snow pressure, but one will suffice. Our kitchen chimney rises from the roof near the eave, and the downward pressure of the snow lying on the roof above it was, on two occasions, so great as to shear away the chimney and land it bodily upon the snow-drift underneath. At present a tall pine tree constitutes a strut which will prevent a recurrence of this disaster. When we arrive early we usually find, here and there, heavy residues of snow. Once, indeed, to obtain entrance to our kitchen we had to cut a staircase of six steps in the drift at the back of the house.

A stream, brawling down the mountain, leaps forth as a cascade thirty paces from our dwelling. We have a cistern beside the stream high enough to command the whole of our cottage. A plentiful water supply for ordinary purposes is always at hand. For drinking purposes there is a spring which bursts, crystal-clear, from the breast of the mountain, a quarter of an hour below us. From this, especially when the flocks and herds are on the alps, the drinking water ought in all cases to be drawn.[1]

An hour ago our little cataract ' leaped in glory.' It is now gone—whither? I climb to the top of it, and follow upwards the forsaken bed of the stream. In half an hour I hear the sound of water, and soon afterwards come to a point where the bank of the stream is

---

[1] During the lifetime of its late proprietor I succeeded in getting this water introduced into the Bel Alp Hotel. It may be useful to remark that impure drinking water is just as obnoxious in a bedroom as at a *table d'hôte.*

broken down, and a barrier of clods built across it, all its water being thus diverted into another channel. The streams are a valuable part of the peasant's property; they are employed to water the sloping meadows, and the laws regulating the distribution of the water are very strict. A burgher has a right to the use of a stream for a certain number of hours. If he exceeds that number he trespasses on his neighbours' rights, and is punished if found out. Now the hotel needs water; and in droughty weather a struggle is carried on between the hotel and the meadows, during which the supply of the former is not unfrequently cut off. Such was the case this morning. When the supply ceases the cause is known, and a peasant provided with pick, spade, or shovel, climbs the mountain and restores the deflected stream.

Within doors we work on the ground floor and sleep aloft. We have two bedrooms there and a servants' bedroom, all lined with smooth pine. The house is covered with pine shingle, prettily cut. Slates, I considered, would be a discord in the landscape, and the right of any man to desecrate a scene of natural beauty by such discord may be questioned. We are sometimes wrapped round by absolutely silent air; more frequently by air in motion, rising sometimes, as already indicated, to the force and sound of raging storms. Over us sail heavy-laden clouds which shower their rain-drops like loud-sounding pellets on the shingly roof. 'The music of the rain' at night is often soothing to a wakeful brain. I walk at times on our terrace under the stars, and trace among them with a clearness not seen elsewhere the whiteness of the 'Milky Way.' Just now the moon is full, gleaming from the glacier, and throwing her light, like a holy robe, over the mountains.

We sometimes quit our Swiss home regretfully, with

a warm sun shining on the newly-fallen snow, which vastly enhances the loveliness of the scenes around us. We moreover regret bidding good-bye to the gorgeous colouring of the trees and undergrowth, which might bear comparison with the beauty of the foliage that I have seen in autumn in the neighbourhood of your own Boston. I have never forgotten the autumn splendour of Mr. Winthrop's trees at Brookline.

Sometimes, however, we depart under difficulties. Last year, for example, on October 16, our porters hoisted on their backs the luggage intended for home, and through a dense fog, with the snow three feet deep, and still heavily falling, we moved downwards. Fog on the mountains is terribly bewildering; and as we descended, one of our men, who had tramped the neighbourhood from his infancy, and had on that account been chosen for our leader, stopped short, and declared that he did not know which way to proceed. There was no danger, but the difficulty was considerable. A thousand feet or so lower down we got entirely clear of the snow.

Towards the end of June the flocks and herds are driven to the upper pastures, private ownership ceasing and communal rights, as to grazing, beginning at an elevation of about four thousand feet above the Rhone, or seven thousand feet above the sea. The peasants and their families accompany their living property, remaining for two or three months in huts built expressly with a view to their annual migration. Nearly the whole of them move into Naters for the winter; but we remain alone, amid the solemn silence of the hills, three weeks or a month after the peasants have disappeared. Their time of disappearance depends upon the exhaustion of the pasturage. Many of them have intermediate huts and bits of land between Naters and their highest dwellings. The possessors of such huts descend by successive steps to the valley.

Snow falls, of course, for the most part, in winter ; but the exact period at which it falls is not to be predicted. A winter may pass with scarcely any snow, while in early spring it may fall in immense quantities. Then follows a time of avalanches, when the snow, detaching itself from the steep mountain-sides, shoots downwards with destructive energy.

I have seen here, in midsummer, snow so heavy that the herds had to be driven a long way down to get a little pasture. Three or four years ago, a fall of unequalled severity began on the night of September 12. There was a brief respite of sunshine, during which the peasants, had they been wise, might have brought down their flocks. But they failed to do so. Snowing recommenced, the sheep were caught upon the mountains, and for a long time they could not be reached by their owners. Parties of men, fourteen or fifteen in number, at length ascended in search of the sheep. My wife and I trudged after one party, and extremely hard work we found it to do so. The leader first broke ground, floundering and ploughing a deep channel in the snow. He was soon exhausted, and fell back, while a fresh man came to the front. Each of them thus took the post of leader in his turn.

At a considerable elevation we parted company with the men. It was a sombre, sunless afternoon, and the scene was desolate in the extreme. Here and there we could discern groups of men, two or three in number, engaged in skinning the dead sheep they had discovered. A joint of meat would have remained sweet for any length of time in the snow, but the warmth preserved by the fleece caused the flesh of the sheep to putrefy.

For thirteen days the chief portion of the flock

remained unaccounted for. During all this time the animals were without food, and, indeed, were given up for lost. Nearly two hundred of them, however, were afterwards driven down to the Bel Alp alive. I saw them arrive after their long fast, and they seemed perfectly brisk and cheerful. Some of them were entirely bare of wool, the covering having been eaten off their backs by their famishing companions. I have been assured that the sheep which indulged in this nutriment all died, balls of undigested wool being found in their stomachs afterwards. Avalanches were frequent at the time here referred to, and by them numbers of sheep on the lower slopes were swept away.

It is only those burghers who are comparatively well off that ascend to the higher grazing grounds. Even they seem to find the struggle for existence a hard one. Two or three cows and a few sheep or goats constitute, in well-to-do cases, the burgher's movable wealth, while the land privately owned is divided into very small parcels.

The peasants' huts, built, for the most part, of pine-logs, richly coloured by the oxidising action of the sun, are not always as wholesome as they might be. The upper part of every hut is divided into two dwelling-rooms, one for sleeping, and the other for cooking and other purposes. The single sleeping-room is sometimes occupied by a numerous family, space being obtained by placing one bed above another, like the berths in a ship. There is no chimney, the smoke escaping through apertures in the roof.

In our neighbourhood, the roofs are usually formed of flags obtained from a rock capable of cleavage. The sleeping-room is always over the cowshed, this position being chosen for the sake of warmth. Through chinks in the floor, the sleepers not only obtain warmth, but often air which has passed through the lungs of the animals underneath. The result, as regards health, is not satisfactory, the women and children suffering most. Were it not that the contaminated respiration of the night is neutralised by outdoor life during the day, the result would be still less satisfactory.

Thanks to a liberal London chemist, I am provided, from time to time, with simple medicines for those requiring medical treatment, and with plaisters for those requiring surgeon's aid. Thanks also to the physicians who visit the Bel Alp Hotel, I am sometimes able to apply these remedies with specially good effect. In the absence of a qualified doctor I do the best I can myself. The peasants come to me in considerable numbers, while I frequently go to them.

I do my best to induce the people to open the windows of their sleeping-rooms during the day. The advice is, in many cases, attended to; while, even where it is neglected, whenever I am seen approaching a hut containing a patient, the windows are thrown open. Justice, firmness, and kindness suffice to make people accept an almost despotic rule; and this, in my own small way, I find to be true of my Alpine neighbours.

As I write, a rush, followed by a heavy thud, informs me that a mass of snow has shot from the southern slope of our roof down upon our terrace. The rush is one of a series, brought down by the strong morning sun. This reminds me to tell you something more about the avalanches which are such frequent destroyers of life in the Alps. Whole villages, imprudently situated, are from time to time overwhelmed. We had an eye to this danger when we chose the rocky prominence on which our cottage is built.

Climbers and their guides are not unfrequently

carried away by avalanches, and many a brave man lies at the present moment undiscovered in their *débris*. Some years ago, a famous guide, and favourite companion of mine, allowed himself to be persuaded to attempt the ascent of a mountain which he considered unsafe, and lost his life in consequence. On the slope of this mountain, with the summit fully in view, a report resembling a pistol shot was heard by the party. It was the cracking of the snow. My friend observed the crack, and saw it widen. Tossing his arms in the air he exclaimed, ' We are all lost!' The fatal rush followed in a moment, and my noble guide, with a Russian gentleman to whom he was roped, were dug, dead, out of the snow some days afterwards. The other members of the party escaped.

I will now describe to you an adventure of my own on one of these avalanches. Five of us, tied together by a rope, were descending a steep slope of ice, covered by a layer of snow, which is always a position of danger. Through inadvertence the snow was detached, an avalanche was formed, and, on it, all five of us were carried down at a furious pace. We were shot over crevasses, and violently tossed about by the inequalities of the surface. The length of the slope down which we rushed in this fashion was about a thousand feet. It was a very grave accident, and within a hair's breadth of being a very calamitous one. A small gold watch, which I then carried, was jerked out of my pocket, and, when we stopped, I found a fragment of the watch-chain hanging around my neck.

I made an excursion into Italy, returned after an absence of nearly three weeks, and, half jestingly, organised a party to go in search of the watch. The proverbial needle in a bundle of straw seemed hardly more hopeless as an object of discovery ; still, I thought it barely possible that the snow which covered the watch might, during my absence, have sufficiently melted away to bring the watch to the surface. An ascent of some hours brought us to the scene of our impetuous *glissade*, and soon afterwards, to our surprise and delight, the watch was found on the surface of the snow. Its case must have fitted water-tight, for on being wound up it began to tick immediately. It is now in the possession of my godson.

Falling stones constitute another serious, and frequently fatal, danger in the Alps. And here the goats, which roam about the upper slopes and gullies, often play a mischievous part. An incident of this kind, witnessed by myself, occurred many years ago about midway between Chamounix and the Montanvert. I was accompanied at the time by a friend and his son. A herd of goats was observed browsing on the heights above us. Their appearance suggested caution. Suddenly an ominous tapping was heard overhead, and, looking up, I saw a stone in the air. Whenever it touched the rock-strewn ground it was deflected, so that from the direction of the stone at any moment it was difficult to infer its final direction. I called out to my friend, ' Beware of the stone ! ' and he, turning towards his son, repeated the warning. It had scarcely quitted his lips, when the missile plunged down upon himself. He fell with a shout, and I was instantly at his side. The stone had struck the calf of his leg, embedding one of its angles in the flesh, and inflicting a very ugly wound.

By good fortune, a spring of pure water was at hand ; a wet compress was rapidly prepared, and the wound was bandaged. Then, hurrying down to Chamounix, which was some two thousand feet below us, I brought up men and means, to carry the patient to his hotel. Perfect quiet would have soon set everything right, but the premature motion of the limb was succeeded by inflammation and other serious consequences.

Slipping in perilous places is the most fruitful cause of Alpine disaster. It is usual for climbers to rope themselves together, and the English Alpine Club has taken every pains to produce ropes of the soundest material and the best workmanship. The rope is tied around the waist, or is fastened to a belt clasping the waist, of the climber. The rope is an indispensable accompaniment of Alpine climbing, and no competent mountaineer will recommend its abandonment. Prudence, however, is necessary in the use of it. The men tied together ought to be few in number. A party of three or four, including the guide, or guides, is in my opinion large enough. In a numerous party, there is a temptation to distribute responsibility, each individual tending to rely too much upon the others; while, in a small party, the mind of each man is more concentrated on the precautions necessary for safety. Besides this, we have the terrible enhancement of the calamity when the slipping of a single individual carries a number of others to destruction. It was a slip —by whom we know not—that caused the disaster on the Matterhorn which so profoundly stirred the public mind some years ago. On that occasion, one of the foremost guides of the Alps, and one of the best gentlemen climbers, lost their lives, in company with two younger colleagues.

The fearful disaster on the Jungfrau this year was, doubtless, due to the same cause. Six strong climbers, all natives of Switzerland, succeeded, without guides, in scaling the mountain from the northern side. From the summit they attempted to descend the southern slope, the danger of which varies with the condition of the snow or ice. I had frequently wondered that no accident had ever previously occurred here ; for, to an experienced eye, the possibility of a fatal accident was plain enough. On this slope the six climbers met their doom. They were roped together, and probably only one of them slipped; but his slip involved the destruction of them all. A few weeks after its occurrence I inspected the scene of the disaster, saw the rocks down which the men had fallen, and the snowfield on which their bodies were found.

Their reaching the summit without guides proved them to be competent men. But they could hardly have accomplished the ascent without fatigue, and tired men sometimes shrink from the labour of hewing safe steps in obdurate ice. Neglect on this score may have been the cause of the accident. But this is mere surmise, the only thing certain being the mournful fact that on the Jungfrau, this year, six men in the very prime of life went simultaneously to destruction.

On the fine October morning when these lines are written, we find ourselves surrounded everywhere by glittering snow. The riven Aletsch glacier and its flanking mountains are dazzling in their whiteness. After a period of superb weather, streaks and wisps of boding cloud made their appearance a few days ago. They spread, became denser, and finally discharged themselves in a heavy fall of snow. But the sunshine rapidly recovered its ascendency, and the peasants, who had already descended some distance with their cows and sheep, hoped that two days of such warmth would again clear their pastures.

They were deceived, for through the whole of yesterday the snow fell steadily. It interrupted the transport of our firewood, on mules' backs, from the pine woods

nearly one thousand feet below us. This morning, however, I opened the glass door of our little sitting-room,
which faces south, and stepped out upon our terrace.
The scene was unspeakably grand. To the right rose
the peak of the Weisshorn, the most perfect embodi-
ment of Alpine majesty, purity, and grace. Next
came the grim Matterhorn, then the noble Mischabel-
hörner, surmounted by the ' Dom.' Right opposite rose
the Fletschhorn, a rugged, honest-looking mass, of true
mountain mould; while to the left of Napoleon's road
over the Simplon Pass stretched the snow-ridge of the
Monte Leone—which, no doubt, derives its name from
its resemblance to a couchant lion. Soft gleaming
clouds wrapped themselves at times grandly round the
mountains, revealing and concealing, as they shifted,
melted, or were re-created, the snow-capped peaks.

About one thousand five hundred feet below us the
white covering came to an end, while, beyond this,
sunny green pastures descended to the valley of the
Rhone. From the chimneys of our cottage, a light
wind carried the smoke in a south-westerly direction ;
the clouds, just referred to, being, therefore, to leeward,
and not in ' the wind's eye,' did not portend bad
weather. To the north, the peaks grouped themselves
round the massive Aletschhorn, the second in height
among these Oberland Mountains. Over the Aletschhorn
the sky was clear, which is one of the surest signs of
fine weather. Once, on a morning as fair and exhilarating as the present one, but earlier in the year, Mrs.
Tyndall and myself, from the top of the Aletschhorn—
a height of fourteen thousand feet—looked *down* upon
the summit of the Jungfrau.

The general aspects of the Alpine atmosphere, and,
more especially, the forms and distribution of the

clouds, are very different in autumn from what they are in early summer. The grandest effects of our mountains are, indeed, displayed when no tourist is here to look at them. To us, who remain, this is not a disadvantage; for, like the poet's 'rapture on a lonely shore,' there is rapture for the lover of Nature in the lonely mountains, and 'a radiance of wisdom in their pine woods.' I well remember, after the tourist season at Niagara Falls had ended, my deep delight in visiting alone the weird region of the 'Whirlpool.' Your countryman, Thoreau, did not love the wilds more than I do.

One striking feature invariably reveals itself here at the end of September and the beginning of October. From the terrace of our cottage we look down upon a basin vast and grand, at the bottom of which stands the town of Brieg. Over Brieg, the line of vision carries us to the Simplon Pass, and the mountains right and left of it. Naters stands in a great gap of the mountains, while meadows and pine-clad knolls stretch, with great variety of contour, up to the higher Alpine pastures. The basin has no regularly rounded rim, but runs into irregular bays and estuaries, continuous with the great valley of the Rhone.

At the period referred to, valley, basin, bays, and estuaries are frequently filled by a cloud, the upper surface of which seems, at times, as level as the unruffled surface of the ocean. A night or two ago I looked down upon such a sea of cloud, as it gleamed in the light of a brilliant moon. Above the shining sea rose the solemn mountains, overarched by the star-gemmed sky. Here your young imaginations must aid me, for my pen fails to pursue any further the description of the scene.

As I write, on a day subsequent to that already mentioned, a firmament of undimmed azure shuts out

the view into stellar space. No trace of cloud is visible;
and yet the substance from which clouds are made is,
at this moment, mixed copiously with the transparent
air. That substance is the vapour of water; and I take
this beautiful day as an illustration, to impress upon
you the fact that water-vapour is not a thing that can
be seen in the air. Were the atmosphere above and
around me at the present moment suddenly chilled,
visible clouds would be formed by the precipitation of
vapour now invisible.

Some years ago, I stood upon the roof of the great
cathedral of Milan. The air over the plains of Lombardy was then as pure and transparent as it is here
to-day. From the cathedral roof the snowy Alps are to
be seen; and on the occasion to which I refer, a light
wind blew towards them. When this air, so pure and
transparent as long as the sunny plains of Lombardy
were underneath to warm it, reached the cold Alps, and
was tilted up their sides, the aqueous vapour it contained was precipitated into clouds of scowling blackness, laden with snow.

If you pour cold water into a tumbler on a fine
summer day, a dimness will be immediately produced
by the conversion into water, on the outside surface of
the glass, of the aqueous vapour of the surrounding
air. Pushing the experiment still further, you may
fill a suitable vessel with a mixture of ice and salt,
which is colder than the coldest water. On the hottest day in summer, a thick fur of hoar frost is
thus readily produced on the chilled surface of the
vessel.

The quantity of vapour which the atmosphere contains varies from day to day. In England, north-
easterly winds bring us dry air, because the wind, before
reaching us, has passed over vast distances of dry

land. South-westerly winds, on the other hand, come charged with the vapour contracted during their passage over vast tracts of ocean. Such winds, in England, produce the heaviest rains.

And now we approach a question of very great interest. The condensed vapour which reaches the lowlands as rain, falls usually upon the summits as snow. To a resident among the Alps it is interesting to observe, the morning after a night's heavy rain, a limit sharply drawn at the same level along the sides of the mountains, above which they are covered with snow, while below it no snow is to be seen. This limit marks the passage from snow to rain.

To the mountain snow all the glaciers of the Alps owe their existence. By ordinary mechanical pressure snow can be converted into solid ice ; and, partly by its own pressure, partly by the freezing of infiltrated water, the snow of the mountains is converted into the ice of the glaciers.

The great glaciers, such as the one now below me, have all large gathering grounds, great basins or branches where the snow collects and becomes gradually compacted to ice. Partly by the yielding of its own mass, and partly by sliding over its bed, this ice moves downwards into a trunk valley, where it forms what De Saussure called 'a glacier of the first order.' Such a glacier resembles a river with its tributaries. We may go further and affirm, with a distinguished writer on this subject, that ' between a glacier and a river there is a resemblance so complete, that it would be impossible to find, in the one, a peculiarity of motion which does not exist in the other.'

Thus, it has been proved that owing to the friction of its sides which holds the ice back, the motion of a

glacier is swiftest at its centre ; that because of the
friction against its bed, the surface of a glacier moves
more rapidly than its bottom ; that when the valley
through which the glacier moves is not straight, but
curved, the point of swiftest motion is shifted from its
centre towards the concave side of the valley. Wide
glaciers, moreover, are sometimes forced through narrow
gorges, after which they widen again. At some distance
below the spot where I now write is the gorge of the
Massa through which, in former ages, the great Aletsch
glacier was forced to pass, widening afterwards, and overspreading a large tract of country in its descent to the
valley of the Rhone. All these facts hold equally good
for a river.

On summer days of cloudless glory, the air is sometimes still, and the heat relaxing upon the mountains.
The glacier is then in the highest degree exhilarating.
Down it constantly rolls a torrent of dry tonic air,
which forms part of a great current of circulation.
From the heated valleys the light air rises, and coming
into contact with the higher snows, is by them chilled
and rendered heavier. This enables it to play the part
of a cataract, and to roll down the glacier to the valley
from which it was originally lifted by the sun. But
the action of the sun upon the ice itself is still more
impressive. Everywhere around you is heard the hum
of streams. Down the melting ice-slopes water trickles
to feed little streamlets at their bases. These meet
and form larger streams, which again, by their union,
form rivulets larger still. Water of exquisite purity
thus flows through channels flanked with azure crystal.
The water, as if rejoicing in its liberty, rushes along in
rapids and tumbles in sounding cascades over cliffs of
ice. The streams pass under frozen arches and are
spanned here and there by slabs of rock which, acting

as natural bridges, render the crossing of the torrent easy from side to side. Sooner or later these torrents plunge with a thunderous sound into clefts or shafts, the latter bearing the name of *moulins* or mills, and thus reach the bottom of the glacier. Here the river produced by the melting of the surface-ice rushes on unseen, coming to the light of day as the Rhone, or the Massa, or the Visp, or the Rhine, at the end of the glacier.

A small dark-coloured stone sinks into the ice, while a large stone protects the ice beneath it. Through the small stone the heat readily passes by conduction to its lower surface and melts the ice underneath ; while the barrier offered by the large stone to the passage of the heat cannot be overcome. Round the large stone, therefore, the exposed ice melts away, leaving the rock supported upon a stalk or pillar of protected ice. Slabs of granite, having a surface of one hundred or two hundred square feet, are to be found at this moment poised upon pillars of ice on the Ober Aletsch glacier. Some of them are nearly horizontal. They are called ' tables,' and right royal tables they are for those privileged to feast upon them. Sand strewn upon the glacier by the streams also protects the ice. The protected parts, being left behind, like the 'tables,' form what are called the ' sand cones ' of the glacier. On the adjacent glacier there are cones from ten to twenty feet in height, and they sometimes reach an even greater elevation. On first seeing them, you would imagine them to be heaps of pure sand. But a stroke of the ice-axe shows the sand to be merely the superficial covering of a cone of specially hard ice. The medial moraine, which stretches like a great flexile serpent along the centre of the glacier now below me, also protects the ice, causing it to rise as a spine which attains in some places a height of fifty feet above the surface of the glacier.

It is easy to understand that with a substance like glacier ice, when some parts of it are held back by friction, while other parts tend to move forward, strains must occur which will crack and tear the ice, forming clefts or fissures, to relieve the strains. The crevasses of glaciers are thus produced.

And here we have another conspicuous danger of the Alps. Crevasses have been the graves of many a gallant mountaineer. They are especially dangerous when concealed by roofs of snow, which is frequently the case in the higher portions of the glacier. Of this danger my own experience furnishes examples not to be forgotten. Passing them by, I may mention that, during the present year, an esteemed English clergyman was lost upon an easy glacier of the Engadine, through the yielding of a snow bridge over which he was passing. The crevasse into which he fell could not have been deep, as he was able to converse with a companion above, and to make the tapping of his ice-axe heard. He did not, as far as I know, complain of being hurt, but desired his companion to hasten to procure a rope. The distance to be passed over, however, before the rope and the necessary help could be obtained, was considerable ; and when rope and help arrived, the clergyman was dead.

A discussion followed in the newspapers as to the amount of blame to be assigned to the gentleman who went for the rope. It was said by one writer that he ought to have tied his clothes together, and, by their aid, to have drawn up his friend. The reader of Mr. Laurence Oliphant's last remarkable volume will remember that Mr. Oliphant was once lifted from a dangerous position by a device of this kind.

I never lifted a man out of a crevasse by a rope of clothes, but the lost guide to whom I have already referred, and myself, were once let down by such a rope into a chasm, from which, by means of a real rope, which had been entombed with himself, we rescued a fellow-traveller. Even with the best of ropes, it would require a very strong single man above, and an extremely expert ice-man below, to effect a rescue from a crevasse of any depth. In most cases it would be impossible. So that I think but little blame was incurred by the omission of the clothes-rope experiment.

If a doubt be at all permitted, it must, I think, be on the ground that, having found rescuers, the gentleman failed to accompany them back to the glacier. He pleaded exhaustion, and it is a valid plea. With wider knowledge, however, he might, perhaps, have had himself carried to the glacier rather than remain behind. To a person dying of cold time is everything; and time may have been lost by the guides in finding the particular crevasse in which the unhappy traveller was entombed. The survivor, however, may have been able to describe with accuracy the position of the fissure. If so, he was in my opinion blameless.

Taking its whole tenor into account, the title of this article, instead of being 'Life in the Alps,' might perhaps, with more appropriateness, have been ·'Life and Death in the Alps.'

SUPPLEMENT, 1890.

There remain two little points—the first of which I have not seen noticed elsewhere—which I should like, before we part, to mention to my 'young companion.' On

the mountain-slopes around my châlet we sometimes observe little valleys, or *couloirs*, as the French call them,
where the stones and boulders lying on the surface are
more crowded together than elsewhere. They follow
each other so as to suggest the notion that they are
moving in a regular stream down the *couloir*. But
when we observe them closely from year to year, we
cannot find the slightest evidence of the sliding of the
stones. Walking over the hills, we soon detect another
fact which proves that though no trace of slipping can
be observed, a movement of the stones downward does
undoubtedly take place. To prove this I take you to a
grassy slope, with a great number of boulders strewn
over it. I select one of the largest of these, a boulder
thirty or forty tons in weight, and ask you to look
closely at the slope behind it. For a considerable distance, say one hundred yards, upwards there is a grass-
covered furrow corresponding in width to the size of the
stone. Look now to the front of the boulder. Sods
and smaller stones are piled up in front. You cannot
resist the conclusion that the furrow behind marks the
track along which the boulder has moved, and that the
sods and stones in front have been pushed up by the
boulder in its descent. Turning to other stones in the
neighbourhood, we find them all, more or less, in this
condition. The whole crowd of stones is, in fact,
moving slowly down the slope of the mountain. And
yet, even by the closest scrutiny, you can never find any
trace of fresh earth which the stone has recently quitted.
The growth of the grass appears to be as quick as the
motion of the boulder, thus destroying all obvious
evidence of mere sliding.

I do not myself doubt that the motion takes place
mainly in the spring, when the melting snows render
the soil upon the slopes particularly soft and yielding ;

yet the displacement in a single year cannot be much more than a hair's-breadth. Looking abroad, we discern still larger evidences of imperceptible motion. We find rocks cracked across, cavities formed, crevasses gaping, and separated from each other by wrinkled soil; the whole leaving no doubt upon the mind as to the slow motion of the whole surface of this mountain downwards. And when we examine the outlines of the hills, we find that this sliding down must, in former ages, have occurred on a vast scale. The torrents which furrow the mountains, and which sometimes cut great chasms in their sides, carry, in the long run, both sliding stones and moving earth to lower levels.

We have now come to our concluding observation. On the 25th of September, 1890, my friend M. E. Sarasin, of Geneva, and myself, had the good fortune to witness a rare and beautiful phenomenon. The sun was sloping to the west, and the valley below us, in which lies the great Aletsch glacier, was filled by a dense fog. Standing in a certain position, with the sun behind us, we saw, swept through the fog in front of us, a grand colourless arch of light. It occupied a position close to that which an ordinary coloured rainbow might have occupied. Twice in England, and twice only, I have seen this wonderful luminous band— the first time, in company with my wife, on the high moorland of Hind Head. The white bow was first described by the Spanish navigator, d'Ulloa, after whom it is named. Its explanation can only be briefly indicated here. Along with the true rainbow, and within it, there are usually produced a number of other bows, by what Dr. Young named the 'interference of light.' They are called supernumerary bows. When the raindrops are all of the same size, and exceedingly small, the colours of these bows expand and overlap each other, producing by their mixture the uncoloured white bow. We sometimes hear of ' fog bows,' but it may be doubted whether true impalpable fog can produce a bow. At all events, we may have very dense fogs with all the conditions of light necessary for producing the bow, without its appearance. A sensible drizzle must be mixed with the fog when the bow is produced.

On the occasion to which I have referred, my friend and I noticed another singular phenomenon, which is usually known as the ' Spectre of the Brocken.' The Brocken, or Blocksberg, is a mountain in Germany rendered famous by the poet Goethe. As I stood with my back to the sinking sun, my shadow was cast on the fog before me. It was surrounded by a coloured halo. When I stood beside my friend, our shadows were seen with an iridescent fringe. We shook our heads; the shadows did the same. We raised our arms and thrust our ice-axes upwards; the shadows did the same. All our motions, indeed, were imitated by the shadows. They appeared like gigantic spectres in the mist, thus justifying the name by which they are usually known. The combination of the Brocken spectre and white bow constituted a most striking phenomenon.

1889.

ABOUT COMMON WATER.[1]

WE have already spent what I trust has proved to you an agreeable and instructive half-hour over water in its solid form. We have conversed about the behaviour of those vast collections of ice which go by the name of glaciers, tracing them to their origin in mountain snow. Closely compacted, but still retaining a certain power of motion, the snow passes from the mountain-slopes and reservoirs where it was first collected into the valleys, through which, becoming more and more compacted, it moves as a river of ice. From the end of this solid river always rushes a liquid one, rendered turbid by the fine matter ground from the rocks during the descent of the glacier. An effect which I thought remarkable when I first saw it may be worth mentioning here. Thirty-two years ago I followed the river Rhone to the place where it enters the Lake of Geneva. The water of the lake is known to be beautifully blue, and I fancied beforehand that the admixture of the water of the Rhone must infallibly render the lake turbid. To my surprise, there was no turbidity observable. A moment's reflection rendered the reason of this obvious. The Rhone water, rushing from its parent glaciers, was colder, and therefore heavier, than the water of the lake. Instead of mixing with the latter, it sank beneath it, disposing of itself along the bottom of the lake, and leaving the surface-water with its delicate azure unimpaired.

I propose now to talk to you for half an hour about

[1] Written for *The Youth's Companion.*

water in its more common and domestic forms. On
the importance of water it is not necessary to dwell,
for it is obvious that upon its presence depends the life
of the world. As an article of human diet, its importance is enormous. Not to speak of fruits and vegetables, and confining ourselves to flesh, every four
pounds of boneless meat purchased at the butcher's shop
contain about three pounds of water. I remember
Mr. Carlyle once describing an author, who was making
a great stir at the time, as 'a weak, watery, insipid
creature.' But, in a literal and physical sense, we are
all 'watery.' The muscles of a man weighing one hundred and fifty pounds weigh, when moist, sixty-four
pounds, but of these nearly fifty pounds are mere
water.

It is not, however, of the water compacted in the
muscles and tissues of a man that I am now going to
speak, but of the ordinary water which we see everywhere around us. Whence comes our drinking-water?
A little reflection might enable you to reply :—' If you
go back far enough you will find that it comes from
the clouds, which send their rain down upon the earth.'
' But how,' it may be asked, 'does the water get up
into the cloud region ?' Your reply will probably be,
' It is carried up by evaporation from the waters of the
earth.'

A great Roman philosopher and poet, named Lucretius, wrote much about *atoms*, which he called 'the
First Beginnings.' When it was objected that nobody
could see the atoms, he reasoned in this way :—' Hang
out a wet towel in the sun, and after some time you
will find that all the water has gone away. But you
cannot see the particles of the water that has thus disappeared. Still, it is perfectly certain that the water
which, when put into the towel, could be seen, and felt,

and tasted, and *weighed*, must have escaped from the towel in this invisible way. How, then, can you expect me to show you the atoms, which, as they are the first beginnings of things, are probably much smaller than your "invisible" particles of water?'

In this invisible state, to which water may be reduced, it is called aqueous *vapour*.

Let it then be admitted that water rises into the air by evaporation; and that in the air it forms the clouds which discharge themselves upon us as rain, hail, and snow. If you look for the source of any great American river, you will find it in some mountain-land, where, in its infancy, it is a mere stream. Added to, gradually, by other tributary streams, it becomes broader and deeper, until finally it reaches the noble magnitude of the Mississippi or the Ohio. A considerable portion of the rain-water sinks into the earth, trickles through its pores and fissures, coming here and there to light as a pellucid spring. We have now to consider how 'spring-water' is affected by the rocks, or gravel, or sand, or soil, through which it passes.

The youths who choose this journal for a 'companion' know already that Mrs. Tyndall and myself are lovers of the highlands. I tried last year to give them some notion of 'Life in the Alps.' Well, here in England, Alpine heights are not attainable; but we have built our house upon the highest available land within two hours' of London. Thousands of acres of heather surround us, and storms visit us more furious than those of the Alps. The reason is, that we are here on the very top of Hind Head, where the wind can sweep over us without impediment.

There is no land above our house, and therefore there are no springs at hand available for our use. But lower down, in the valleys, the springs burst forth, providing the people who live near them with the brightest and purest water. These happy people have all my land, and all the high surrounding land, as a collecting-ground, on which the rain falls, and from which it trickles through the body of the hill, to appear at lower levels.

What, then, am I obliged to do? It stands to reason that if I could bore down to a depth lower than the springs, the water, instead of flowing to them, would come to me. This is what I have done. I have sunk a well two hundred and twenty-five feet deep, and am thereby provided with an unfailing supply of the most delicious water.

The water drawn from this well comes from what geologists call the greensand. Within sight of my balcony rise the well-known South Downs, which are hills of chalk covered with verdure. Now, if a bucket of water were taken from my well, and a similar bucket from a well in the South Downs, and if both buckets were handed over to a laundress, she would have no difficulty in telling you which she would prefer. With my well-water it would be easy to produce a beautiful lather. With the South Downs well-water it would be very difficult to do so. In common language, the one water is *soft*, like rain-water, while the other is *hard*.

We have now to analyse and understand the meaning of 'hard water,' and to examine some of its effects. Suppose, then, three porcelain basins to be filled, the first with pure rain-water, the second with greensand-water, and the third with chalk-water; all three waters at first being equally bright and transparent. Suppose the three basins placed on a warm hob, or even exposed to the open air, until the water of each basin has wholly evaporated. In evaporation the water only disappears; the mineral matter remains. What, then, is the result?

In the rain-water basin you have nothing left behind; in the greensand-water basin you have a small residue of solid mineral matter; in the chalk-water basin you have a comparatively large residue. The reason of this is that chalk is soluble in rain-water, and dissolves in it, like sugar or salt, though to a far less extent; while the water of my well, coming from the greensand, which is hardly soluble at all, is almost as soft as rain-water.

The simple boiling of water is sufficient to precipitate a considerable portion of the mineral matter dissolved in it. One familiar consequence of this is, that kettles and boilers in which hard water is used become rapidly incrusted within, while no such incrustation is formed by soft water. Hot-water pipes are sometimes choked by such incrustation; and the boilers of steamers have been known to be so thickly coated as to prevent the access of heat to the water within them. Not only was their coal thus wasted, but it has been found necessary in some cases to burn the very spars in order to bring the steamers into port.

There is no test of the presence of suspended matter in water or air so searching and powerful as a beam of light. An old English writer touched this point when he said :—' The sun discovers atomes, though they be invisible by candle-light, and makes them dance in his beams.' In the purest water—it may be filtered water; it may be artificially-distilled water; it may be water obtained by the melting of the purest ice—a sufficiently strong searching beam reveals suspended matter. I have done my best to get rid of it, but can hardly say that I have completely succeeded.

Differences in quantity are, however, very strikingly revealed. When, in a darkened study, I send a concentrated beam through our well-water, after boiling, it appears turbid ; sent through the South Downs well-water, it appears *muddy*, so great is the quantity of chalk precipitated by the boiling. The mere exposure of hard water to the open air, where it can evaporate, softens it considerably, by the partial precipitation of the mineral matter which it held in solution.

This last observation is important, because it enables us to explain many interesting and beautiful effects. In chalybeate springs, iron is dissolved in the water. Round about such springs, and along the rivulets which flow from them, red oxide of iron—iron rust —is precipitated by the partial evaporation of the water. In Iceland, the water of the Great Geyser holds a considerable quantity of flint or silica in solution. By a most curious process of evaporation this silica, as shown by Bunsen, has been so deposited as to enable what was at first a simple spring to build up, gradually, the wonderful tube of the Geyser, which is seventy-four feet deep and ten feet across, with a smooth basin, sixty feet wide, at the top.

Again, the great majority of our grottos and caves are in limestone rock, which, in the course of ages, has been dissolved away by a stream. To the present hour are to be found, in most of these caves, the streams which made them. I have been through many of them, but through none which can compare in beauty with St. Michael's Cave in the Rock of Gibraltar. From the roof hang tapering stalactites, like pointed spears. From the floor rise columnar stalagmites. The stalactites gradually lengthen, while the stalagmites gradually rise. In numerous cases stalactite and stalagmite meet, the sharp point of the former resting upon the broad top of the latter. Columns of singular beauty, reaching from floor to roof, are thus formed. Stalactites and stalagmites are to be seen in all phases of their approach towards each other; from the little spear, beginning like a small icicle in the roof, and the little mound of stalagmite on the floor, exactly underneath, up to the actual contact of both. The pillars and spears, the arches and corridors, the fantastic stone drapery, the fretted figures on the walls—all contribute to produce an effect of extraordinary magnificence.

What is the cause of the wonderful architecture and decoration of St. Michael's Cave? Probably some of my clever readers will have anticipated both this question nd its answer. The rai , charged with its modicum of carbonic acid by the air, falls upon the limestone rock overhead percolates through it, dissolves it, and, thus 1 den, reaches th roof of the cave. Here it is expose to eva oration. The dissolved solid is, in part, deposited, and the base of the stalactite is planted against the root. The charged water continues to drip, and the stalactite to lengthen. Escaping from the point of the stalactite, the drop falls upon the floor, where evaporation continues, and mineral matter is deposited. The stalagmite rises; the mound becomes a pillar, towards which the spear overhead accurately points, until, in course of time, they unite to form a column.

A similar process goes on over the fretted walls. They shine with the water passing over them. Each water-film deposits its infinitesimal load, the quantity deposited here and there depending on the inequalities of the surface, which cause the water to linger longer, and to deposit more at some places than at others.

The substance most concerned in the production of all this beauty is called by chemists carbonate of lime. It is formed by the union of carbonic acid and lime. What lime is, of course, you already know; its companion, carbonic acid, is, at ordinary temperatures, a very heavy gas. It effervesces in soda-water, and it constitutes a portion of the breath exhaled from the lungs. The weight of the gas, as compared with air, may be accurately determined by the chemist's balance.

But its weight may also be shown in the following way. Let a wide glass shade be turned upside down, and filled with carbonic-acid gas. This is readily done, though when done you do not see the gas. Well, iron sinks in water, because it is heavier than water; it swims on mercury, because it is lighter than mercury. For the same reason, if you blow a soap-bubble and dexterously shake it off, so that it shall fall into the glass shade, it is stopped at the top of the shade, bobbing up and down, as if upon an invisible elastic cushion. The light air floats on the heavy gas. Almost any other acid, poured upon chalk or marble, liberates the carbonic acid. Its grasp of the lime is feeble, and easily overcome. When we dissolve and mix a common soda-powder, the tartaric acid turns the weaker carbonic acid out of doors.

Many natural springs of carbonic acid have been discovered, one of which I should like to introduce to your notice. In the neighbourhood of the city of Naples there is a cave called the *Grotto del Cane*, a name given to it for a curious and culpable reason. During one of the eruptions of Vesuvius I paid a visit, in company with two friends, to Naples, and went to see, among the other sights of that wonderful region, the Grotto of the Dog. At a place adjacent we met a guide and some other visitors. At the heels of the guide was a timid little quadruped, which, for the time being, was the victim that gave the cave its name. We could walk into the cave without inconvenience, knowing, at the same time, from the descriptions we had heard and read, that our feet were plunged in a stream of heavy carbonic acid flowing along the bottom of the cave. The poor little dog, much against its will, was brought into the grotto. The stream of carbonic acid was not deep enough to cover the animal; its master, accordingly, pressed its head under the suffocating gas. It struggled for a time, but soon became motionless —apparently lifeless. Taken into the air outside, through a series of convulsions painful to look upon, it returned to life.

The experiment is a barbarous one, and ought not to be tolerated. There are many ways of satisfying the curious, without cruelty to the dog. I made the following experiment, which seemed to surprise the by-standers. Placing a burning candle near the bottom of my hat, in the open air outside the cave, I borrowed a cap, and by means of it ladled up the heavy gas. Pouring it from the cap into the hat, the light was quenched as effectually as if water had been poured upon it. Made with glass jars instead of hats, this is a familiar laboratory experiment.

We must now proceed slowly forward, making our footing sure as we advance. Lime is sparingly soluble in water, giving it a strong acrid taste. Lime-water is as clear as ordinary water; the eye discerns no differ-ence between them. And now I want to point out to you one of the ways in which the carbonate of lime, which we have been speaking of, may be formed.

I suppose you to have before you a tumbler, or beaker, filled with clear lime-water. By means of a pair of bellows, to the nozzle of which a glass tube is attached, you can cause pure air to bubble through the lime-water. It continues clear. You have been just informed that the breath exhaled from the lungs con-tains carbonic acid, and if this acid be brought into

contact with lime, carbonate of lime will be formed.
Knowing this, you can make the following experiment :—Drawing your breath inward so as to fill your
lungs, you breathe, by means of a glass tube, through
the lime-water. Before you have emptied your lungs the
clear lime-water will have become quite milky, the milki-
ness being due to fine particles of carbonate of lime—
otherwise chalk—formed by the union of the carbonic
acid of your breath with the lime of the water.

Take a well-corked champagne-bottle, from which
the wine has been half removed, but which still retains, above the remaining wine, a quantity of carbonic-
acid gas. It is easy to devise a means of causing this
gas to bubble through lime-water. A heavy white precipitate of chalk is immediately formed.

We now come to a point of great practical importance. The carbonate of lime exists in two forms :
the simple carbonate, of which chalk is an example,
which embraces a certain amount of carbonic acid ; and
the bicarbonate, which contains twice as much. But
the bicarbonate is far more soluble in water than the
simple carbonate. Pure water dissolves only an extremely small quantity of the simple carbonate of lime.
But carbonic acid is sparingly diffused everywhere
throughout our atmosphere, and rain-water always
carries with it, from the air, an amount of carbonic
acid, which converts the simple carbonate of the chalk
into the bicarbonate, of which it can dissolve a considerable quantity. Every gallon of water, for example,
taken from the chalk contains more than twenty grains
of the dissolved mineral.

By boiling, or by evaporation, this bicarbonate is
re-converted into the insoluble carbonate, which renders
our flasks of boiled chalk-water turbid, forms incrusta-
tions in our kettles, and deposits itself as stalactites and

stalagmites in our limestone caves. But there is another way of converting the bicarbonate into the carbonate, which is well worthy of our attention. It will show how a man of science thinks before he experiments, and how, by experiment, he afterwards verifies his thought. Bearing in mind that the chalk-springs hold lime in solution as bicarbonate, it is plain that if we could rob this bicarbonate of half its carbonic acid, we should reduce it to the simple carbonate, which is almost wholly insoluble.

Think the matter over a little. What we have to combat is an excess of carbonic acid. Lime-water, without any carbonic acid, is easily prepared. Suppose, then, that we add to our chalk-water, with its double dose of carbonic acid, some pure lime-water; what would you expect? You would, at all events, think it probable that the bicarbonate of the chalk-water would give up its excess of carbonic acid to the lime, and assume the condition of the simple carbonate, which, because of its insolubility, would be precipitated as a white powder in the water. And, because chalk is heavier than water, you would conclude that the powder would sink to the bottom, leaving a clear, softened water overhead. Thus reasoned Dr. Clark, of Aberdeen, when he invented his beautiful process of softening water on a large scale. I have myself seen the process applied with success in various chalk-districts in England.

Let us make a calculation. Every pound of chalk contains nine ounces of lime and seven ounces of carbonic acid. Dissolved by rain-water, this simple carbonate becomes bicarbonate, where every nine ounces of lime combine with fourteen ounces of carbonic acid. If, then, a quantity of pure lime-water containing nine ounces of lime be added to these

twenty-three ounces of bicarbonate solution, the lime
will seize upon seven ounces of the fourteen, and form
two pounds of the nearly insoluble carbonate. In other
words, nine ounces of lime can precipitate thirty-two
ounces of chalk. Counting thus on a large scale, we
find that a single ton of lime, dissolved in lime-water,
suffices to precipitate three and a-half tons of the simple
carbonate.

Let me now describe to you what I saw at Canter-
bury, where works for the softening of water were
constructed by the late Mr. Homersham, civil engineer.
I found there three reservoirs, each capable of containing one hundred and twenty thousand gallons of water.
There was also a fourth, smaller cistern, containing water
and lime in that state of fine division which is called
" cream of lime." The mixture of water and lime is
violently stirred up by currents of air driven through
it. Brought thus into intimate contact with every
particle, the water soon takes up all the lime it can
dissolve. The mixture is then allowed to stand ; the
solid lime falls to the bottom, and the pure lime-water
collects overhead.

The softening process begins by introducing a
measured quantity of this lime-water into one of the
larger cisterns. The hard water, pumped directly from
the chalk, is then permitted to fill the cistern. When
they come together, the two clear liquids form a kind
of thin whitewash, which is permitted to remain quiet
for twelve or, still better, for twenty-four hours. The
carbonate of lime sinks to the bottom of the reservoir,
covering it as a fine white powder ; while above it is a
water of extreme softness and transparency, and of the
most delicate blue colour. This water harbours no
organisms. Properly conducted to our homes, no infectious fever could ever be propagated by such water.

Blue is the natural colour of both water and ice.
On the glaciers of Switzerland are found deep shafts
and lakes of beautifully blue water. The most striking
example of the colour of water is probably that furnished
by the Blue Grotto of Capri, in the Bay of Naples.
Capri is one of the islands of the Bay. At the bottom
of one of its sea-cliffs there is a small arch, barely
sufficient to admit a boat in fine weather, and through
this arch you pass into a spacious cavern, the walls and
water of which shimmer forth a magical blue light.
This light has caught its colour from the water through
which it has passed. The entrance, as just stated, is
very small; so that the illumination of the cave is
almost entirely due to light which has plunged to the
bottom of the sea, and returned thence to the cave.
Hence the exquisite azure. The white body of a diver
who plunges into the water for the amusement of
visitors is also strikingly affected by the coloured
liquid through which he moves.

Water yields so freely to the hand that you might
suppose it to be easily squeezed into a smaller space.
That this is not the case was proved more than two
hundred and sixty years ago by Lord Bacon. He filled
a hollow globe of lead with the liquid, and, soldering up
the aperture, tried to flatten the globe by the blows of
a heavy hammer. He continued hammering ' till the
water, impatient of further pressure, exuded through
the solid lead like a fine dew.' Water was thus proved
to offer an immense resistance to compression. Nearly
fifty years afterwards, a similar experiment, with the
same result, was made by the members of the Academy
Del Cimento in Florence. They, however, used a globe
of silver instead of a globe of lead. This experiment is
everywhere known as ' the Florentine experiment ';

but Ellis and Spedding, the eminent biographers of Bacon, have clearly shown that it ought to be called 'the Baconian experiment.'

This stubbornness of water in the liquid condition has a parallel in its irresistible force when passing from the liquid into the solid state. Water expands in solidifying; and ice floats on water in consequence of this expansion. The wreck of rocks upon the summits of some mountains is extraordinary. Scawfell Pike in England, and the Eggischhorn and Sparrenhorn in Switzerland, are cases in point. Under the guise of freezing water, a giant stone-breaker has been at work upon these heights. By his remorseless power, even the great and fatal pyramid of the Matterhorn is smashed and riven from top to bottom. I once lay in a tent for a night near a gully of the Matterhorn, and heard all night long the thunderous roar of the stone-avalanches which sweep incessantly down this mountain.

On the slopes surrounding our Alpine home we find heaps and mounds, where slabs and blocks are piled together in apparent confusion. But we soon come to the sure and certain conclusion that these severed pieces are but parts of a once coherent rock, which has been shattered by the freezing of water in its fissures and its pores.

When the severed masses are large, they are sometimes left poised as 'rocking-stones.' A favourite excursion of ours in Switzerland takes us along a noble glacier, to the base of the great final pyramid of the Aletschhorn. There, a few years ago, was to be found a huge rock, with a horizontal upper surface so spacious that twenty of us have sometimes lunched upon it together. Literally as well as technically, it was a noble 'glacier-table.' That great boulder, of apparently iron strength, is now reduced to fragments by the universal pulveriser—freezing water. I say pulveriser; for, over and above its work of destruction upon the mountains, has it not disintegrated the bare rocks of the ancient earth, and thus produced the soils which constitute the bases of the whole vegetable world?

When water passes from the liquid to the solid condition, it is usually by a process of architecture so refined as to baffle our most powerful microscopes. I never observe without wonder this crystalline architecture. Look at it on the window-panes, or on the flags over which you walk on a frosty morning. Nothing can exceed the beauty of the branching forms that overspread the chilled surfaces. Look at the feathery plumes that sometimes sprout from wood, or cloth, or porous stones. The reflecting mind cannot help receiving from this definite grouping and ordering of the ultimate particles of matter suggestions of the most profound significance.

Many months ago I read a stanza from your delightful poet, Bryant, wherein he refers to the 'stars' of snow. Those stellar forms of falling snow repeat themselves incessantly. I have seen the Alps in midwinter laden with these fallen stars; and three or four days ago, they showered their beauty down upon me in England. Dr. Scoresby observed them in the Arctic regions, and Mr. Glashier has made drawings of them nearer home.

The ice-crystal is hexagonal in form, and the snow-stars invariably shoot forth six rays. The hexagonal architecture is carried on in the formation of common ice. Some years ago I set a large lens in the sun, and brought the solar rays to a focus in the air. I then placed a slab of pure ice across the convergent beam.

Sparks of light, apparently generated by the beam, immediately appeared along its track.

Examining the ice afterwards with a magnifying-lens, I found that every one of those brilliant points constituted the centre, or nucleus, of a beautiful liquid flower of six petals. There was no deviation from this number, because it was inexorably bound up with the crystalline form of the ice.

Thus, in a region withdrawn from the inattentive eye, we find ourselves surprised and fascinated by the methods of Nature.

1890.

PERSONAL RECOLLECTIONS OF THOMAS CARLYLE.[1]

IT is an age of 'Reminiscences'; known to me, in great part, through extracts and reviews. Pleasant reading, in their fulness, many of these records must surely be. Carlyle has given us his 'Reminiscences'—written, alas! when he was but the hull of the true Carlyle. Still, methinks the indignation thereby aroused was out of proportion to the offence. It is not, however, my task or duty to defend the 'Reminiscences.' In clearer and happier moments, Carlyle himself would have recoiled from publishing their few offending pages. When they were written, all things were seen by him through the medium of personal suffering, physical and mental. This lurid atmosphere defaced, blurred, and sometimes inverted like mirage, his coast-line of memory. The figure of himself, standing on that quivering and delusive shore, has suffered more from the false refraction than anything else. With the piercing insight which belonged to him all this, in healthier hours, would have been seen, weighed, and rectified by Carlyle himself.

Vast is the literature which has grown around the memory of this man. It is not my desire, or intention, to sensibly augment its volume. I wish merely to contribute a few memorial notes which I am unwilling to let die, but which, in presence of what has gone before, are but as a pebble dropped upon the summit of a tor.

[1] Written for the most part from memory in the Alps, 1889, and published in the *Fortnightly Review*, January, 1890.

There are amongst us eminent men who knew Carlyle
longer, and who saw him oftener, than I did—whose
store of memories is therefore far larger than mine.
But it was my fortune, during some of the most impressive phases of his life, to be very close to him ; and
though my visits to his home in Chelsea, and our common rambles in London and elsewhere, were, to my
present keen regret, far fewer than either of us wished
them to be, they gave me some knowledge of his inner life
and character. Better however than in any formal record,
that life is to be sought and found in his imperishable
works.   There we best see the storm of his passion, the
depth of his pity, the vastness of his knowledge—his
humour, his tenderness, his wisdom, his strength.   As
long as men continue capable of appreciating what is
highest in literary achievement, these works must hold
their own.

When, before a group of distinguished and steadfast friends, the statue of Carlyle was unveiled on the
Thames Embankment, I briefly referred to my first
acquaintance with his works.   'Past and Present,' the
astonishing product of seven weeks' fierce labour in the
early part of the year, was published in 1843 ; and
soon after its publication I met some extracts from the
work in the Preston newspapers.   I chanced, indeed,
to be an eye-witness of the misery which at that time
so profoundly moved Carlyle.   With their hands in
their pockets, with nothing in their stomachs, but
with silent despair fermenting in their hearts, the
'hunger-stricken, pallid, yellow-coloured' weavers of
Preston and the neighbourhood stalked moodily through
the streets.   Their discontent rose at length to riot,
and some of them were shot down.   Such were the
circumstances under which Carlyle appealed to Exeter

Hall, with its schemes of beneficence for aborigines
far away. ' These yellow-coloured for the present
absorb all my sympathies. If I had a twenty millions
with model farms and Niger expeditions, it is to them
I would give it.' Under the same circumstances he
warned his ' Corn-lawing friends ' that they were driving
into the frenzy of Socialism ' every thinking man in
England.' With my memory of the Preston riots still
vivid, I procured ' Past and Present,' and read it per-
severingly. It was far from easy reading ; but I found
in it strokes of descriptive power unequalled in my
experience, and thrills of electric splendour which
carried me enthusiastically on. I found in it, moreover, in political matters, a morality so righteous, a
radicalism so high, reasonable, and humane, as to make
it clear to me that without truckling to the ape and
tiger of the mob, a man might hold the views of a
radical.

The first perusal of the work gave me but broken
gleams of its scope and aim. I therefore read it a
second time, and a third. At each successive reading
my grasp of the writer's views became stronger and
my vision clearer. But even three readings did not
satisfy me. After the last of them, I collected economically some old sheets of foolscap, and wrote out
thereupon an analytical summary of every chapter.
When the work was finished I tied the loose sheets
together with a bit of twine and stowed them away.

For many years they remained hidden from me. I
had passed through the railway madness of the ' forties,'
emerging sane from the delirium. I had studied in
Germany, had lectured at the Royal Institution, and
in 1853 had been appointed its Professor of Natural
Philosophy. For fifteen years I had enjoyed the
friendship of Faraday, whose noble and illustrious life

came to an end in 1867, on Hampton Court Green. Reverently, but reluctantly, I took his place as Superintendent of the Royal Institution, vastly preferring, if it could have been so arranged, to leave Mrs. Faraday in undisturbed possession of the rooms which had been her happy home for six-and-forty years. The thing, however, could not be. On returning from one of my Alpine expeditions I found at the entrance of the place which had been occupied successively by Davy and Faraday, my name upon the wall. It was to me more of a shock than a satisfaction.

The change, however, brought me nearer to Carlyle; and to Albemarle Street from time to time he wended his way to see me. Once he found me occupied, not with a problem of physics, but with a question of biology of fundamental import. The origin of life was, is, and ever will be, a question of profoundest interest to thoughtful men. In the early ' seventies ' I was busy experimenting on this question, my desire being to bring to bear upon it physical methods which should make known the unmistakable verdict of science regarding it, and thus abolish the doubt and confusion then existing. Permitting air to purify itself by the subsidence of all floating motes, so that the track through it of a sunbeam, even when powerfully concentrated, was invisible, infusions of meat, fish, fowl, and vegetables were exposed to such air and found incapable of putrefaction. The vital oxygen was still there; but with the floating motes, the seminal matter of the atmosphere had vanished, and with it the power of generating putrefactive organisms. The organisms, in other words, required the antecedent seed—*there was no spontaneous generation*. By means of gas stoves rooms had been raised to the proper temperature, and into one of these rooms, which was stocked with my

moteless chambers, I took Mr. Carlyle. He listened
with profound attention to the explanation of the experiments. They were quite new to him; for *microbes*,
bacilli, and *bacteria* were not then the household words
which they are now. I could notice amazement in his
eyes as we passed from putrefaction to antiseptic
surgery, and from it to the germ theory of communicable disease. To Carlyle life was wholly mystical—
incapable of explanation—and the conclusion to which
the experiments pointed, that life was derived from
antecedent life, and was not generated from dead
matter, fell in with his notions of the fitness of things.
Instead, therefore, of repelling him, the experiments
gave him pleasure.

After quitting the laboratory I took my guest upstairs, and placed him. in an armchair in front of a
cheerful fire. The weather was cold, and I therefore
prepared for him a tumbler of mulled claret. And
now we arrive at the cause which induces me to speak
thus early of a late event. About a fortnight prior to
this visit, while rummaging through a mass of ancient
papers, I had come upon the long-lost sheets of foolscap
which contained my analysis and summary of the various
chapters of 'Past and Present.' The packet, tied with
twine as aforesaid, and bearing the yellow tints of age,
lay in an adjacent drawer. At length I said to him,
' Now you shall see something that will interest and
amuse you.' I took the ragged sheets from the drawer,
told him what they were and how they had originated,
and read aloud some of the passages which had kindled
me when young. He listened, sometimes clinching a
paragraph by a supplement or ratification, but frequently breaking forth into loud and mellow laughter
at his own audacity. It would require gifts greater
even than those of Boswell to reproduce Carlyle. I

think it was my sagacious friend, Lady Stanley of Alderley, who once remarked to me that in the reported utterances of Carlyle we miss the deep peal which rounded off and frequently gave significance to all that had gone before.[1] Our fun over the eviscerated 'Past and Present' continued for some time, after which it ceased, and an expression of solemn earnestness overspread the features of the old man. 'Well,' he said at length, in a voice touched with emotion, 'what greater reward could I have than to find an ardent young soul, unknown to me, and to whom I was personally unknown, thus influenced by my words.' We continued our chat in a spirit of deeper earnestness, and after he had exhausted his goblet we walked together down Albemarle Street to Piccadilly, his tough old arm encircling mine. There I saw. him safely seated in a Brompton omnibus, which was his usual mode of locomotion. When he was inside every conductor knew that he carried a great man.

All this was late in the day of my acquaintance with Carlyle. I first saw him, and heard his voice, in the picture gallery of Bath House, Piccadilly. I noticed the Scottish accent, not harsh or crabbed as it sometimes is, but rich and pleasant, which clung to him throughout his life, as it did also to Mrs. Somerville. I first became really acquainted with him at the 'Grange,' the Hampshire residence of the accomplished and high-minded Lord Ashburton. Sitting beside him at luncheon, I spoke to him, and he answered me bluntly. James Spedding was present, and to render myself sure of his identity I asked Carlyle, in a low voice, whether the gentleman opposite was not Spedding? 'Yes,' he

[1] From Dr. Garnet's excellent *Life of Carlyle* I learn that Mrs. Allingham had also drawn attention to this point.

replied aloud, 'that's Spedding.' He had no notion of tolerating a confidential whisper. The subject of homœopathy was introduced. Carlyle's appreciation of the relation of cause and effect was as sharp and clear as that of any physicist; and he thought homœopathy an outrageous defiance of the proportion which must subsist between them. I sought to offer an explanation of the alleged effects of 'infinitesimals,' by reference to the asserted power of the Alpine muleteer's bell to bring down an avalanche. If the snow could be loosened by a force so small, it was because it was already upon the verge of slipping. And if homœopathic globules had any sensible effect, it must be because the patient was on the brink of a change which they merely precipitated. Carlyle, however, would listen to neither defence nor explanation. He deemed homœopathy a delusion, and those who practised it professionally impostors. He raised his voice so as to drown remonstrance; while a 'tsh!' with which Mrs. Carlyle sometimes sought to quiet him, was here interposed. Casting homœopathy overboard, he spoke appreciatively of George III. The capacity of the King was small, but he paid out conscientiously the modicum of knowledge he possessed. This was illustrated by the way in which he collected his library, always seeking the best advice and purchasing the best books. Carlyle's respect for conscientiousness and earnestness extended to all things. We once went together to an exhibition of portraits at South Kensington. Pausing before the portrait of Queen Mary (Bloody Mary, as we had been taught to call her), he musingly said, ' A well-abused woman, but by no means a bad woman—rather, I should say, a good woman—acting according to her lights.' He ought, perhaps, to have extended the same tolerance to Ignatius Loyola, whom he abhorred and scathed. In the evening, while we stood before the drawing-room fire, I spoke to him of Emerson. There was something lofty in the tone of Carlyle's own voice as he spoke of the 'loftiness' of his great American friend.[1] I mentioned Lewes's life of Goethe, which I had been just reading, and ventured to express a doubt whether Lewes, as a man, was strong enough to grapple with his subject. He was disposed to commend the Life as the best we had, but he was far from regarding it as adequate. Carlyle was a bold rider, and during this visit to the Grange he indulged in some wild galloping. Professor Hofmann was his companion, and he humorously described their motion as tantamount to being shot like a projectile through space. Brookfield was one of the guests, a clergyman of grace and culture, who might have been a great actor, and who entertained a high notion of the actor's vocation. One evening he gave us an illustration of his dramatic gifts—extemporising, and drawing by oblique references, the principal personages round him into his performance. It was then I first heard the resonant laugh of Carlyle. Himself a humourist on a high plane, he keenly enjoyed humour in others. Lady Ashburton, with fine voice and expression, read for us one of Browning's poems. It was obvious from his ejaculatory remarks that Carlyle enjoyed and admired Browning.

As time went on I drew more closely to Carlyle,

[1] Their friendship continued unimpaired to the end. Not long before Carlyle's death, I noticed two volumes of the same shape and binding on the table of his sitting-room. Opening one of them, I found written on the fly-leaf :—

'To Thomas Carlyle

'with unchangeable affection

'from Ralph Waldo Emerson.'

The two volumes were Emerson's own collected works. 'That,' I said, 'is as it ought to be : you and Emerson must remain friends to the last.' 'Aye,' he responded, 'you are quite right; take the volumes with you, but return them punctually' : which I did.

seeking, among other things, to remove all prejudice
by making clear to him the spirit in which the highest
scientific minds pursued their work. They could not
detach themselves from their fellow-men, but history
showed that they thought less of worldly profit and
applause, and practised more of self-denial than any
other class of intellectual workers. Carlyle had been
to the Royal Society, but found the meetings he attended flat, stale, and unprofitable. Not knowing how
the communications were related to the general body
of research, they, of course, lacked the sap which their
roots might have supplied to them. He was surprised to
find me fairly well acquainted with 'Wilhelm Meister's
Wanderjahre,' declaring that, as far as his knowledge
went, the persons were few and far between who showed
the least acquaintance with Goethe's 'Three Reve-
rences'—reverence for what is above us, reverence for
what is around us, reverence for what is beneath us.
To this feature of Goethe's ethics Carlyle always attached great importance. Among the spoken and
written words of our age the utterances of Goethe were,
in his estimation, the highest and weightiest. Of
Fichte and Schiller he sometimes spoke with qualified
admiration—of Goethe never. He may have been
indebted to the great German for a portion of his
spiritual freedom, and such indebtedness men do not
readily forget. Unswerving in his loyalty, Carlyle,
towards the end of his life, would have ratified by
re-subscription the ardent outburst of 1831. 'And
knowest thou no prophet, even in the vesture, environment, and dialect of this age? None to whom the
Godlike has revealed itself, and by him been again
prophetically revealed; in whose inspired melody, even
in these rag-gathering and rag-burning days, Man's
life again begins, were it but afar off, to be divine?

Knowest thou none such? I know him and name him—Goethe.'[1] The majesty of Goethe's character seemed, in Carlyle's estimate of him, to dissolve all his errors, both of intellect and conduct. The standards of the homiletic market-place were scornfully brushed aside; drawbacks and qualifications were blown away like chaff, 'the golden grain' of the mighty German husbandman being alone garnered and preserved.

I had various talks with him about Goethe's mistaken appreciation of the 'Farbenlehre' as the greatest of his works. To Carlyle this was a most pathetic fact. The poet thought he had reached the adamant of natural truth, and alas! he was mistaken. But, after all, was he mistaken? Over German artists the 'Farbenlehre' had exercised a dominant influence. Could it be all moonshine? Thus he mused. While holding firmly to the verdict that with regard to theory Goethe was hopelessly wrong, I dwelt with pleasure on the wealth of facts which his skill and industry had accumulated. This to a certain extent gratified Carlyle, but he sighed for the supplement necessary to the scientific completeness of his hero. He was intimately acquainted with every nook and corner of Goethe's work—sometimes more intimately than the poet's own countrymen. I once had occasion to quote the poem 'Mason Lodge,' translated and published in 'Past and Present.'[2] The article in which it was quoted was

---

[1] *Sartor Resartus*, Library Edition, p. 244.

[2] Book III. chap. xv. A very noble song, and a great favourite of Carlyle's. With it he wound up his Rectorial address at Edinburgh. The reciting of two of its verses, under peculiar circumstances, had an important influence on my own destiny.

'Solemn before us,	'Here eyes do regard you,
Veiled, the dark Portal,	In Eternity's stillness;
Goal of all mortal:	Here is all fulness,
Stars silent o'er us,	Ye brave, to reward you,
Graves under us silent!	Work and despair not.'

afterwards translated into German ; the original poem, therefore, required hunting up. None of my friends in Berlin knew anything about it. On learning this I went down to Chelsea, where, in answer to my inquiry, Carlyle promptly crossed his sitting-room and took from a shelf the required volume.

Thus, through the years, I kept myself in touch with this teacher and inspirer of my youth. The ' Life of Frederick ' drew heavily upon his health and patience. His labours were intensified by his conscientiousness. He proved all things, with the view and aim of holding fast that which was historically good. Never to err would have been superhuman ; but if he erred, it was not through indolence or lack of care. The facts of history were as sacred in his eyes as the ' constants ' of gravitation in the eyes of Newton ; hence the severity of his work. The ' Life of Frederick,' moreover, worried him ; it was not a labour into which he could throw his whole soul. He was continually pulled up by sayings and doings on the part of his hero which took all enthusiasm out of him. ' Frederick was the greatest administrator this world has seen, but I could never really love the man.' Such were his words. While engaged on this formidable task, he was invited to stand for the Rectorship of Edinburgh University. For the moment he declined, promising, however, to consider the proposal when his labours on Frederick were ended. The time came, and he accepted the invitation. Disraeli was pitted against him, but he won the election by an overwhelming majority. His transport to Edinburgh had then to be considered. After many talks with him and his wife, the simplest and safest solution of the difficulty seemed to be that I should take charge of him myself.

It was arranged that he should go first to Freystone, in Yorkshire, and pay a short visit to Lord
Houghton. On the morning of March 29, 1866, I drove
to Cheyne Row, and found him punctually ready at
the appointed hour. Order was Carlyle's first law, and
punctuality was one of the chief factors of order. He
was therefore punctual. On a table in a small back
parlour below-stairs stood a 'siphon,' protected by
wickerwork. Carlyle was conservative in habit, and in
his old age he held on to the brown brandy which was
in vogue in his younger days. Into a tumbler Mrs.
Carlyle poured a moderate quantity of this brandy, and
filled it up with the foaming water from the siphon.
He drank it off, and they kissed each other—for the
last time. At the door she suddenly said to me, ' For
God's sake send me one line by telegraph when all
is over.' This said, and the promise given, we drove
away.

In due time we reached Freystone, where the
warmest of welcomes greeted Carlyle. A beautiful
feature in the record of Carlyle's relations to his friends
is the loving loyalty of Lord Houghton. Not long
prior to his lamented death he sent me an extract from
a letter written by Carlyle to his wife on the occasion,
I believe, of his first visit to Freystone. It had been
purchased by Lord Houghton from some collector of
letters, into whose hands it had fallen. It showed how
long-standing Carlyle's malady of sleeplessness had
been. It spoke of the weary unrest of the previous
night—the ceaseless tossing to and fro—and of the
comfort he experienced in thinking of her, as he
smoked his morning cigar in the sunshine. On the
first night of his last visit to Freystone, the unrest was
not only renewed but intensified. Railways had multiplied ; they clasped Freystone as in a ring, and their

whistles were energetically active all night. I feared
the result, and my fears proved only too well grounded.
In the morning I found Carlyle in his bedroom, wild
with his sufferings. He had not slept a wink. It
ought to be noted that the day previous he had dined
two or three hours later than was his wont, and had
engaged in a vigorous discussion after dinner. Looking at me despairingly, he said, ' I can stay no longer
at Freystone, another such night would kill me.'
' You shall do exactly as you please,' was my reply. ' I
will explain matters to Lord Houghton, and he, I am
persuaded, wi'l comply with all your wishes.' I spoke
to Lord Houghton, who, though sorely disappointed,
agreed that it was best to allow his guest complete
freedom of action. It was accordingly arranged that
we should push on to Edinburgh. Carlyle's breakfast
was prepared. He partially filled a bowl with strong
tea, added milk, and an egg beaten up. Rendered
thus nutritive, the tea seemed to soothe and strengthen
him. As he breakfasted our projects were discussed.
Once, after a pause, he exclaimed, ' How ungrateful it
is on my part, after so much kindness, to quit Freystone in this fashion.' Taking prompt advantage of
this moment of relenting, I said, ' Do not quit it, but
stay. We will take a pair of horses and gallop over
the country for five or six hours. When you return
you shall have a dinner like what you are accustomed to at home, and I will take care that there
shall be no discussions afterwards.' He laughed, which
was a good sign. I stood to my guns, and he at
length yielded. Lord Houghton joyfully ratified the
programme, and two horses were immediately got
ready.

The animal bestrode by Carlyle was a large bony
grey, with a terribly hard mouth. He seemed dis-

posed to bolt, and obviously required a strong wrist to rein him in. Carlyle was no longer young : *paralysis agitans* had enfeebled his right hand—for some time my anxiety was great. But after sundry imprecations and strenuous backward pulls, the horse was at length clearly mastered by its rider, and we fleetly sped along. Through lanes, over fields, along high-roads, past turnpike gates where I paid the toll. This continued for at least five hours, at the end of which we returned, and handed the bespattered horses over to the groom. The roads and lanes had been abominable, mud to the fetlocks, not to speak of the slimy fields. Had the groom's feelings been allowed open vent, we should have had imprecations on his part also. We heard only a surface murmur, but the storm, I doubt not, discharged itself behind our backs in the stable. Carlyle went to his room, donned his slippers and his respectable grey dressing-gown. Carrying with him one of the long 'churchwardens' which he always obtained from Glasgow, he stuffed it full of tobacco. Choosing a position on the carpet by the hall fire which enabled him to send the products of combustion up the chimney, to the obvious astonishment of the passing servants he began to smoke. Having with me at the time a flask of choice pale brandy, of this, mixed with soda-water, I gave him a stiff tumbler. The ride had healthily tired him, and he looked the picture of content. At six o'clock his simple dinner was set before him, and he was warned against discussion. It was the traditional warning of the war-horse to be quiet when he hears the bugle sound. In the evening discussion began with one of the guests, and I could see that Carlyle was ready to dash into it as impetuously as he had done the night before. I laid my hand upon his arm and said sternly, ' We must have

no more of this.' He arched his brows good-humouredly,
burst into laughter, and ended the discussion. I accompanied him to his bedroom, every chink and fissure
of which had been closed to stop out both light and
sound. ' I have no hope of sleep,' he said, ' and I will
come to your room at seven in the morning.' My
reply was, ' I think you *will* sleep, and if so, I will
come to your room instead of your coming to mine.'
My hopes were mainly founded on the vigorous exercise
he had taken ; but the next day being Good Friday, I
also hoped for a mitigation of the whistle nuisance.

At seven o'clock, accordingly, I stood at his door.
There was no sound. Returning at eight, I found the
same dead silence. At nine, hearing a rustle, I opened
his door and found him dressing. The change from
the previous morning was astonishing. Never before
or afterwards did I see Carlyle's countenance glow with
such happiness. It was seraphic. I have often thought
of it since. How in the case of a man possessing a
range of life wide enough to embrace the demoniac and
the godlike, a few hours' sound sleep can lift him from
the grovelling hell of the one into the serene heaven of
the other! This question of sleep or sleeplessness hides
many a tragedy. He looked at me with boundless
blessedness in his eyes and voice. ' My dear friend, I
am a totally new man; I have slept nine hours without once awaking.' That night's rest proved the prelude and guarantee of his subsequent triumph at Edinburgh.

We had been joined at Freystone by Huxley,[1] and
in due time started, all three together, for the beautiful

---

[1] And by the able and lamented Mr. Maclennan. Dr. Hirst also
paid a brief visit to Freystone, and was afterwards one of Carlyle's
hearers in Edinburgh.

metropolis of the North. There Carlyle was lodged in the house of his gentle and devoted friend, Erskine of Linlathen. He was placed as far from the noises of the street, in other words as near the roof, as possible. I saw him occasionally in his skyey dormitory, where, though his sleep did not reach the perfection once attained at Freystone, it was never wholly bad. There was considerable excitement in Edinburgh at the time —copious talking and hospitable feasting. The evening before the eventful day I dined at Kinellan with my well-beloved friends, Sir James and Lady Coxe, whose permanent guest I was at the time. Sir David and Lady Brewster were there, and Russell of the *Scotsman*. The good Sir David looked forward with fear and trembling to what he was persuaded must prove a *fiasco*. ' Why,' he said to me, ' Carlyle has not written a word of his Address ; and no Rector of this University ever appeared before his audience without this needful preparation.' In regard to the writing I did not share Sir David's fear, being well aware of Carlyle's marvellous powers of utterance when he had fair play. *There*, however, was the rub. Would he have fair play? Would he come to his task fresh and strong, or with the pliancy of his brain destroyed by sleeplessness? This surely is the tragic side of insomnia, and of the dyspepsia which frequently generates it. ' It takes all heart out of me, so that I cannot speak to my people as I ought.' Such were the words of a worthy Welsh clergyman whom I met in 1854 among his native hills, and whose unrest at night was similar to that of Carlyle. Time would soon deliver its verdict.

The eventful day came, and we assembled in the anteroom of the hall in which the address was to be delivered —Carlyle in his rector's robe, Huxley, Ramsay, Erskine, and myself in more sober gowns. We were all four to be doctored. The great man of the occasion had declined the honour, pleading humorously that in heaven there might be some confusion between him and his brother John, if they both bore the title of doctor. I went up to Carlyle, and earnestly scanning his face, asked : ' How do you feel?' He returned my gaze, curved his lip, shook his head, and answered not a word. ' Now,' I said, ' you have to practise what you have been preaching all your life, and prove yourself a hero.' He again shook his head, but said nothing. A procession was formed, and we moved, amid the plaudits of the students, towards the platform. Carlyle took his place in the rector's chair, and the ceremony of conferring degrees began. Looking at the sea of faces below me—young, eager, expectant, waiting to be lifted up by the words of the prophet they had chosen —I forgot all about the degrees. Suddenly I found an elbow among my ribs—' Tyndall, they are calling for you.' I promptly stood at ' 'tention ' and underwent the process of baptism. The degrees conferred, a fine tall young fellow rose and proclaimed with ringing voice from the platform the honour that had been confered on ' the foremost of living Scotchmen.' The cheers were loud and long.

Carlyle stood up, threw off his robe, like an ancient David declining the unproved armour of Saul, and in his carefully-brushed brown morning-coat came forward to the table. With nervous fingers he grasped the leaf, and stooping over it looked earnestly down upon the audience. ' They tell me,' he said, ' that I ought to have written this address, and out of deference to the counsel I tried to do so, once, twice, thrice. But what I wrote was only fit for the fire, and to the fire it was compendiously committed. You must therefore listen to and accept what I say to you as coming straight

from the heart.' He began, and the world already
knows what he said. I attended more to the aspect of
the audience than to the speech of the orator, which
contained nothing new to me. I could, however, mark
its influence on the palpitating crowd below. They
were stirred as if by subterranean fire. For an hour
and a half he held them spellbound, and when he ended
the emotion previously pent up burst forth in a roar of
acclamation. With a joyful heart and clear conscience
I could redeem my promise to Mrs. Carlyle. From the
nearest telegraph-office I sent her a despatch of three
words : 'A perfect triumph,' and returned towards the
hall. Noticing a commotion in the street, I came up
with the crowd. It was no street brawl ; it was not the
settlement of a quarrel, but a consensus of acclamation,
cheers and ' bravos,' and a general shying of caps into
the air ! Looking ahead I saw two venerable old men
walking slowly arm-in-arm in advance of the crowd.
They were Carlyle and Erskine. The rector's audience
had turned out to do honour to their hero. Nothing in
the whole ceremony affected Carlyle so deeply as this
display of fervour in the open air.

All this was communicated by letter to Mrs. Carlyle ;
and as I shared the general warmth of the time, it
is to be assumed that my letters were of the proper
temperature. She, at all events, wrote warmly enough
about me afterwards. Wound up, as she had been, to
such an intense pitch of anxiety, the thin-spun life was
almost ' slit ' by the telegram. Her joy was hysterical.
But after a little time, aided by the loving care of
friends, she shook away all that was abnormal in her
happiness. She dined that evening with John Forster.
Dickens and Wilkie Collins were of the party. She
entered the drawing-room exultant, waving the telegram
in the air. Warm felicitations were not wanting, and

probably on that occasion her cup of bliss was fuller
than it had been for years before.

Carlyle's great task having ended thus happily, he
joined in festivities, public and private. Meat and wine
I have forgotten, but I have not forgotten the jocund
after-dinner songs. They were sung by their composers.
Dry science became plastic in the hands of these artists,
and the forms it assumed must have astonished Carlyle.
He joined heartily in the fun. Two banquets dwell
specially in my memory—a *Symposium Academicum*,
got up in Carlyle's honour, and a dinner at the house of
his steadfast friend, Professor Masson. At both hilarity
ran high. The figure of Dr. Maclagan, with eyes
directed piteously upwards, with body bent, and hands
clasped in agony over some excruciating medical
absurdity, has left an unfading photograph upon my
brain. Till then I had thought the dinners of our
Royal Society Club in London the most genial in the
world ; but they could not hold a candle to this Edinburgh Symposium. The dinner at Masson's was equally
jovial. Lord Neaves was there—one of the most pleasant
personages I had ever met. He was charged with his
own bright ditties, which he sang with infective anima-
tion. Some time previously John Stuart Mill had
written his ' Examination of the Philosophy of Sir
William Hamilton,' wherein he had reduced the external world to 'a series of possibilities of sensation.'
Lord Neaves had thrown this theory into lyric rhyme.
The refrain of his song was ' Stuart Mill on Mind and
Matter.' The whole table joined in the refrain, Carlyle,
with voice-accompaniment, swaying his knife to and
fro, like the bâton of a ' conductor.' If, afterwards, in
a fit of depression, he described the time he spent in
Edinburgh as ' a miserable time,' he must have been
the victim of self-delusion. It was a time of joy and

gladness, which he amply shared ; but he seemed unable, subsequently, to shoot the rays of memory through the heavy atmosphere which immediately surrounded him. Like light-rays in a fog, they were quenched by re-percussion from his own melancholy broodings. In Edinburgh all the necessary elements combined to render him happy. In the background slumbered the consciousness of success. In the same region lay thoughts of his wife, whose pride in his triumph would reverberate its glow upon him. Clinging to her image were memories of a time when her union with him was deemed a *mésalliance.* Who could think so now ? He stood consciously there as a victor over difficulties which would have broken to pieces not the feeble only, but the strong—a victor in the chief city of his country, which he had entered fifty-seven years previously as a wayworn peasant-boy. Such, during his actual stay in Edinburgh, were Carlyle's pleasant musings—swept, alas! into practical oblivion by calamity soon afterwards.

Huxley and I had 'proposed to ourselves an excursion in the Highlands ; but snow had fallen, covering the hills and rendering them unfit for exercise. Our thoughts turned homewards, and our bodies soon followed our thoughts. Before coming away I visited Carlyle in his bedroom. He was correcting the proofs of his Address. ' Now,' he said, ' the tollgates at Freystone are to be settled for.' I made light of them, and urged him to say ' Good-bye.' But he would not. ' The thought of them clings to me like unwashed hands.' He recognised as mean the cause of the discomfort, and used a congruous metaphor to express it. I still refused to make out a bill, so he put down all the items he remembered, added them together, and said, ' I owe you so much.' Looking over the account I retorted, with mock sternness, ' I beg your pardon, you owe me

fourpence-halfpenny more.' He laughed heartily, produced the fourpence-halfpenny, which, with an air of
business-like gravity, I pocketed, and bade him ' Good-
bye.'

Immediately after my arrival in London I called
upon Mrs. Carlyle. It was a bright welcome that she
gave me. A deep and settled happiness had taken
possession of her mind ; though she still could afford a
flash of sarcasm for one of the Edinburgh audience who
had visited her the day before. The glow of pride in
her husband was obvious enough. Not before a select
few, but before the world at large, he had won for himself renown, and for her choice of him, justification.
She wrote to him, ' I have not been so fond of everybody
since I was a girl.' We chatted long over the occur-
rences in the North, which I thought would give her
a new lease of happy life. Referring to her anxiety
about the Address, she said she had never entertained
the thought of his breaking down. As long as he had
life there was no fear of that. But she thought it quite
possible that life itself might snap, and that he might
fall down dead before the people. It must have been
her lithe fingers, and her high-strung nerves, that gave
to the pressure of her hand an elastic intensity which I
have not noticed elsewhere. Such warmth of pressure
had been always mine. As might be surmised, it was
not relaxed on this occasion, when, all unconscious of
impending disaster, I stood up and bade her 'Good-bye.'

I went to the Isle of Wight, which was my usual
refuge when tired, made Freshwater Gate my head-
quarters, and was refreshed as I had often been before
by the broad-blown, brotherly voice of Tennyson. Two
walks in the island have always had a special charm
for me ; one along ' the ridge of a noble down ' which

stretches from Freshwater Gate to the Needles; the
other along the spine of the island from Freshwater
Gate to Carisbrook, past ancient Barrows, with the
Solent on the one side and the ocean on the other.
From Carisbrook it was an easy walk to Cowes, whence
steamers plied to Southampton. Returning from the
island on the occasion now referred to, I chose this
latter route, and on reaching the railway-station at
Southampton, went straight to the bookstall to pick up
a copy of the *Times.* On opening the paper I was
stunned. Before me stood in prominent letters, 'Sudden
death of Mrs. Carlyle.' I sped to London, and on my
writing-table found a note from Miss Jewsbury. Carlyle
had arrived in Chelsea. 'For Heaven's sake,' said my
correspondent, 'come and see the old man! he is utterly
heart-broken.' In a few pathetic words Leslie Stephen
has told the story of her death: 'Mrs. Carlyle had asked
some friends to tea on Saturday, April 21. She had
gone out for a drive with a little dog; she let it out
for a run, when a carriage knocked it down. She sprang
out and lifted it into the carriage. The carriage went
on, and presently she was found sitting with folded
arms in the carriage, dead.'

I drove forthwith to Chelsea. The door was opened
by Carlyle's old servant, Mrs. Warren, who informed me
that her master was in the garden. I joined him there,
and we immediately went upstairs together. It would
be idle, perhaps sacrilegious on my part, to attempt
any repetition of his language. In words, the flow of
which might be compared to a molten torrent, he referred to the early days of his wife and himself—to
their struggles against poverty and obstruction; to her
valiant encouragement in hours of depression; to their
life on the moors, in Edinburgh, and in London—how
lovingly and loyally she had made of herself a soft

cushion to protect him from the rude collisions of the world. The late Mr. Venables, whose judgment on such a point may be trusted, often spoke to me of Carlyle's extraordinary power of conversation. In his noon of life it was without a parallel. And now, with the floodgates of grief fully opened, that power rose to a height which it had probably never attained before. Three or four times during the narrative he utterly broke down. I could see the approach of the crisis, and prepare for it. After thus giving way, a few sympathetic words would cause him to rapidly pull himself together, and resume the flow of his discourse. I subsequently tried to write down what he said, but I will not try to reproduce it here. While he thus spoke to me, all that remained of his wife lay silent in an adjoining room.

His house was left unto him desolate. Sympathy from all quarters flowed towards him, but it seemed to do him little good. His whole life was wrapped in mourning. I think it probable that in the lamentations which have reached the public through the ' Reminiscences,' he did himself wrong. His was a temper very likely to exaggerate his shortcomings ; very likely to blame himself to excess for his over-absorption in his work, and his too great forgetfulness of his wife. The figure of Johnson standing bareheaded in the market-place of Lichfield, to atone for some failure of duty to his father, fascinated Carlyle ; and now in his hour of woe he imitated Johnson, not by baring his head, but by lacerating his heart. They had had their differences — due probably more to her vivid and fanciful imaginings than to anything else. He, however, took the whole blame upon himself. It was loving and chivalrous, but I doubt whether it was entirely just. I think it likely that in her later years she would have condemned some of the utterances of her earlier ones. As time passed she grew more and more mellow and tender—more and more into the form and texture of the wife needed by Carlyle. Had she lived a little longer his self-reproaches would never have been heard.[1] Let me, however, forsake surmises and return to facts. He had laid his wife in Haddington Churchyard. The summer had passed, and harsh, dark winter was approaching. To spend the winter in Cheyne Row with all its associations was more than he could be expected to bear. But what was to be done? A loving answer to this question

[1] There was a fund of tenderness and liberality in Mrs. Carlyle; but her sarcasm could, on occasion, bite like nitric acid. Like her husband, she could hit off a character or peculiarity with a simple stroke of the tongue. Her stories sparkled with wit and humour. It may be an old yarn, but she caused me to shake with laughter by her inimitable way of telling the story of an old French priest, who discoursed to his peasant congregation on Samson's feat of tying the foxes' tails together, and sending them with burning brands through the standing corn. The ruin to agricultural produce was described so vividly, and with such local and domestic applications, that the people burst into weeping. Their sobs and tears reacted on the old priest himself. He also fell to weeping, but tried to assuage the general grief by calling out, ' Ne pleurez pas, mes enfants. Ne pleurez pas ; ce n'est pas vrai ! ' Her voice was exquisitely comic as she told this story. The only intimation that I ever had of past unhappiness on her part was given during an evening visit when I found her alone. She then told me that some years previously she had kept a journal, in which, to relieve her mind, she wrote down her most secret thoughts and feelings. She condemned, as she spoke to me, this habit of introspection. One day she had left the book upon her desk, and on returning to her room, found there a visitor actually looking into the journal. He probably regarded it as a mere library book ; but her wrath and rage, on finding sayings and sentiments intended for her eye alone, and kept secret even from Carlyle, thus pried into, were uncontrollable. As she spoke to me her anger seemed to revive, and its potency could not be doubted. When I quitted her, I carried away the impression that her maturer judgment had caused her to regard these journal entries as the foolish utterances of a too sensitive past.

came to him in his hour of need. The first Lady Ashburton had been Carlyle's friend, and the second, with a more fervent nature, was no less so. She had taken at Mentone a beautiful villa, the Villa Madonna, and thither she pressed Carlyle to come. I saw him frequently at this mournful time, and talked much with him about his plans. The Mentone scheme he deemed at first clearly impracticable; but the more it was thought over, the more evident it became that it was the only really practicable course open to him. As the gloom of December set in, the necessity of getting him away from London became more and more apparent. Counting the days at my disposal, I found that it was within my power to convey him to Mentone, deposit him there, and return in time for my personal duties in the Royal Institution. Lectures would begin, but men were there whose friendship had never failed me, and on whom I could rely that all things would be well conducted during my absence. Seeing the possibility, my action was prompt. I offered to take charge of him, cutting short hesitation and discussion by pointing to the inexorable march of time. Over the packing of his pipes we had a wrangle. It was clearly evident that his mode of packing would bring the 'churchwardens' to grief, and I emphatically told him so. But he would have his way. He knew how to pack pipes, and would be answerable for their safety. Out of fifty thus packed at Cheyne Row, three only reached Mentone unbroken. I afterwards enjoyed the triumph of sending him fifty without a single fracture.

But I anticipate. Rime was in the air, sucking the vital warmth out of every living thing when we started on the morning of December 22. A raw breeze blew in our faces as we crossed the Channel, or rather a breeze created by the vessel's motion, for the air was still. I

tried to muffle him up ; but immediately resigned my
attempted task to a young lady, who wound and pinned
his comforter in a manner unattainable by me. Carlyle
was interested to learn that his kind protectress was the
daughter of Sir John Herschel. She was then Miss
Amelia Herschel, she is now Lady Wade. In Paris we
spent the night at the Grand Hôtel de St. James,
Rue St. Honoré. A bad sleeper myself, I had long
before chosen this hotel, because its bedrooms opened
into a garden. We were well lodged ; but some slight
creak or clatter of a loose window roused Carlyle, who
became vocal. Noise at night was a terror and a
torture to him. I rose, reproved and corrected the
peccant window, the night afterwards passing quietly.
Next morning we started. At the Gare de Lyon we
were met by my lamented friend Jamin, a Member of
the Institute, who helped us with the railway officials,
and sent us on our journey with a hearty God-speed.

In England, as stated, the weather was harsh ; it
continued so in France. We had the good luck to
secure a *coupé* in the Marseilles train. Throughout
the day the landscape was cut off by freezing mist, and
at the Lyons station the outlook was specially dismal ;
due precautions however had been taken against cold.
In view of my winter expedition to the Mer de Glace
in 1859 I had purchased a sheepskin bag, lined with
its own wool, and provided with straps to attach it
comfortably to the waist. Swathed with this to the
hips, such heat as he could generate was preserved for
his feet and limbs. At Lyons food, wine, and a bottle
of water for the night were secured. The water-bottle
stood on a shelf in front of us. ' Observe it,' I said to
my companion. He did so with attention. At times
the water would appear quite tranquil ; then it would
begin to oscillate, the motion augmenting till the liquid

splashed violently to and fro up the sides of the bottle ;
then the motion would subside, almost perfect stillness
setting in. In due time this would be again disturbed,
the oscillations setting in as before. Carlyle was well
acquainted with the effects of synchronism in periodic
motion, but he was charmed to recognise in the water-
bottle an analyst of the vibrations of the train. It
told us when vibrations of its own special period were
present in, and when they were absent from, the confused and multitudinous rumble which appealed to our
ears. This was monotonous and permitted us to have
some sleep. On opening our eyes in the morning we
found a deep-blue sky above us, and a genial sun shining on the world. The change was surprising ; we had
obviously reached ' the Sunny South.'

We rested at Marseilles, and walked through the
sunlit city. Carlyle seated himself on a bench in the
shade of trees, while I went back to our hotel. On
returning I found him in conversation with a paralysed
beggar boy, from whom he had extracted the sad story
of his life. ' The poor we have always with us,' may be
truly said of all kindreds and tongues. In Marseilles
we had them singing in the streets for eleemosynary
sous. Carlyle contributed liberally. At the proper
time we took our tickets for Nice. In his later years
the factory smoke which pollutes our air, the dyer's
chemistry which pollutes our rivers, the defacement of
natural beauty which many of our industries have
brought in their train, were hateful to him. The railway whistle, rather than the grand roar of the rushing
locomotive, was his abomination. Tumult and confusion, especially when mixed with the stupidity of men
and women, he detested. Such confusion we found at
the Marseilles railway-station, and his disgust thereat
was registered in his voice and written on his counte-

nance. At Nice the railway came to an end, and a carriage was needed to take us over the hills to Mentone. We had a vigorous altercation at a cab-stand, where gross extortion was attempted. We retired to a respectable hotel, the courteous proprietor of which, after some waiting, provided us with the required vehicle. The lights of Monaco shone below us as we slowly crept over the hills. From the summit we trotted down to Mentone, reaching it at two o'clock in the morning. He was expected, and a loving friend was on the alert to welcome him. The reception was such as a younger man might envy. It was indeed plain to me that the storm-tossed barque had reached a haven in which it could safely rest.

I allowed myself a few pleasant excursions in the neighbourhood. We all ascended to the high-perched village of Sant' Agnese, whence, though strenuously opposed by Carlyle, I continued the ascent to the summit of the 'Aiguille.' This is the highest peak of the region. The sun was setting as I reached the top, flooding the Maritime Alps and the bays and promontories of the Mediterranean with blood-red light. It was a grand scene. We dined with the accomplished Lady Marian Alford. The present Lord Brownlow, as Mr. Cust, was there at the time, and a finer specimen of physical manhood I thought I had never before seen. After dinner a discussion arose about the sun as the physical basis of life. Carlyle's usual dislike to anything savouring of materialism showed itself, while I, with my usual freedom, told him that he was sure to come to grief if he questioned the sun's capacity as regards either light or life. In the morning, at an early hour, I found him vigorously marching along the fringe of the Mediterranean. In the afternoon we had a long drive on the Corniche Road. The zenithal

firmament, as we returned, was a deep blue, the western
sky a fiery crimson. Newton's suggestion—it could
hardly be called a theory—as to the cause of the
heavenly azure was mentioned. Carlyle had learned a
good deal of natural philosophy from Leslie, of whom
he preserved a grateful remembrance. From Leslie he
had learnt Newton's view of the colour of the sky, and
he now stood up for it. Leslie, he contended, was a
high and trustworthy authority. 'An excellent man,'
I admitted, 'in his own line, but not an authority on
the point now under discussion.' Carlyle continued to
press his point, while I continued to resist. He became
silent, and remained so for some time. A 'dépendance'
of the Villa Madonna had been placed at his sole disposal, and in it his fire was blazing pleasantly when we
returned from our drive. I helped him to put on his
dressing-gown. Throwing himself into a chair, and
pointing to another at the opposite side of the fire, he
said : ' I didn't mean to contradict you. Sit down
there and tell me all about it.' I sat down, and he
listened with perfect patience to a lengthy dissertation
on the undulatory theory, the laws of interference, and
the colours of thin plates. As in all similar cases, his
questions showed wonderful penetration. The power
which made his pictures so vivid and so true enabled
him to seize physical imagery with ease and accuracy.
Discussions ending in this way were not unfrequent
between us, and, in matters of science, I was always
able, in the long run, to make prejudice yield to reason.
On the day of my departure we all drove to Monaco—
our warm-hearted hostess, Carlyle, and a young lady
who was then a lovely child, and who is now a charming mother. On the little pier I bade them good-bye
and went on board the steamer for Nice. Almost at
the point where we had quitted the rime the train

plunged into it again. It had clung to its clime per-
sistently, while sunshine covered the Mentone hills.

After Carlyle's return from Mentone in the spring
we had various excursions together. I accompanied
him to Melchet, the beautiful seat of Lady Ashburton,
and rode with him through the adjacent New Forest.
We drove to Lyndhurst to see Leighton's frescoes. We
frequently walked together. One day, the storm being
wild and rude, a refuge from its buffets was thought
desirable. He said he knew of one. I accordingly
followed his lead to a wood at some distance. We
skirted it for a time, and finally struck into it. In the
heart of the wood we found a clearing. The trees had
been cut down and removed, their low stumps, with
smooth transverse sections, remaining behind. It was
a solemn spot, perfectly calm, while round the wood
sounded the storm. Dry dead fern abounded. Of this
I formed a cushion, and placing it on one of the tree-
stumps, set him down upon it. I filled his pipe and
lighted it, and while he puffed, conversation went on.
Early in the day, as we roamed over the pastures, he
had been complaining of the collapse of religious feeling in England, and I had said to him, ' As regards the
most earnest and the most capable of the men of a
generation younger than your own, if one writer more
than another has been influential in loosing them from
their theological moorings, thou art the man!' Our
talk was resumed and continued as he sat upon the
stump and smoked his placid pipe within hearing of
the storm. I said to him, ' Despite all the losses you
deplore, there is one great gain. We have extinguished
that horrible spectre which darkened with its death-
wings so many brave and pious lives. It is something
to have abolished Hell fire!' ' Yes,' he replied, ' that

is a distinct and an enormous gain. My own father was a brave man, and, though poor, unaccustomed to cower before the face of man ; but the Almighty God was a different matter. You and I do not believe that Melchet Court exists, and that we shall return thither, more firmly than he believed that, after his death, he would have to face a judge who would lift him into everlasting bliss or doom him to eternal woe. I could notice that for three years before he died this rugged, honest soul trembled to its depths at even the possible prospect of hell-fire. It surely is a great gain to have abolished this Terror.'

Sir Benjamin Brodie, the great surgeon, a man of highly philosophic mind, whose intimate friendship I enjoyed for many years before his death, always held and insisted that a good memory was essential to the making of a great man. That Carlyle's memory was astonishing numerous proofs could be given. One instance, associated with a fact of some interest, occurs to me as I write. When, struck down by the malady which has shorn away before their time so many precious lives, the gifted Clifford was approaching his end, I called one evening to see him in Quebec Street, and found Professor Croom Robertson at his bedside. Clifford had been reading a work on Germany ' by Thomas Carlyle, Barrister-at-Law,' and conjecture was set afloat to determine at what period of his career Carlyle had donned this designation. It was known that he once had thoughts of becoming a lawyer, but it was not known that he had ever used the title of a lawyer. Clifford said, ' The subject is one which Carlyle might be expected to handle ; the style is, to some extent, that with which we are so well acquainted, still, the book is one which nobody, knowing Carlyle, could suppose him to have written at any period of his life.' I went down to Chelsea next day, and made inquiries about the authorship of the volume. ' Oh,' said Carlyle with a laugh, ' that was " the Miracle." ' There was in Annandale a second Thomas Carlyle, whose cleverness, when a youth, caused him to be looked upon as a prodigy. Both he and the other Thomas sent from time to time mathematical questions to a local newspaper, and answered them mutually. Here Carlyle's extraordinary memory and narrative power came into play. He ran some centuries back, struck into ' the Miracle's ' family history, and traced it to that hour. While studying at the University of Marburg, I had been one morning startled by the intelligence that Thomas Carlyle, *der Engländer*, had arrived in that historic town. On inquiry, however, I found that it was not my Carlyle, but Carlyle the Irvingite, who had come on a visit to Professor Thiersch. It was, in fact, ' the Miracle.' The Professor, a very distinguished Greek scholar and a pious man, had just joined the Irvingites , hence the visit of ' the Miracle.' Carlyle spoke with feeling regarding what he considered to be the decadence and spiritual waste of his namesake and competitor, who, when he came to Marburg, had, I was told, the rank and function of an ' Apostle.'

An event, important in its relation to Carlyle's memory, is to be noted here. Meeting one day in the Athenæum Club Mr. (now Sir Mountstuart) Grant-Duff, he informed me that an accomplished American friend of his was very anxious to know Carlyle, but that he was held back by the notion that Carlyle disliked Americans. I was able to say upon the spot that this was an error. From my own direct questionings I had learned that the feelings of the old man were those of gratitude rather than of dislike. At a time when

his own countrymen, failing to recognise his need of a
form of expression suited to his genius, had set him
down as merely eccentric and wayward—meting out to
him the wages of eccentricity and waywardness, and
describing the work in which he had invested his highest
faculty as ' a heap of clotted nonsense '—America,
through her noblest son, had opened to him her mind,
her heart, her purse. Still, to make assurance doubly
sure, I told Grant-Duff that I would go down to Chelsea
and make myself acquainted with Carlyle's present
feelings. I went, and mentioned this conjectural dislike of Americans. ' What nonsense ! ' he exclaimed ;
' bring him down here immediately.' The gentleman
here referred to was, and is, Mr. Charles Norton, of
Harvard College. He came to Carlyle, and his visit
was the starting-point of a friendship which proved its
steadfastness after Carlyle was dead and gone. With
chivalrous firmness of purpose Mr. Norton has sought,
and I am told successfully sought, to stem and roll
back the foul wave of detraction and abuse, whereby
inconsiderate England threatened to overwhelm the
memory of a man to whom her best and bravest owe a
debt never to be cancelled. On this sad subject, however, it is not my intention to dwell ; but many
patriotic men regard it as a calamity of unspeakable
magnitude, that Carlyle's opinions on the grave questions which now agitate us should be reduced to nullity.
Were he amongst us he could point for our instruction
to certain apposite phases of the French Revolution,
which he—incomparable limner that he was !—has
thrown upon the canvas of History. The manifold
coiling of fraternal arms ; the friendships sworn and
re-sworn at the ' Feast of Pikes ' ; the pathetic ' *Souper
fraternel*,' with citizens ' hobnobbing in the streets to
the reign of Liberty, Equality, Brotherhood ' ; and

then, ah me! the law of gravity illustrated by the
incessant fall of the guillotine; the hackings, strang-
lings, fusillades, and noyades; cargoes of men, women,
and children sunk by their sworn brothers in the Loire
and the Rhone! One can fancy his presageful countenance were be to witness the revival, in our own day,
of this ghastly farce of ' fraternity '—unsexed, it is true,
and converted into ' sisterly embraces.' When the
manhood of England has departed, this nauseous
sentimentalism may go down with the electorate—*not
before.*

My recollection here reaches back to two powerful
and important letters published by Carlyle, one in the
Examiner and another in the *Spectator*, during a
former Repeal agitation. Each of them bore the
initial ' C.' as signature. His bold outspokenness and
fiery eloquence had endeared him to the enthusiastic
Young Irelanders, and it was thought that a word from
him would, at the time, be a word in season. These
letters had been read by me with profound interest
when they first appeared, and I notified their existence
to more than one able editor, when Carlyle's name was
mentioned a year or two ago in the House of Commons.
Standing recently beside the bookstall at Godalming
railway-station, I took up a quaint little book, with a
quaintly-printed title on its cover—' A Pearl of English
Rhetoric. Thomas Carlyle on the Repeal of the
Union.' It was a reprint of one of the letters signed
' C.,' to which I have just referred. After long burial
it had been unearthed, and thus restored to the public.
I give here a sample of its arguments against
Repeal :—

' Consider,' says the pearl-diver, ' whether, on any
terms, England can have her house cut in two and a
foreign nation lodged in her back parlour itself? Not

in any measure conceivable by the liveliest imagination that will be candid ! England's heavy job of work, inexorably needful to be done, cannot go on at all unless her back parlour, too, belong to herself. With foreign controversies, parliamentary eloquences, with American sympathisers, Parisian *émeutiers*, Ledru Rollins, and a world just now [1848] fallen into bottomless anarchy, parading incessantly through her back parlour, no nation can go on with any work. . . . Let Irish patriots seek some other remedy than repealing the Union ; let all men cease to talk or speculate on that, since once for all it cannot be done. In no conceivable circumstances could or durst a British Minister propose to concede such a thing : the British Minister that proposed it would deserve to be impeached as a traitor to his high post, and to lose his worthless head. Nay, if, in the present cowardly humour of most Ministers and governing persons, and loud, insane babble of anarchic men, a traitorous Minister did consent to help himself over the evil hour by yielding to it and conceding its mad demand—even he, whether he saved his traitorous head or lost it, would have done nothing towards the Repeal of the Union. While a British citizen is left, there is left a protestor against our country being occupied by foreigners, a repealer of the Repeal.'

Carlyle's mind was not of a texture to be greatly flurried by the prospect of confusion and bloodshed which the Repeal of the Union would infallibly carry in its train. He would have grimly accepted this result. But he would have been moved to the depths of his nature by the Liberal palinode of 1886, and the consequent spread of untruth among a straightforward and truth-loving people. ' A national wound,' he would have said, ' may be healed by the healthy surgery of the sword, but not when it is accompanied by national putrefaction.' He would have made his own observations on the fell potency of that party virus which has brought men whom he regarded and loved as younger brothers into partnership with so much that is mean and mendacious in political life. They have, I doubt not, their hours of misgiving, if not of self-accusation.

A word or two may here be thrown in as to Carlyle's relation to the 'Nigger question.' He undoubtedly rated the white man above the black. The capacity of rising to a higher blessedness, and of suffering a deeper woe, he deemed the prerogative and doom of the white. Hence his sympathy with the yellow-coloured weavers of Lancashire, as against ' black Quashee over the seas.' Even among ourselves he insisted on indelible differences. Wise culture could make the cabbage a good cabbage and the oak a good oak ; but culture could not transform the one into the other. It is interesting to observe how Locke's image of a sheet of white paper, on which education could write everything at will, laid hold of even powerful minds. I had many discussions with the late Mr. Babbage upon this subject. His belief in the all-potency of education, as applied to the individual, I could not share. Brains differ, like voices ; and as the voice-organ of a great singer must be the gift of Nature, so the brain-organ of the great man must also be a natural gift. Nobody who knew Carlyle could dream for a moment that he meant to be unfair, much less cruel, towards the blacks. ' Do I then hate the Negro ? No ; Except when the soul is killed out of him I decidedly like poor Quashee. A swift, supple fellow ; a merry-hearted affectionate kind of creature, with a great deal of melody and amenability in his composition.' It was not the guilt of ' a skin not coloured like his own,' but the demoralising idleness of the negro amid his pumpkins, that drew down the condemnation of Carlyle. His feelings towards the idle, pampered white man were more contemptuous and unsparing than towards the black. ' A poor negro overworked on the Cuba sugar grounds, he is sad to look upon ; yet he inspires me with sacred pity, and a kind of human respect is not denied him. But with what feelings can I look upon an over-fed white flunkey, if I know his ways ? Pity is not for him, or not a soft kind of it ; nor is any remedy visible except abolition at no distant date.' In ' Sartor ' he writes : ' Two men I honour, and no third. First, the toil-worn craftsman that, with earth-made implement, laboriously conquers the earth, and makes her man's. A second man I honour, and still more highly : Him who is seen toiling for the spiritually indispensable ; not daily bread, but the bread of life.'

Still, it must be admitted that Carlyle estimated the whites as of greater value than the blacks ; and he deprecated the diversion towards the African of power which might find a more profitable field of action at home. Perhaps he saw too vividly, and resented too warmly, the mistakes sometimes made by philanthropists, whereby their mercies are converted into cruelties. We see at the present moment a philanthropy, which would be better named an *insanity*, acting in violent opposition to the wise and true philanthropists, who are aiming at the extinction of rabies among dogs, and of its horrible equivalent, hydrophobia, among men. Reason is lost on such people, and instead of reason Carlyle gave them scorn. Perhaps he was too scornful. History had revealed to him the unspeakable horrors of a black insurrection. Hence his action, as regards

Governor Eyre, after the outbreak at Morant Bay.[1]
' Hell had broken loose, and the fire must be quenched
at any cost.' Perhaps he was right; perhaps he was
wrong. The question at the time produced an extraordinary cleavage among intimate friends ; but not, to
my knowledge, did it produce any permanent estrange-
ment. Huxley and Spencer fought like brothers under
a common flag ; Hooker and myself, equally fraternal,
under the opposite one. We surely did not love each
other less afterwards because of this temporary divergence of judgment. I fervently trust that all our
differences may have a similar end.

' It is related,' says Dr. Garnet, ' that, fascinated by
the grand figure of Michael Angelo, he [Carlyle] once
announced his intention of writing his life.' He would
have thus added to his picture-gallery ' The Hero as
Artist.' Carlyle would have found ' The Hero as Man
of Science ' a more fitting theme. He had mastered
the ' Principia,' and was well aware of the vast revolu-
tionary change wrought, not in Science only, but in the
whole world of thought, by the theory ef gravitation.
The apparently innocent statement, that every particle
of matter attracted every other particle with a force
which was a function of the distance between them,
carried the mind away from the merely *falling* atoms
of Epicurus and Lucretius to conceptions of *molecular*
forces. By their aid we look intellectually into the
architecture of crystals. But the inquiring spirit of man
cannot stop there. It now recognises, with what ulti-

[1] I may here say that when speaking to Governor Eyre upon the
subject, he declared to me that he knew as little, at the time, about
the floggings of women and other cruelties, as I did. But though
he might have mitigated the severity of the verdict against himself
by shifting the odium on to his subordinates, he refused to do so,
and accepted all the blame.

mate results we know not, the all-potent play of molecular forces in the animal and vegetable organisms. Without, however, trenching upon these points, which Carlyle saw as in a glass darkly, he would have found in Newton or Boyle an appropriate subject. Had he taken either of them in hand, he would undoubtedly have turned out an impressive figure. Boyle especially would, I imagine, have appealed to his sympathies and love.

The mistake, not unfrequently made, of supposing Carlyle's mind to be unscientific, may be further glanced at here. The scientific reader of his works must have noticed the surprising accuracy of the metaphors he derived from Science. Without sound knowledge such uniform exactitude would not have been possible. He laid the whole body of the sciences under contribution—Astronomy, from the nebular theory onwards ; mathematics, physics, chemistry, geology, natural history—drawing illustrations from all of them, grinding the appropriate parts of each of them into paint for his marvellous pictures. Quite as clearly as the professed physicist he grasped the principle of Continuity, and saw the interdependence of ' parts ' in the ' stupendous Whole.' To him the Universe was not a Mechanism, but an Organism—each part of it thrilling and responding sympathetically with all other parts. Igdrasil, ' the Tree of Existence,' was his favourite image :—' Considering how human things circulate each inextricably in communication with all, I find no similitude so true as this of a tree. Beautiful ; altogether beautiful, and great. The " *Machine* of the Universe,"—alas, do but think of that in contrast ! ' [1] Other penetrative minds have made us familiar with the

[1] *Heroes and Hero-Worship*, Library Edition, p. 25.

'Social Organism,' but Carlyle saw early and utilised nobly the beauty and the truth of the metaphor.

In the month of May, 1840, the foregoing words were spoken. Harking back to 1831, we find him at Craigenputtock, drawing this picture :—' As I rode through the Schwarzwald I said to myself: That little fire which glows star-like across the dark-growing moor, where the sooty smith bends over his anvil, and thou hopest to replace thy lost horseshoe—is it a detached, separated speck, cut off from the whole universe ; or is it indissolubly joined to the whole ? Thou fool, that smithy-fire was primarily kindled at the sun.' (Joule and Mayer were scientifically unborn when these words were written.) He continues :—' Detached, separated ! I say there is no such separation ; nothing hitherto was ever stranded, cast aside ; but all, were it only a withered leaf, works together with all, and lives through perpetual metamorphoses.' With its parts in ' æterne alterna-tion ' the universe presented itself to the mind of Car-lyle. ' The drop which thou shakest from thy wet hand rests not where it falls, but to-morrow thou findest it swept away; already on the wings of the north-wind it is nearing the Tropic of Cancer. How came it to evaporate and not lie motionless ? Thinkest thou there is ought motionless ; without force and utterly dead ? ' [1] Such passages—and they abound in his writings—might justify us in giving Carlyle the credit of poetically, but accurately, foreshadowing the doctrine of the Conservation of Energy. As a physiologist describes the relation of nerve to muscle, he hits off the function, and the fate, of demagogues in revolu-tionary times :—' Record of their thought remains not ; death and darkness have swept it out utterly. Nay,

---

[1] *Sartor Resartus*, Library Edition, pp. 68, 69.

if we had their thought, all that they could have articu-
lately spoken to us, how insignificant a fraction were
that of the Thing which realised itself, which decreed
itself, on signal given by them!' Thus, a howling
Marat, or a sea-green Robespierre was able to unlock
forces infinitely in excess of his own.

It was not the absence of scientific power and precision, so much as the overwhelming importance which
Carlyle ascribed to ethical considerations and influences,
that determined his attitude towards natural science.
The fear that moral strength might be diminished by
Darwin's doctrine accounts for such hostility as he
showed to the 'Origin of Species.' We had many calm
and reasonable conversations on this and kindred subjects; and I could see that his real protest was against
being hemmed in. He demanded a larger area than
that offered by science for speculative action and its
associated emotion. 'Yes, Friends,' he says in 'Sartor,'
'not our Logical Mensurative faculty, but our Imaginative one is King over us.'[1] Worship he defined as
'transcendent wonder'; and the lifting of the heart by
worship was a safeguard against moral putrefaction.
Science, he feared, tended to destroy this sentiment.
I may remark here that, as a corrective of superstition, science, even when it acts thus, is altogether
salutary. But preoccupation alone could close the eyes
of the student of natural science to the fact that the
long line of his researches is, in reality, a line of
wonders. There are freethinkers who imagine themselves able to sound with their penny twine-balls the
ocean of immensity. With such Carlyle had little
sympathy. He was a freethinker of wiser and nobler
mould. The miracles of orthodoxy were to him, as to

---

[1] Book III., Symbols.

his friend Emerson, ' Monsters.' To both of them ' the blowing clover and the falling rain ' were the true miracles. Napoleon gazing at the stars, and gravelling his *savants* with the question : ' Gentlemen, who made all that ? ' commended itself to their common sympathy. It was the illegitimate science which, in its claims, overstepped its warrant—professing to explain everything, and to sweep the universe clear of mystery—that was really repugnant to Carlyle.

Here a personal recollection comes into view which, as it throws a pleasant light on the relations of Carlyle and Darwin, may be worth recording. Like many other noble ladies, Lady Derby was a warm friend of Carlyle ; and once, during an entire summer, Keston Lodge was placed by Lord Derby at Carlyle's disposal. From the seat of our common friend, Sir John Lubbock, where we had been staying, the much-mourned William Spottiswoode and myself once walked over to the Lodge to see Carlyle. He was absent ; but as we returned we met him and his niece, the present Mrs. Alexander Carlyle,[1] driving home in a pony-carriage. I had often expressed to him the wish that he and Darwin might meet ; for it could not be doubted that the nobly candid character of the great naturalist would make its due impression. The wish was fulfilled. He met us with the exclamation : ' Well, I have been to see Darwin.' He paused, and I expressed my delight. ' Yes,' he added, ' I have been to see him, and a more charming man I have never met in my life.'

The sad years rolled on, and I began at length to notice a lowering of his power of conversation, and a

---

[1] To whom he was indebted not only for her affectionate care of his health, but occasionally, in later years, for wise counsel where his own faltering judgment might have led him wrong.

tendency to somnolence, which contrasted strongly
with the brisk and fierce alacrity of former times. On
one occasion when I called, this was specially noticeable. He was seated before the fire, with Mr. Browning [1] for his companion. We entered into conversation,
which, in Carlyle's case, was limited to the answering
of a question addressed to him now and then. I was
aware of the poet's habit of early rising, and of his
hard work, and I wished to know something of the
antecedents of so strenuous and so illustrious a life.
Mr. Browning's father and grandfather came thus to be
spoken of. Carlyle seemed at length to rouse himself.
'Browning,' he said, 'it was your ancestor that broke
the boom stretched across the Foyle, and relieved
Derry, when the city was besieged by James's army.'
He named the ship. 'Surely not,' I said; 'it was the
Dartmouth.' In saying this, I relied more upon songs
committed to memory in boyhood,[2] than upon historical knowledge. Carlyle was right. The relief of
Derry is described by Macaulay, who has given honour
to whom honour is due.

One other trivial item, almost the last, may be
here set down. In his days of visible sinking, I took
down to him a small supply of extremely old pale
brandy from the stores of Justerini & Brooks, together with a few of the best cigars that I could find.
On visiting him subsequently, I found that he had
hardly touched either the one or the other. Thinking

[1] Vigorous when this page was written; now, alas! no more.
The reverent affection with which the poet spoke of, and to, Carlyle
was a delightful feature of this interview.

[2] The strophe on which my opinion was founded runs thus:—

'The *Dartmouth* spreads her snow-white sail,
 Her purple pendant flying O,
While we the dauntless heroes hail,
 Who saved us all from dying O.'

them worth a trial, I mixed some brandy and water in a tumbler, and placing a cigar between his fingers, gave him a light. The vigour of his puffs astonished me ; his strength as a smoker seemed unimpaired. With the view of supporting him, I placed myself on the sofa behind him. After a time, putting aside the half-consumed cigar, he drank off the brandy-and-water, and with a smile gleaming in his eye,[1] remarked ' That's well over.' Soon afterwards he fell asleep. Quietly relinquishing my position as pillow, I left him in slumber. This, to the best of my recollection, was the last time I saw Thomas Carlyle.

The disintegration of the firm masonry went rapidly on, and at length the noble tower fell. Carlyle died on February 5, 1881.

Immediately afterwards I was visited by Mr. Froude, who came to inform me of the arrangements made for the funeral. In touching language he described the placid beauty of the dead man's face, contrasting it with the stern grandeur of Mrs. Carlyle's countenance in her last sleep. The brave and sympathetic Stanley wished to have him in Westminster Abbey, but this Carlyle had steadily declined. Troops of friends from all accessible places would have reverently made their way to the burial-ground of Ecclefechan, but it was thought desirable to make the funeral as quiet and as simple as possible. Lecky, Froude, and myself formed a small delegation from London. We journeyed together northwards, halting at Carlisle for the night. Snow was on

---

[1] I think it was the late Mr. Donne who once remarked to me that Carlyle's beard, by hiding the grimly-set mouth, greatly improved his aspect. ' His eye was tender and sweet.' A comparison of the frontispiece of *Heroes and Hero-Worship* with that of *Sartor Resartus* (Library Editions) will illustrate Mr. Donne's meaning and justify his observation.

the ground next morning as we proceeded by rail to the station of Ecclefechan. Here we found the hearse powdered over by the frozen shower of the preceding night. Through the snow-slop we walked to Mainhill, the farmhouse where Carlyle, in 1824, completed the translation of 'Wilhelm Meister.' It may have been the state of the weather, but Mainhill seemed to me narrow, cold, humid, uncomfortable. We returned to Ecclefechan, I taking shelter for a time in the signal-room of the station. Here I conversed with the signalman, an intelligent fellow, who wished me to know that Mr. James Carlyle, who was still amongst them, was fit to take rank in point of intellect with his illustrious brother. At the appointed hour we joined the carriage procession to the churchyard. There, without funeral rite or prayer, we saw the coffin which contained the body of Carlyle lowered to its last resting-place. So passed away one of the glories of the world.

ON UNVEILING THE STATUE OF THOMAS CARLYLE.

(26th October, 1882.)

AMID scenes well calculated to tinge the mind with solemnity, if not with awe, I have lately thought a good deal of the hour that is now come, and of the man in loving memory of whom we are here assembled. And with my thoughts sometimes mingled the very genuine wish that the honourable but trying task now before me had been committed to worthier hands. Without conscious disloyalty, however, I could not decline the request of the Carlyle Memorial Committee; and so, without further preface or apology, here I am.

You have heard much this year of the bi-decennial festival known as the Preston Guild. Two Guilds ago, that is to say in 1842, I was a youth in Preston, being attached to a division of the Ordnance Survey then stationed there. It was a period of gloom and suffering in the manufacturing districts. Some time prior to the Guild, processions of another kind filled the streets—crowds of shiftless and hungry men who had been discharged from the silent mills. In their helplessness and misery they had turned out, so that their condition might be seen of all. Well, in Lune Street, down which we could look from our office, the tumult one day became unmanageable. Heated by its own interaction and attrition, the crowd blazed out into open riot, and attacked the bakers' shops. Soldiers had been summoned to meet this contingency. Acting under orders, they fired upon the people, and the riot was quelled at the cost of blood.

At the very time when these things were occurring in Lancashire, Thomas Carlyle was at work on 'Past and Present' at No. 5 Cheyne Row, Chelsea. The cry of the famishing weavers came up to him from the North, and drew from him his memorable appeal to Exeter Hall :—' In thee too is a kind of instinct towards justice, and I will complain of nothing. Only, Quashee over the seas being once provided for, wilt thou not open thine eyes to the hunger-stricken, pallid, yellow-coloured " Free Labourers " of Lancashire, Yorkshire, Buckinghamshire, and all other shires ? These yellow-coloured, for the present, absorb all my sympathies. If I had a twenty millions, with model farms and Niger expeditions, it is to them that I would give it. Why, in one of these Lancashire weavers, dying of hunger, there is more thought and heart, a greater arithmetical amount of misery and desperation, than in whole gangs of Quashees.' Copied into the Preston newspapers at the time, these were the first words of Carlyle that I ever read. After the rattle of musketry and spatter of bullets, among the weavers and spinners in Lune Street, they rang, I confess, with strange impressiveness in my ears.

Carlyle's defects of feeling—if such there were—could only have reference to the distribution of his sympathy, not to its amount. His pity was vast, and only his division of it between black and white could be called in question. The condition of his toiling fellow-countrymen oppressed him like a nightmare. Day and night for years he had brooded upon this subject, if haply a gleam might be discerned showing the way towards amelioration. Braver or wiser words were never addressed to the aristocracy of England than those addressed to them by Carlyle. Braver or wiser words were never addressed to the Radicalism of England than those uttered by the same strenuous spirit. He saw clearly the iniquity of the Corn Laws, and his condemnation fell upon them, like the stone of Scripture, grinding them to powder. With equal clearness he saw the vanity of expecting political wisdom from intellectual ignorance, however backed by numbers. It was like digging for diamonds in Thames mud. Hence the pressing need of public education, and hence his powerful advocacy, in advance, of what his friend Forster has, in these later days, in great part realised. He urged the necessity of an organised system of emigration, and it might have been well had his prevision been translated long ago into action. But though, as regards these and other matters, he uttered his views with a strength and clearness peculiar to himself, his aim, politically, was rather to elevate and ennoble public life generally than to enunciate special measures. His influence went far beyond the sphere of politics. No man of his day and generation threw so much of resolution and moral elevation into the hearts and lives of the young. Concerning the claims of duty and the dignity of work, never man spake like this man. A friend and I agreed some time ago to describe him as 'dynamic,' not 'didactic'—a spiritual force, which warmed, moved, and invigorated, but which refused to be clipped into precepts. He desired truth in the inward parts. To the Sham, however highly placed and run after, his language was : ' Depart hence, in the Devil's name, unworshipped by at least one man, and leave the thoroughfare clear.' But his spirit leaped to recognise true merit and manfulness in all their phases and spheres of action. Braidwood amid the flames of Tooley Street, and the riddled *Vengeur* sinking to the cry of ' *Vive la République!* ' found in his strong soul sympathetic admiration. He, however, prized courage less than truth ; and when he found the story of the *Vengeur* to be a lie, he transfixed it, and hung it up as an historic scarecrow. The summer lightning of his humour, and the splendour of an imagination perhaps without a parallel in literature, served only to irradiate and vivify labours marked by a thoroughness in searching, and a patience in sifting, never yet surpassed. The bias of his mind was certainly towards what might be called the military virtues ; thinking, as he did, that they could not be dispensed with in the present temper of the world. But, though he bore about him the image and superscription of a great military commander, had he been a statesman, as he might well have been, he would at any fit and proper moment have joyfully accepted as the weapons of his warfare, instead of the sword and spear, the ploughshare and pruning-hook of peaceful civic life.

One point, touching Carlyle's ethics, may be referred to here. Taking all that science has done in the past, all that she nas achieved in the present, and all that she is likely to compass in the future—will she at length have told us everything, rendering our knowledge of this universe rounded and complete ? The answer is clear. After science has completed her mission upon earth, the finite known will still be embraced by the infinite unknown. And this ' boundless contiguity of shade,' by which our knowledge is hemmed in, will always tempt the exercise of belief and imagination. The human mind, in its structural and poetic capacity, can never be prevented from building its

castles—on the rock or in the air, as the case may be—
in this ultra-scientific region. Certainly the mind of
Carlyle could not have been prevented from doing so.
Out of pure Unintelligence he held that Intelligence
never could have sprung, and so, at the heart of things,
he placed an Intelligence—an Energy which, ' to avoid
circuitous periphrasis, we call God.' I am here repeating his own words to myself. Every reader of his works
will have recognised the burning intensity of his conviction that this universe is ruled by veracity and justice,
which are sure in the end to scorch and dissipate all
falsehood and wrong.

And now I come to the charge so frequently made
against him, that he was the apostle of Might. He felt,
perhaps more deeply than his assailants, the radical and
ineffaceable difference which often subsists between
might and right. But

> His faith was large in time,
> And that which shapes it to some perfect end,

His own words, which are to be found in the 8th chapter
of ' Chartism,' are these :—' Might and Right do differ
frightfully from hour to hour ; but give them centuries
to try it in, and they are found in the end to be identical.'
Viewed in the light of this utterance, the advocacy of
Might is not, in the abstract, offensive ; for it meant at
bottom the assertion that, in the end, that only is
mighty which has the ' Law of the Universe ' on its
side. With Carlyle, as with Empedocles, Lucretius,
and Darwin, the Fit survives. His doctrine is the
doctrine of science, not ' touched ' but saturated with
religious emotion. For the operation of Force—the
scientific agent—his deep and yearning soul substituted
the operation of the Energy before referred to, which,
to avoid periphrasis, we call God.

The ' Abbey ' would have opened its doors to

welcome him, but he chose to lie down among his own, in the humble burial-ground of Ecclefechan, where many a reverent pilgrim of the future will look upon his grave. Since his death we have had misjudgment and misapprehension manifold regarding him and his; but these are essentially evanescent, and I therefore pass them by with a simple comparison to mark their value. In Switzerland I live in the immediate presence of a mountain, noble alike in form and mass. A bucket or two of water, whipped into a cloud, can obscure, if not efface, that lordly peak. You would almost say that no peak could be there. But the cloud passes away, and the mountain, in its solid grandeur, remains. Thus, when all temporary dust is laid, will stand out, erect and clear, the massive figure of Carlyle.

It now becomes my duty to unveil and present to the British public, and to the strangers within our gates who can appreciate greatness, the statue of a great man. Might I append to these brief remarks the expression of a wish, personal perhaps in its warmth, but more than personal in its aim, that somewhere upon this Thames Embankment could be raised a companion memorial to a man who loved our hero, and was by him beloved to the end? I refer to the loftiest, purest, and most penetrating spirit that has ever shone in American literature—to Ralph Waldo Emerson, the life-long friend of Thomas Carlyle.

1891.

ON THE ORIGIN,
PROPAGATION, AND PREVENTION OF PHTHISIS.

IT is now a little over nine years since I received here, at Hind Head, a memoir by Professor Koch on the 'Etiology of Tuberculosis.' Taking it in all its bearings, the memoir seemed to me of extraordinary interest and importance, not only to the medical men of England, but to the community at large. I therefore drew up and sent an account of it to the *Times*. The discovery of the tubercle bacillus was therein announced for the first time, and by experiments of the most definite and varied character the propagation and action of this terrible organism were demonstrated.

With regard to his recent labours, Professor Koch may or may not have been hasty in the publication of his remedies for consumption. On this point it would be out of place, on my part, to say a word. But the investigations which first rendered his name famous, and which, I believe, were introduced to the English public by myself, are irrefragable. His renowned inquiry on anthrax caused him to be transferred from a modest position near Breslau, to the directorship of the Imperial Sanitary Institute of Berlin, where he was soon surrounded by able colleagues and assistants. Conspicuous among these was Dr. Georg Cornet, whose labours on the diffusion of tuberculosis constitute the subject of this article.

After the investigation of Koch, various questions of moment pushed themselves imperiously to the front: How is phthisis generated ? How is it propagated ? What is the part played by the air as the vehicle of tubercle bacilli ? How are healthy lungs to be protected from their ravages ? What value is to be assigned to the hypothesis of predisposition and hereditary transmission ? Cornet describes the attempts made to answer these and other questions. The results were conflicting, and when subjected to critical examination they were proved, for the most part, inadequate and inconclusive. The art of experiment is different from that of observation ; so much so, that good observers frequently prove but indifferent experimenters. It was his education as an experimenter that gave Pasteur such immense advantage over Pouchet in their celebrated controversy on ' spontaneous generation ' ; and it is on the score of experiment that the writers examined by Cornet were found most wanting. One evil result of this conflict of opinions as to the propagation and prevention of phthisis, was the unwarrantable indifference which it generated among medical men.

The researches referred to and criticised by Cornet are too voluminous to be mentioned in detail. Valuable information was, to some extent, yielded by these researches, but they nevertheless left the subject in a state of vagueness and uncertainty. Cornet, in fact, when he began his inquiry, found himself confronted by a practically untrodden domain. He entered it with a full knowledge of the gravity of his task. The result of his investigation is a memoir of 140 pages, the importance of which, and the vast amount of labour involved in it, can be appreciated by those only who have read it and studied it from beginning to end.

That the matter expectorated by phthisical patients
is infectious had been placed by previous investigations
beyond doubt. The principal question set before himself by Cornet had reference to the part played by the
air in the propagation of lung disease :—Is the breath
of persons suffering from phthisis charged, as assumed
by some, with bacilli ? or is it, as assumed by others,
free from the organism ? The drawing of the air
through media able to intercept its floating particles,
and the examination of the media afterwards, might, at
first sight, appear the most simple way of answering
this question. But to examine a thousand litres of air
would require a considerable time, and this is only one-
twelfth of the volume which a man breathing quietly
expires every day. If the air were only sparingly
charged with bacilli, the amount necessary for a
thorough examination might prove overwhelming.
Instead of the air, therefore, Cornet chose for examination the *precipitate* from the air ; that is to say, the
dust of the sick-room, which must contain the bacilli
in greater numbers than the air itself.

He chose for his field of operations seven distinct
hospitals (Krankenhäuser), three lunatic asylums (Ir-
renanstalten), fifty-three private houses, and various
other localities, including private asylums, lecture-
rooms, surgical wards, public buildings, and the open
street. The smallness of the bacilli has given currency
to erroneous notions regarding their power of floating
in the air. The bacilli are not only living bodies, but
heavy bodies, which sink in water and pus, and fall
more rapidly in calm air. Cornet gathered his dust
from places inaccessible to the sputum issuing directly
from the coughing patient. He rubbed it off high-
hung pictures, clock-cases, the boards and rails at the
back of the patient's bed, and also off the walls behind

it. The enormous care necessary in such experiments, and, indeed, in the use of instruments generally, has not yet, I fear, been universally realised by medical men. With a care worthy of imitation, Cornet sterilised the instruments with which his dust was collected, and also the vessels in which it was placed.

The cultivation of the tubercle bacilli directly from the dust proved impracticable. Their extraordinary slowness of development enabled other organisms—weeds of the pathogenic garden—which were always present, to overpower and practically stifle them. Cornet, therefore, resorted to the infection of guinea-pigs with his dust. If tuberculosis followed from such inoculation, a proof of virulence would be obtained which the microscope could never furnish The dust, after being intimately mixed with a suitable liquid, was injected into the abdomen of the guinea-pig. For every sample of dust, two, three, four, or more animals were employed. In numerous cases the infected animal died a day or two after inoculation. Such rapid deaths, however, were not due to the tubercle bacillus, which, as already stated, is extremely slow of development, but to organisms which set up peritonitis and other fatal disorders. Usually, however, some of the group of guinea-pigs escaped this quick mortality, and, to permit of the development of the bacilli, they were allowed to live on thirty, forty, or fifty days. The survivors were then killed and examined. In some cases the animals were found charged with tubercle bacilli, the virulence of the inoculated matter being thus established. In other cases the organs of the guinea-pigs were found healthy, thus proving the harmlessness of the dust.

It must here be borne in mind that the bacilli mixed with Cornet's dust must have first floated in the

air, and have been deposited by it. Considering the
number of persons who suffer from phthisis, and the
billions of bacilli expectorated by each of them, it
would seem a fair *à priori* deduction that wherever
people with their normal proportion of consumptive
subjects aggregate, the tubercle bacillus must be present everywhere. Hence the doctrine of 'ubiquity,'
enunciated and defended by many writers on this question. Common observation throws doubt upon the
doctrine, while the experiments of Cornet are distinctly
opposed to it. Tested by the dust deposited on their
furniture or rubbed from their walls, the wards of some
hospitals were found entirely free from bacilli, while
others were found to be richly and fatally endowed with
the organism. Cornet, it may be remarked, does not
contend that his negative results possess demonstrative
force. He is quite ready to admit that, where he failed
to find them, bacilli may have escaped him. But he
justly remarks that, until we have discovered a bac-
terium magnet, capable of drawing every bacillus from
its hiding-place, experiment must remain more or less
open to this criticism. Cornet's object is a practical
one. He has to consider the *probability*, rather than
the remote *possibility*, of infection. The possibility,
even in places where no bacilli show themselves, may
be admitted, while the probability is denied. Such
places, Cornet contends, are practically free from
danger.

In the differences as to infectiousness here pointed
out, we have an illustration of wisely-applied knowledge,
care, and control, as contrasted with negligence, or
ignorance, on the part of hospital authorities. And
this may be a fitting place to refer to a most impressive
example of what can be accomplished by resolute
supervision on the part of hospital doctors and nurses.

A glance at the state of things existing some years ago will enable us to realise more fully the ameliorations of to-day. I once had occasion to ask Professor Klebs, of Prague, for his opinion of the antiseptic system of surgery. He replied, 'You in England are not in a position to appreciate the magnitude of the advance made by Lister. English surgeons were long ago led to recognise the connection between mortality and dirt, and they spared no pains in rendering their wards as clean as it was possible to make them. Wards thus purified showed a mortality almost as low as other wards in which the antiseptic system was employed. The condition of things in *our* hospitals is totally different ; and it is only amongst us, on the Continent, that the vast amelioration introduced by Lister can be properly apprehended.' I may say that Lister himself once described hospitals in his own country which, in regard to uncleanness and consequent mortality, might have vied with those on the Continent. Klebs's letter was written many years ago. Later on the authorities in German hospitals bestirred themselves, with the splendid result disclosed by Cornet, that institutions which were formerly the chief breeding-grounds of pathogenic organisms are now raised to a pitch of salubrity surpassing that of the open street.

Cornet thus grapples with the grave question which here occupies us. How, he asks, does the tubercle bacillus reach the lungs, and how is it transported thence into the air ? Is it the sputum alone that carries the organism, or do the bacilli mingle with the breath ? This is the problem of problems, the answer to which will show whether we are able to protect ourselves against tuberculosis, whether we can impose limits on the scourge, or whether, with hands tied, we have to surrender ourselves to its malignant sway. If the

tubercle bacilli are carried outwards by the breath,
then nothing remains for us but to wait till an infected
puff of expired air conveys to us our doom. A kind of
fatalism, sometimes dominant in relation to this question, would thus have its justification. There is no
inhabited place without its proportion of phthisical
subjects, who, if the foregoing supposition were correct,
would be condemned to infect their neighbours. Terrible in this case would be the doom of the sufferer,
whom we should be forced to avoid, as, in earlier ages,
the plague-stricken were avoided. Terrible, moreover,
to the invalid would be the consciousness that with
every discharge from his lungs he was spreading death
among those around him. ‘ Such a state of things,’
says Cornet, ‘ would soon loosen the bonds of the family
and of society.’ Happily, the facts of the case are very
different from those here set forth.

‘ I would not,’ says our author, ‘ go into this subject so fully, I would not here repeat what is already
known, were I not convinced that, in regard to this
special point, the most erroneous notions are prevalent,
not only amongst the general public, but even among
highly-cultivated medical men. Misled by such notions, precautions are adopted which are simply calculated to defeat the end in view. Thus it is that while
one physician anxiously guards against the expired
breath of the phthisical patient, another is careful to
have his spittoon so covered up that no bacilli can
escape into the air by evaporation. Neither of them
makes any inquiry about the really crucial point—
whether the patient has deposited *all* his sputum in
the spittoon, thus avoiding the possibility of the expectorated matter becoming dry, and reduced afterwards to
a powder capable of being inhaled.

‘ While a positive phthisiophobia appears to have

taken possession of some minds, others ignore almost completely the possibility of infection. The fact that investigations have been published of late, with the object of discovering tubercle bacilli in the breath, sufficiently indicates that the conclusive researches of earlier investigators have not received the proper amount of attention.

'We must regard it,' says Cornet, 'as firmly established that, under no circumstance, can the bacteria contained in a liquid, or strewn upon a wet surface, escape by evaporation or be carried away by currents of air. By an irrefragable series of experiments Nägeli has placed this beyond doubt.'

The evidence that the sputum is the real source of tuberculous infection is conclusive ; and here Cornet earnestly directs attention to the fact that in the houses of the poor the patient commonly spits upon the floor, where the sputum dries and is rubbed into infectious dust by the feet of persons passing over it. The danger becomes greatest when the dry floor is swept by brush or broom. There is a still graver danger connected with the habits of well-to-do people who occupy clean and salubrious houses. This is the common practice of spitting into pocket-handkerchiefs. Here the sputum is soon dried by the warmth of the pocket, the subsequent use of the handkerchief causing it to be rubbed into virulent dust. This constitutes a danger of the highest consequence, both to the individual using the handkerchief and to persons in his immediate neighbourhood.

It is a primary doctrine with both Koch and Cornet that tuberculosis arises from infection by the tubercle bacillus. Predisposition, or hereditary tendency, as a *cause* of phthisis, is rejected by both of them. Facts, however, are not wanting which suggest the notion of predisposition. Cornet once attended, in a hotel, an actress far advanced in phthisis. A guest taking possession of her room after her death, or removal, might undoubtedly become infected. The antecedents of the room being unknown, the case of such a guest would, in all probability, be referred to predisposition. It might be declared, with perfect sincerity, that for years he had had no communication with phthisical persons. There is very little doubt that numbers of cases of tuberculosis, which have been referred to predisposition or inheritance, are to be really accounted for by infection in some such obscure way.

Cornet draws attention to hotels and lodging-houses at, and on the way to, health-resorts. He regards them as sources of danger, and he insists on the necessity of disinfecting the rooms and effects after the death or removal of tuberculous patients. He recommends physicians, before sending patients abroad, or to health-resorts at home, to inform themselves, by strict inquiry, regarding the precautions taken to avoid infectious diseases, tuberculosis among the number. The attention of those responsible for the sanitary arrangements in the health-resorts of England may be invited to the following observation of Cornet:—' On a promenade, amidst a hundred phthisical persons who are careful to expectorate into spittoons, the visitor is far safer than among a hundred men, taken at random, and embracing only the usual proportion of phthisical persons who spit upon the ground.'

With regard to the *permanence* of the tubercle contagium, the following facts are illustrative. A woman, who had for two years suffered from a phthisical cough, and who had been in the habit of spitting first upon the ground, and afterwards into a glass or a pocket-handkerchief, was visited by Cornet. During her life-time he proved the dust of her room to be infectious. Six weeks after her death he again visited the dwelling. Rubbing the dust from a square meter of the wall on which he had formerly found his infectious matter, and which had not been cleansed after the woman's death, he inoculated with it three of his guinea-pigs. Examined forty days after the inoculation, two of the three were found tuberculous. Cornet reasons thus :— ' No doubt the dust which had thus proved its virulence would have retained it for a longer time. Schill and Fischer, indeed, have proved that, after six months' preservation, dried sputum may retain its virulence. During this period, therefore, the possibility of infection by this dust is obviously open. When, moreover, the quantity of infectious matter inhaled is very small, a considerable time elapses before the development of the bacilli renders the malady distinct. Even if a year should elapse after the death of a phthisical patient before another member of the same household shows symptoms of lung disease, we are not entitled to assume a hereditary tendency without further proof. Aware of the facts above mentioned, we ought rather to ascribe the disease to infection by the dwelling, not to mention its possible derivation from other sources.'

On January 14, 1888, Cornet visited a patient who, for three-quarters of a year, had suffered from tuberculosis of the lung and larynx. The dust of the room occupied by this man was proved to contain virulent infective matter. A brother of the patient who, at the time of the examination of the dwelling, was alleged to be in perfect health, exhibited phthisis of the larynx four months afterwards. ' We are surely,' says Cornet, ' warranted in ascribing this result, not to heredity, or any other hypothetical cause, but to the naked fact that the dust of this dwelling contained tubercle bacilli which were capable of infecting the lungs and larynx of a man, as they did the peritoneum of a guinea-pig.'

On December 31, 1887, Cornet visited a man who for two years had suffered from phthisis. He lived in the same room with two brothers who were very robust, one of whom, however, had begun to cough, though without any further evidence of serious disorder. The patient had been at home for eight days, while previously he had acted as foreman in a tailoring establishment. It was proved, to a certainty, that this patient had taken the place of a colleague who had died from phthisis of the throat, and who had been in the habit of expectorating copiously upon the floor. In the workroom, moreover, the present sufferer had occupied a place next to the man who died. Cornet called upon the proprietor of the establishment, who allowed him every opportunity of examining the room, in which eight or ten workmen were engaged. With dust rubbed from about two square meters of the wall, near the spot where the patient now works, Cornet infected guinea-pigs and produced tuberculosis. He ridicules the notion of ascribing this man's malady to any hereditary endowment or predisposition, derived, say, from a phthisical mother, which, after sleeping for twenty years, woke up to action at the precise time when he was surrounded by infective matter. Our author regards this, and other similar cases which he adduces, as of special interest. The tuberculous virus was here found in rooms containing several workmen, who had thus an opportunity of infecting each other. The infection, moreover, occurred among tailors, who are known to be special sufferers from phthisis.

The general belief some time ago, which, to some

extent, may hold its ground to the present hour, was that this wasting malady arose from some peculiarity in the individual constitution, independent of infection from without. Enormous mischief has been done through exaggerated and incorrect notions regarding the influence of predisposition and inheritance. Members of the same family were observed to fall victims to this scourge, but each was regarded as an independent source of the disease, to the exclusion of the thought that the one had infected the other. Two or three days ago an old man here at Hind Head told me that he had lost three children in succession through phthisis; and he mentioned another case where five or six robust brothers had fallen, successively, victims to the same disease. 'I am sure,' said the man, with a flash of intelligence across his usually unintelligent countenance, ' *it must be catching.*' Cornet describes some cases which irresistibly suggest family infection. In 1887 he visited a patient, the father of a family, who, six years previously, had lost by consumption a little girl fourteen years old. A year and a half afterwards a daughter of the same man, twenty-one years old, fell a victim to the disease. One or two years later a robust son succumbed, while, a fortnight before Cornet's visit, a child a year and a half old had been carried away. Without doing violence to the evidence, as Cornet remarks, these cases may be justly regarded as due to family infection. For many years the father had suffered from a phthisical cough, and directly or indirectly he, in all probability, infected his children.

In connection with this subject, I may be permitted to relate a sad experience of my own. It is an easy excursion from my cottage in the Alps to the remarkable promontory called ' The Nessel,' on which stands a cluster of huts, occupied by peasants during the summer months. On visiting The Nessel three years ago, I was requested to look into a hut occupied by a man suffering from a racking cough, accompanied by copious expectoration. I did so. It was easy to see that the poor fellow was the victim of advanced lung disease. In the same hut lived his daughter, who, when I first saw her, presented the appearance of blooming health and vigour. Acquainted as I was with Koch's discoveries, I remarked to a friend who accompanied me, that the girl lived in the midst of peril. We had here the precise conditions notified by Cornet : spitting on the floor, drying of the sputum, and the subsequent treading of the infectious matter into dust. Whenever the hut was swept, this dust mingled freely with the air, and was of course inhaled.

I warned the girl against the danger to which she was exposed. But it is sometimes difficult to make even cultivated people comprehend the magnitude of this danger, or take the necessary precautions. A year afterwards I visited the same hut. The father was standing in the midst of the room—a well-built man, nearly six feet high, and as straight as an arrow. He was wheezing heavily, being at intervals bowed down by the violence of his cough. On a stool in the same room sat his daughter, who, a year previously, had presented such a picture of Alpine strength and beauty. Her appearance shocked me. The light had gone out of her eyes, while the pallor of her face and her panting breath showed only too plainly that she also had been grasped by the destroyer. There are thousands at this moment in England in the position I then occupied— standing helpless in the presence of a calamity that might have been avoided. All that could be done was to send the sufferers wine and such little delicacies as I could command. Last summer I learned that both father and daughter were dead, the daughter having been the first to succumb.

In opposition to those who consider that they have found bacilli in the breath of phthisical patients, Cornet adduces a number of very definite results. Patients have been caused to breathe against plates of glass coated with glycerine, which would undoubtedly have held the bacilli fast. Water has been examined, through which the air expired by phthisical lungs had been caused to pass. In this case the bacilli, being moist, would have been infallibly intercepted by the water. The aqueous vapour exhaled by consumptive lungs has been carefully condensed by ice ; but no bacilli has, in any of these cases, been detected. It behoves those who have arrived at an opposite result to repeat their experiments with the most scrupulous care, so that no doubt should be suffered to rest upon a point of such supreme importance. The lungs, air-passages, throat, and mouth all present wet surfaces, and it has been proved that even with sputum rich in bacilli, over which a current of air of considerable force had been driven, the air was found perfectly free from the organism.

The immunity as regards infection which to so great an extent is observed, is ascribed by Cornet in part to the intensely viscous character of the sputum when wet. Even after it has been subjected to a drying process its complete desiccation is opposed by its hygroscopic character. Cornet calls other investigators to bear him witness that the task of reducing well-dried sputum to a fine powder, even in a mortar, is by no means an easy one. It is difficult to produce, in this way, a dust fine enough to remain suspended in the air. It would be an error to suppose that dry tuberculous phlegm, when trodden upon in the streets, sends a cloud

of infected dust upwards. Its hygroscopic qualities in
great part prevent this. When dried sputum is reduced to powder in a humid place, it attracts to itself
moisture, and collects into little balls. The streets in
which phthisical persons expectorate are rendered in-
nocuous by rain, or by the artificial watering common
in towns. Cornet regards this watering as an enormous
sanitary advantage. No doubt when dry east winds
prevail for a sufficient time, infectious dust will mingle
with the air. During easterly winds infectious diseases
are known to be particularly prevalent. Our sufferings
from influenza during the present year have been connected in my mind with the long-continued easterly
and north-easterly winds, which, sweeping over vast
areas of dry land, brought with them the contagium that
produced the malady. Besides the difficulty encountered
before the sputum reaches the state of very fine powder,
other difficulties are presented by the numberless angles
and obstacles of the respiratory tract, and by the
integrity of the ciliary-epithelium, to the more or less
vigorous action of which is due the fact that amid
thousands of opportunities we have only here and there
a case of infection.

The action of the tubercle bacillus is determined by
the state of the surface with which it comes into contact. Wounds or lesions, caused by previous diseases,
such as measles, whooping cough, and scarlatina, may
exist along the respiratory canal. By illness, moreover,
the epithelium may be impaired, the inhaled bacilli
being thus offered a convenient domicile. If it be
thought desirable to call such a state of things 'predisposition,' Cornet will raise no objection. Wherever
a wounded or decaying tissue exists the bacillus will
find, unopposed, sufficient nutriment to enable it to increase in number, and to augment in vigour, before it

comes into contact, and conflict, with the living cells underneath. It is not any such predisposition, but predisposition by inheritance as a *source* of phthisis that is contended against by Cornet. That Koch entertained a different opinion is declared to be absolutely erroneous. The admission that a disease may be favoured, or promoted, by this or that circumstance is not tantamount to the assertion that in all, or nearly all cases, this circumstance is the cause, concomitant, or necessary precursor of the disease. This is the view generally entertained regarding ' predisposition.'

Cornet's further reasoning on this subject reveals his views so clearly that I will endeavour, in substance, to reproduce it here. Let a box be imagined filled with finely-divided bacillus dust, and let a certain number of guinea-pigs be caused, for a very short time, to inhale this dust. A few of them will be infected, while the great majority will escape. If the inhalation be prolonged, the number of animals infected will increase, until at length only one or two remain. With an exposure still more prolonged the surviving ones would undoubtedly succumb. Why, then, in the first instance, does one animal contract tuberculosis and another not ? Have they not all inhaled the same air, under the same conditions ? Are the animals that have escaped the first contagion less ' disposed ' than the survivors to the disease ? Assuming the animals to be all perfectly healthy, such differences will be observed. But, supposing them to be weakened in different degrees by previous disorders, the differences revealed in the case of healthy animals would be more pronounced. This, with human beings, is the normal state of things.

Take the case of a veteran who has been to the front in fifty different battles, who, right and left of him, has seen his comrades fall, until haply he remains the sole survivor of his regiment, without scratch or contusion. Shall we call him bullet-proof? Will his safety be ascribed to an absence of ' predisposition ' to attract the bullets—thus enjoying an immunity which the superstition of former ages would have ascribed to him? Is he more bullet-proof or less vulnerable than the comrade who by the first volley in the first battle was shot down? ' How often,' says Cornet, ' do such cases repeat themself in life? and are we able to do more than describe them as accidents? Unscientific as this word may appear, it is more in harmony with the truth than any artificial hypothesis.'

The opportunities for incorrect reasoning in regard to phthisis are manifold. It is observed, for example, that a hospital attendant, who has had for years, even for decades, consumptive patients in his charge, has, nevertheless, escaped infection. The popular conclusion finds vent in the words, ' It cannot be so dangerous after all!' Here, however, attention is fixed on a single fortunate individual, while the hundreds who, during the same time, have succumbed are forgotten. The danger of infection in different hospitals is a variable danger. In some we find bacilli, while in others we do not find them. It is no wonder, then, that among attendants who are thus exposed to different degrees of danger, some should be infected and others not. When, in cases of diphtheria, typhus, cholera, small-pox, which are undeniably infectious diseases, an attendant escapes infection, we do not exclaim, ' They are not so dangerous after all!' But this is the favourite expression when pulmonary consumption is in question. ' When,' adds Cornet, with a dash of indignation, ' we observe the enormous increase of phthisis among the natives of Mentone, and find this ascribed to the abandonment of land labour, instead of to intercommunication with the consumptive patients who spend their winter at that health-resort, it would seem as if some people shut their eyes wilfully against the truth.'

Again and again our author insists on the necessity of the most searching oversight on the part of physicians who have consumptive patients in charge. ' I cannot,' he says, ' accept as valid the assertion that in well-ordered hospitals provision is invariably made for expectoration into proper vessels, the conversion of the sputum into infectious dust being thereby rendered impossible. Take a case in point. One of the physicians to whose kindness I owe the possibility of carrying on my investigation, assured me in the most positive manner that the patients in his hospital invariably used spittoons. A few minutes after this assurance had been given, and under the eyes of the director himself, I drew from the bed of a patient a pocket-handkerchief filled with half-dried phlegm. I rubbed from the wall of the room, at a distance of half a meter from the bed of this patient, a quantity of dust, with which, as I predicted, tuberculosis was produced. If, therefore, physicians, attendants, and patients do not work in unison, if the patient and his attendants be not accurately instructed and strictly controlled, the presence of the spittoon will not diminish the danger.'

In the dwellings of private patients the perils here glanced at were most impressively brought home to the inquirer. In fifteen out of twenty-one sick-rooms, that is to say, in more than two-thirds of them, Cornet found in the dust of the walls and bed-furniture virulent tubercle bacilli. He refers to his published tables to prove that in no ward or room where the organism was found did the patients confine themselves to expectoration into spittoons, but were in the habit of spitting

either upon the floors or into pocket-handkerchiefs. In
no single case, on the other hand, where spitting on the
floor or into pocket-handkerchiefs was strictly and
effectually prohibited, did he find himself able to produce tuberculosis from the collected dust.

A point of considerable importance, more specially
dealt with by Cornet in a further investigation, has
reference to the allegation that physicians who attend
tuberculous patients do not show among themselves
the frightful mortality from phthisis that might be
expected. This is often adduced as proof of the comparative harmlessness of the tubercle bacillus. No investigation, however, has proved that the mortality
among physicians by phthisis does not far exceed the
average. And even should this mortality show no great
preponderance, it is to be borne in mind that the
number of physicians who, thanks to their education,
are able to discern the first approaches of the malady,
and to master it in time, is by no means inconsiderable.
In the health resorts of Germany, Italy, France, and
Africa, we find numbers of physicians who have been
compelled, by their own condition, to establish their
practice in such places.

The memorable paper of which I have here given a
concentrated abstract concludes with a chapter on
' Preventive Measures,' which are assuredly worthy of
grave attention on the part of Governments, of hospital
authorities, and of the public at large. The character
of these measures may be, in great part, gathered from
the foregoing pages. It is more than once enunciated
in Cornet's memoir that the first and greatest danger
to which the phthisical patient is exposed is *himself*.
If he is careless in the disposal of his phlegm, if he
suffers it to become dry and converted into dust, then,

by the inhalation of a contagium derived from the diseased portions of his own lung, he may infect the healthy portions. ' If, therefore,' says Cornet, ' the phthisical patient, to avoid the guilt of self-murder, is compelled to exercise the utmost caution, he is equally bound to do so for the sake of his family, his children, and his servants and attendants. He must bestow the most anxious care upon the disposal of his sputum. Within doors he must never, under any circumstances, spit upon the floor, or employ his pocket-handkerchief to receive his phlegm, but always and everywhere must use a proper spittoon. If he is absolutely faithful in the carrying out of these precautions, he may accept the tranquillising assurance that he will neither injure himself nor prove a source of peril to those around him.'

Though mindful of the danger of interfering with social arrangements, Cornet follows out his preventive measures in considerable detail. Hand-spittoons, with a cover, he recommends, not with the view of preventing evaporation, but because flies have been known to carry infection from open vessels. Without condemning the practice, he does not favour the disinfection of sputum by carbolic acid and other chemicals. He deprecates the use of sand or sawdust in spittoons. On æsthetic grounds, he would have the spittoons of those who can afford it made ornamental, but earthenware saucers, such as those placed under flower-pots, are recommended for the use of the poor. The consumptive patient must take care that not only in his own house, but also in the offices and workshops where he may be engaged, he is supplied with a proper spittoon. In public buildings, as in private houses, the corridors and staircases ought to be well supplied with these necessaries. The ascent of the stairs often provokes coughing and expectoration,

and the means of disposing of phlegm ought to be at
hand. The directors of factories, and the masters of
workshops, as well as the workmen themselves, ought
to make sure that, under no circumstances, shall
spitting on the floor or into a pocket-handkerchief be
tolerated.

One final word is still to be spoken. If we are to
fight this enemy with success, the public must make
common cause with the physician. The fear of spreading panic among the community, and more particularly
among hospital nurses, must be dismissed. Unless
nurses, patients, and public, realise with clear intelligence the dangers to which they are exposed, they will
not resort to the measures necessary for their protection.
Should the sources of infection be only partially removed, the marked diminution of a malady, which now
destroys .more human beings than all other infective
diseases taken together, will, as pointed out by Cornet,
be ' our exceeding great reward.'

Dr. Cornet's great investigation, of which some account is given above, is entitled, ' The Diffusion of
Tubercle Bacilli exterior to the Body.' It was published
in 1888. A shorter, though not less important inquiry,
on ' The Mortality of the Nursing Orders,' was published
in 1889. These two memoirs will be found permanently
embodied in the fifth and sixth volumes of the *Zeit-
schrift für Hygiene*. From a former paragraph it will
be seen that Cornet's attention had been directed to
those who, more than others, come closely into contact
with infectious diseases, and that he throws doubt upon
the notion that neither physicians nor nurses suffer from
this proximity. No definite and thorough inquiry had,
however, been made into this grave question. In face

of the vague and contradictory statements which issued from the authorities of different hospitals, the problem cried aloud for solution. For aid and data, under these circumstances, Cornet resorted to Herr von Gossler, the Prussian Minister of State, who, at that time, had medical matters under his control. From him he received the most hearty furtherance and encouragement. Dr. von Gossler has recently resigned his post in the Prussian Ministry, but his readiness to forward the momentous inquiry on which Cornet was engaged merits the grateful recognition of the public and the praise of scientific men.

The number of female nurses in Prussia, as shown by the statistics of the Royal Bureau of Berlin for 1885, was 11,048. Of these, the Catholic Sisters of Mercy numbered 5,470, or 49·51 per cent.; Evangelical nurses, 2,496, or 22·59 per cent. ; nurses belonging to other societies and associations, 352, or 3·19 per cent.; while of unclassified nurses there were 2,730, or 24·71 per cent. of the whole. The male attendants, at the same time, numbered 3,162. Of these, 383 were Brothers of Mercy, 205 were deacons, while of unclassified attendants there were 2,574.

The sifting of these numbers was a labour of anxious care to Dr. Cornet. It had already been remarked by Guttstadt that the commercial attractions of hospital service were insufficient, without the help of some ideal motive, to secure a permanent staff. This motive was found in devotion through a sense of religious duty to the service of the sick. The sifting of his material made it clear to Cornet that, to secure a safe basis of generalisation, by causing it to embrace a sufficient number of years, he must confine himself solely to the nurses of the Catholic orders. The greater freedom enjoyed and practised by Protestants, in changing their

occupation, in entering the married state, or through
other modes of free action, rendered them unsuitable
for the purpose he had in view. Cornet's inquiry extended over a quarter of a century. The returns furnished by thirty-eight hospitals served by Catholic
sisters and brethren, and embracing a yearly average of
4,020 attendants, showed the number of deaths during
the period mentioned to be 2,099. Of these, 1,32 0
were caused by tuberculosis. In the State, as a whole,
the proportion of deaths from this malady to the total
number of deaths is known to be very high, reaching
from one-fifth to one-seventh of the whole. In the
hospitals this proportion was enormously increased. It
rose on the average to almost two-thirds, or close upon
63 per cent. of the total number of deaths. In nearly
half the hospitals even this high proportion was surpassed, the deaths in these amounting to three-fourths
of the whole. Scarcely any other occupation, however
injurious to health, shows a mortality equal to that
found in these hospitals.

The following statistics furnish a picture of the
state of things prevalent during the five-and-twenty
years referred to. A healthy girl of 17, devoting herself to hospital nursing, dies on the average $21\frac{1}{2}$ years
sooner than a girl of the same age moving among the
general population. A hospital nurse of the age of 25
has the same expectation of life as a person of the age
of 58 in the general community. The age of 33 years
in the hospital is of the same value as the age of 62 in
common life. The difference between life-value in the
hospital and life-value in the State increases from the
age of 17 to the age of 24 ; nurses of this latter age
dying 22 years sooner than girls of the same age in the
outside population. The difference afterwards becomes
less. In the fifties it amounts to only six or seven

years, while later on it vanishes altogether. The reason of this is, that the older nurses are gradually withdrawn from the heavier duties of their position and the attendant danger of infection.

In these hospitals, deaths from typhus and other infectious disorders exhibit a frequency far beyond the normal ; but the enormous total augmentation is mainly to be ascribed to the frequency of deaths from tuberculosis. The excess of mortality is to be referred to the vocation of nursing, and the chances of infection involved in it. Cornet examines other assumptions that might be made to account for the mortality, and gives cogent reasons for dismissing them all. The tranquil lives led by the nurses, the freedom from all anxiety in regard to subsistence, the moderation observed in food and drink, all tend to the preservation of health. They live in peace, free from the irregularities of outside life, and their contentment and circumstances generally are calculated rather to prolong their days than to shorten them.

Cornet is very warm in his recognition of the devotion of these Catholic nurses, two-thirds of whom are sacrificed in the service which they render to suffering humanity. And they are sacrificed for the most part in the blossom of their years; for it is the younger nurses, engaged in the work of sweeping and dusting, whose occupation charges the air they breathe with virulent bacilli. The statistics of their mortality Cornet regards as a monumental record of their lofty self-denial, their noble, beneficent, and modest fidelity to what they regard as the religious duty of their lives.

But, he asks, is it necessary that this sacrifice should continue ? His answer is an emphatic negative, to establish which he again sums up the results which

we have learnt from his first memoir :—It is universally recognised that tuberculosis is caused by tubercle
bacilli, which reach the lungs through the inhalation
of air in which the bacilli are diffused. They come
almost exclusively from the dried sputum of consumptive persons. The moist sputum, as also the expired
breath of the consumptive patient, is, for this mode of
infection, without danger. If we can prevent the drying of the expectorated matter, we prevent in the same
degree the possibility of infection. It is not, however,
sufficient to place a spittoon at the disposal of the
patient. The strictest surveillance must be exercised
by both physicians and attendants, to enforce the proper use of the spittoon, and to prevent the reckless disposal of the infective phlegm. Spitting on the floor
or into pocket-handkerchiefs is the main source of
peril. To this must be added the soiling of the bed-
clothes and the wiping of the patient's mouth. The
handkerchiefs used for this purpose must be handled
with care, and boiled without delay. Various other
sources of danger, kissing among them, will occur to
the physician. A phthisical mother, by kissing her
healthy child, may seal its doom. Notices, impressing
on the patients the danger of not attending to the precautions laid down in the hospital, ought to be posted
up in every sick-room, while all wilful infringements of
the rules ought to be sternly punished. Thus may the
terrible mortality of hospital nurses be diminished, if
not abolished ; the wards where they are occupied being
rendered as salubrious as those surgical wards in which
no bacilli could be found.

Reflecting on the two investigations which I have
here endeavoured to bring, lucidly if briefly, before
my readers, the question arises—'What, under the

circumstances, is the duty of the English public and the English Government?' Will the former suffer themselves to be deluded, and the latter frightened, by a number of loud-tongued sentimentalists, who, in view of the researches they oppose, and the fatal effects of their opposition, might be fairly described as a crew of well-meaning homicides. The only way of combating this terrible scourge of tuberculosis and, indeed, of abolishing all other infectious diseases, is experimental investigation ; and the most effectual mode of furthering such investigation, just now in England, is the establishment of that ' Institute of Preventive Medicine ' which, I am rejoiced to learn, has, after due consideration, been licensed by the President of the Board of Trade. Whatever my illustrious friend, the late Mr. Carlyle, may have said to the contrary, the English public, in its relation to the question now before us, are *not* ' mostly fools ' ; and if scientific men only exhibit the courage and industry of their opponents, they will make clear to that public the beneficence of their aims, and the fatal delusions to which a narrow and perverted view of a great question has committed the anti-vivisectionist.

The letter to the ' Times ' of April 22nd, 1882, describing Koch's epoch-making discovery of the tubercle bacillus, is here introduced :—

To the Editor of the ' Times.'

SIR,—On the 24th of March, 1882, an address of very serious public import was delivered by Dr. Koch before the Physiological Society of Berlin. It touches a question in which we are all at present interested— that of experimental physiology—and I may, therefore,

be permitted to give some account of it in the ' Times.'
The address, a copy of which has been courteously sent
to me by its author, is entitled ' The Etiology of
Tubercular Disease.' Koch first made himself known,
and famous, by the penetration, skill, and thoroughness
of his researches on the contagium of anthrax, or splenic
fever. By a process of inoculation and infection he
traced this terrible parasite through all its stages of
development and through its various modes of action.
This masterly investigation caused the young physician
to be transferred from a modest country practice in the
neighbourhood of Breslau to the post of Government Adviser in the Imperial Health Department of
Berlin.

From this department has lately issued a most important series of investigations on the etiology of infective disorders. Koch's last inquiry deals with a
disease which, in point of mortality, stands at the
head of them all. ' If,' he says, ' the seriousness of a
malady be measured by the number of its victims,
then the most dreaded pests which have hitherto
ravaged the world—plague and cholera included—
must stand far behind the one now under consideration.' Then follows the startling statement that one-
seventh of the deaths. of the human race are due to
tubercular disease. Prior to Koch it had been placed
beyond doubt that the disease was *communicable* ; and
the aim of the Berlin physician has been to determine
the precise character of the contagium which previous
experiments on inoculation and inhalation had proved
to be capable of indefinite transfer and reproduction.
He subjected the diseased organs of a great number of
men and animals to microscopic examination, and found,
in all cases, the tubercles infested by a minute, rod-
shaped parasite, which, by means of a special dye, he

differentiated from the surrounding tissue. ' It was,' he says, ' in the highest degree impressive to observe in the centre of the tubercle-cell the minute organism which had created it.' Transferring directly, by inoculation, the tuberculous matter from diseased animals to healthy ones, he in every instance reproduced the disease. To meet the objection that it was not the parasite itself, but some virus in which it was imbedded in the diseased organ, that was the real contagium, he cultivated his bacilli artificially for long periods of time and through many successive generations. With a speck of matter, for example, from a tuberculous human lung, he infected a substance prepared, after much trial, by himself, with the view of affording nutriment to the parasite. In this medium he permitted it to grow and multiply. From the new generation he took a minute sample, and infected therewith fresh nutritive matter, thus producing another brood. Generation after generation of bacilli were developed in this way, without the intervention of disease. At the end of the process, which sometimes embraced successive cultivations extending over half a year, the purified bacilli were introduced into the circulation of healthy animals of various kinds. In every case inoculation was followed by the reproduction and spread of the parasite, and the generation of the original disease.

Permit me to give, a little more in detail, an account of Koch's experiments. Of six healthy guinea-pigs, four were inoculated with bacilli derived originally from a human lung, which, in fifty-four days, had produced five successive generations. Two of the six animals were not infected. In every one of the infected cases the guinea-pig sickened and lost flesh. After thirty-two days one of them died, and after thirty-

five days the remaining five were killed and examined. In the guinea-pig that died, and in the
three remaining infected ones, strongly-pronounced
tubercular disease had set in. Spleen, liver, and
lungs were found filled with tubercles; while in
the two uninfected animals no trace of the disease
was observed. In a second experiment, six out
of eight guinea-pigs were inoculated with cultivated
bacilli, derived originally from the tuberculous lung
of a monkey, bred and re-bred for ninety-five
days, until eight generations had been produced.
Every one of these animals was attacked, while
the two uninfected guinea-pigs remained perfectly
healthy. Similar experiments were made with cats,
rabbits, rats, mice, and other animals, and, without
exception, it was found that the injection of the parasite into the animal system was followed by decided
and, in most cases, virulent tubercular disease.

In the cases thus far mentioned inoculation had
been effected in the abdomen. The place of inoculation
was afterwards changed to the aqueous humour of the
eye. Three rabbits received each a speck of bacillus-
culture, derived originally from a human lung affected
with phthisis. Eighty-nine days had been devoted
to the culture of the organism. The infected rabbits
rapidly lost flesh, and after twenty-five days were killed
and examined. The lungs of every one of them were
found charged with tubercles. Of three other rabbits,
one received an injection of pure blood-serum in the
aqueous humour of the eye, while the other two were
infected in a similar way, with the same serum, containing bacilli derived originally from a diseased lung,
and subjected to ninety-one days' cultivation. After
twenty-eight days the rabbits were killed. The one
which had received an injection of pure serum was

found perfectly healthy, while the lungs of the two others were found overspread with tubercles.

Other experiments are recorded in this admirable essay, from which the weightiest practical conclusions may be drawn. Koch determines the limits of temperature between which the tubercle bacillus can develop and multiply. The *minimum* temperature he finds to be 86° Fahr., and the *maximum*, 104°. He concludes that, unlike the *Bacillus anthracis* of splenic fever, which can flourish freely outside the animal body, in the temperate zone animal warmth is necessary for the propagation of the newly discovered organism. In a vast number of cases Koch has examined the matter expectorated from the lungs of persons affected with phthisis, and found in it swarms of bacilli, while in matter discharged from the lungs of persons not thus afflicted he has never found the organism. The sputum in the former cases was highly infective, nor did *drying* destroy its virulence. Guinea-pigs infected with expectorated matter which had been kept dry for two, four, and eight weeks respectively, were smitten with tubercular disease quite as virulent as that produced by fresh expectoration. Koch points to the grave danger of inhaling air in which particles of the dried sputa of consumptive patients mingle with dust of other kinds.

The moral of these experiments is obvious. In no other conceivable way than that pursued by Koch could the true character of the most destructive malady by which humanity is now assailed be determined. And however noisy the fanaticism of the moment may be, the common-sense of Englishmen will not, in the long run, permit it to enact cruelty in the name of tenderness, or to debar us from the light and leading of such investigations as that which is here so imperfectly described.

Your obedient servant,

JOHN TYNDALL.

HIND HEAD, *April* 20, 1882.

NOTE.

Twenty years ago I received letters describing to me the grief and ruin introduced into families through the notion, then prevalent, that typhoid fever is non-contagious. When Dr. William Budd published his researches on this subject, showing by facts and reasonings as cogent as it was in the power of science to supply, the infectiousness of the fever, certain writers discerned in that important work a proof of the decadence of Budd's intellect, and gave the public the benefit of their conclusions.

In regard to the contagiousness of phthisis, we have now, it seems, to face the same danger. It may not, therefore, be out of place to cite an illustration of the recklessness stimulated by the assertion that 'consumption is not an infectious disease.' While occupied with experiments on the inhalation of tuberculous air by dogs, Tappeiner was assisted by a robust man of forty, who was specially warned never to tarry in the locality where the dogs were confined. He, however, seemed bent on proving the doctrine of tubercle-contagion to be a delusion, and recklessly exposed himself to the infective air. This strong man, who was free from any suspicion of hereditary taint, was smitten by tuberculosis of exactly the same kind as that exhibited by the dogs, and in fourteen weeks he was a corpse. Examination after death proved the identity of the disease which killed him with that which affected the dogs; the only difference being that he, having lived longer, exhibited the malady in a more advanced stage. (October, 1891.)   J. T.

OLD ALPINE JOTTINGS,

TWENTY years ago, at the instance of Mr. Macmillan, I threw these 'jottings' together for his excellent magazine. In surveying them I notice, what is confirmed by a larger survey, that my life, in regard to working power and consequent enjoyment, has been one of ups and downs. Intellectual work has its delights and drawbacks. Strain and worry of mind are admitted causes of physical disturbance, and of them I have had my share. 'Materialism' is also better understood than it used to be; and no man subject to a weak digestion with periodic loss of sleep will be indisposed to assign to material things a transcendental value. They act upon body and mind; predisposing the organ of intellect and imagination to give to current events, especially on wakeful nights, an over-brilliant colouring. For such derangements I know nothing better than a dose of the glaciers—under the condition, however, that they have an organism needing purification and repair, but otherwise sound all round, to act upon. The reader would err if he imagined that the 'lowness of spirits' revealed here and there in these 'jottings' was a permanent lowness. The contemplated abandonment of the Alps was all moonshine. Seven years after my 'leave-taking' I built my aerie upon the heights, where the snow, which falls as I write, heightens, instead of lowering, the inner temperature of the old mountaineer.

August 24, 1889.

PART I.

SINCE the publication, seven years ago, of a little tract entitled 'Mountaineering in 1861,' I have contributed hardly anything to the literature of the Alps. I have gone to them every year, and found among them refuge and recovery from the work and the worry—which acts with far deadlier corrosion on the brain than real work— of London. Herein consisted the fascination of the Alps for me : they appealed at once to thought and feeling, offering their problems to the one and their grandeurs to the other, while conferring upon the body the soundness and the purity necessary to the healthful exercise of both. There is, however, a natural end to Alpine discipline, and henceforth mine will probably be to me a memory. The last piece of work requiring performance on my part was executed last summer; and, unless temptation of unexpected strength assail me, this must be my last considerable climb. With soberness of mind, but without any approach to regret, I take my leave of the higher Alpine peaks.

And this is why it has occurred to me to throw together these odds and ends of Alpine experience into a kind of cairn to the memory of a life well loved. Previous to the year 1860, I knew the Matterhorn as others did, merely as a mountain wonder, for up to that time no human foot had ever been placed on its repel-

lent crags. It is but right to state that the man who
first really examined the Matterhorn, in company with
a celebrated guide, who came to the conclusion that
it was assailable if not accessible, was Mr. Vaughan
Hawkins. It was at his suggestion that in August 1860
I took part in the earliest assault upon this formidable
peak. We halted midway, stopped less by difficulty,
though that was great, than by want of time. In 1862
I made a more determined attack upon the mountain,
but was forced to recoil from its final precipice; for
time, the great reducer of Alpine difficulties, was again
wanting. On that occasion 1 was accompanied by two
Swiss guides and two Italian porters. Three of these
four men pronounced flatly against the final precipice.
Indeed, they had to be urged by degrees along the sharp
and jagged ridge—the most savage, in my opinion, on
the whole Matterhorn—which led up to its base. The
only man of the four who never uttered the word
' impossible ' was Johann Joseph Bennen, the bravest of
brave guides, who now lies in the graveyard of Ernan,
in the higher valley of the Rhône. We were not only
defeated by the Matterhorn, but were pelted down its
crags by pitiless hail.

On the day subsequent to this defeat, while crossing
the Cimes Blanches with Bennen, we halted to have a
last look at the mountain. Previous to quitting Breuil
I had proposed to him to make another attempt. He
was adverse to it, and my habit was never to persuade
him. On the Cimes Blanches I turned to him and used
these words : ' I leave Breuil dissatisfied with what we
have done. We ought never to have quitted the Matterhorn without trying yonder arête.' The ridge to
which Bennen's attention was then directed certainly
seemed practicable, and it led straight to the summit.
There was moisture in the strong man's eyes as he re-

plied, falling into the *patois* which he employed when
his feelings were stirred : ' What could I do, sir ? not
one of them would accompany me.' It was the accurate
truth.

To reach the point where we halted in 1862 one
particularly formidable precipice had to be scaled. It
had also to be descended on our return, and to get down
would be much more hazardous than to climb. At the
top of the precipice we therefore fastened a rope, and
by it reached in succession the bottom. This rope had
been specially manufactured for the Matterhorn by Mr.
Good, of King William Street, City, to whom I had
been recommended by his landlord, Appold, the famous
mechanician. In the summer of 1865, the early part
of which was particularly favourable to the attempt, one
of the Italians (*Carrel dit le Bersaglier*) who accompanied me in 1862, and who proved himself on that
occasion a first-rate cragsman, again tried his fortune
on the Matterhorn. He reached my rope, and found it
bleached to snowy whiteness. It had been exposed for
three years to all kinds of weather, and to the fraying
action of the storms which assail the Matterhorn ; but
it bore, on being tested, the united weights of three
men.[1] By this rope the summit of the precipice which
had given us so much trouble in 1862 was easily and
rapidly attained. A higher *Nachtlager* was thus
secured, and more time was gained for the examination
of the mountain. Every climber knows the value of
time in a case of the kind. The result of the scrutiny
was that a way was found up the Matterhorn from the
Italian side, that way being the ridge referred to in my
conversation with Bennen three years before.

Committed thus and in other ways to the Matterhorn, the condition of my mind regarding it might be

[1] A yard of this rope is now in my possession.

fitly compared to one of those uncheerful tenements
often seen in the neighbourhood of London, where an
adventurous contractor has laid the foundations, run up
the walls, fixed the rafters, but stopped short, through
bankruptcy, without completing the roof. As long as
the Matterhorn remained unscaled, my Alpine life could
hardly be said to be covered in, and the admonitions of
my friends were premature. But now that the work is
done, they will have more reason to blame me if I fail
to profit by their prudent advice.

Another defeat of a different character was also
inflicted on me in 1862. Wishing to give my friend
Mr. (now Sir John) Lubbock a taste of mountain life, I
went with him up the Galenstock. This pleased him so
much that Bennen and I, desiring to make his cup of
pleasure full, decided on taking him up the Jungfrau.
We sent two porters, laden with coverlets and provisions,
from the Æggischhorn to the Faulberg, but on our
arrival there found one of the porters in the body of the
Aletsch glacier. He had recklessly sought to cross a
snow-bridge which spanned a broad and profound chasm.
The bridge broke under him, he fell in, and was deeply
covered by the frozen *débris* which followed him. He
had been there for an hour when we arrived, and it
required nearly another hour to dig him out. We
carried him more dead than alive to the Faulberg cave,
and by great care restored him. As I lay there wet,
through the long hours of that dismal night, I almost
registered a vow never to tread upon a glacier again.
But, like the forces in the physical world, human
emotions vary with the distance from their origin, and a
year afterwards I was again upon the ice. Towards the
close of 1862 Bennen and myself made 'the tour of
Monte Rosa,' halting for a day or two at the excellent
hostelry of Delapierre, in the magnificent Val du Lys.

We scrambled up the Grauhaupt, a point exceedingly favourable to the study of the conformation of the Alps. We also halted at Alagna and Macugnaga. But, notwithstanding their admitted glory, the Italian valleys of the Alps did not suit either Bennen or me. We longed for the more tonic air of the northern slopes, and were glad to change the valley of Ansasca for that of Saas.

The first days of the vacation of 1863 were spent in the company of Mr. Philip Lutley Sclater. On July 19 we reached Reichenbach, and on the following day sauntered up the valley of Hasli, turning to the left at Imhof into Gadmenthal. Our destination was Stein, which we reached by a grass-grown road through fine scenery. The goatherds were milking when we arrived. At the heels of one quadruped, supported by the ordinary one-legged stool of the *Senner*, bent a particularly wild and dirty-looking individual, who, our guide informed us, was the proprietor of the inn. ' He is but a rough Bauer,' said our guide Jaun, ' but he has engaged a pretty maiden to keep house for him.' While he thus spoke a light-footed creature glided from the door towards us, and bade us welcome. She led us upstairs, provided us with baths, took our orders for dinner, helped us by her suggestions, and answered all our questions with the utmost propriety and grace. She had been two years in England, and spoke English with a particularly winning accent. How she came to be associated with the unkempt brute outside was a puzzle to both of us. It is Emerson, I think, who remarks on the benefit which a beautiful face, without trouble to itself, confers upon him who looks at it. And, though the splendour of actual beauty could hardly be claimed for our young hostess, she was handsome enough and graceful enough to brighten a tired traveller's thoughts, and to raise by her presence the modest comforts she dispensed to the level of luxuries.[1]

It rained all night, and at 3.30 A.M., when we were called, it still fell heavily. At five, however, the clouds began to break, and half an hour afterwards the heavens were swept quite clear of them. At six we bade our pretty blossom of the Alps good-bye. She had previously to bring her gentle influence to bear upon her master to moderate the extortion of some of his charges. We were soon upon the Stein glacier, and after some time reached a col from which we looked down upon the lower portion of the nobler and more instructive Trift glacier. Brown bands were drawn across the ice-stream, forming graceful loops with their convexities turned downwards. The higher portions of the glacier were not in view, still those bands rendered the inference secure that an ice-fall existed higher up, at the base of which the bands originated. We shot down a shingly couloir to the Trift, and looking up the glacier, the anticipated cascade came into view. At its bottom the ice, by pressure, underwent that notable change, analogous to slaty cleavage, which caused the glacier to weather in parallel grooves, and thus mark by the dirt upon its surface the direction of its interior lamination.

The ice-cascade being itself impracticable, we

[1] Thackeray, in his 'Peg of Limarady,' is perhaps more to the point than Emerson :—

> ' Presently a maid
>   Enters with the liquor —
> Half-a-pint of ale
>   Frothing in a beaker ;
> As she came she smiled,
>   And the smile bewitching,
> On my word and honour
>   Lighted all the kitchen.'

scaled the flanking rocks, and were soon in presence of
the far-stretching snow-fields from which the lower
glacier derives nutriment. With a view to hidden
crevasses, we here roped ourselves together. The sun
was strong, its direct and reflected blaze combining
against us. The scorching warmth experienced at
times by cheeks, lips, and neck, indicated, in my case,
that mischief was brewing. But the eyes being well
protected by dark spectacles, I was comparatively indifferent to the prospective disfigurement of my face.
Mr. Sclater was sheltered by a veil, a mode of defence
which the habit of going into places requiring the
unimpeded eyesight has caused me to neglect. There
seems to be some specific quality in the sun's rays
which produces the irritation of the skin experienced
in the Alps. The solar heat may be compared, in
point of *quantity*, with that radiated from a furnace ;
and the heat which the mountaineer experiences on
Alpine snows is certainly less intense than that encountered by workmen in many of our technical
operations. But terrestrial heat appears to lack the
quality which gives the sun's rays their power. The
sun is incomparably richer in what are called chemical
rays than are our fires, and to these chemical rays the
irritation may be due.[1] The keen air of the heights
may also have something to do with it. As a remedy
for sunburn I have tried glycerine, and found it a
failure. The ordinary lip-salve of the druggists' shops
is also worse than useless, but pure cold cream, for a
supply of which I have had on more than one occasion
to thank a friend, is an excellent ameliorative.

After considerable labour we reached the ridge—
a very glorious one as regards the view—which forms

[1] I might have said 'is certainly due.' A powerful 'arc-light'
produces, in a sheltered room, substantially the same effect as the
sun.

the common boundary of the Rhône and Trift glaciers.[1] Before us and behind us for many a mile fell the dazzling névés, down to the points where the grey ice emerging from its white coverlet declared the junction of snow-field and glacier. We had plodded on for hours soddened by the solar heat and parched with thirst. There was

> Water, water everywhere,
> But not a drop to drink.

For, when placed in the mouth, the liquefaction of the ice was so slow, and the loss of heat from the surrounding tissues so painful, that sucking it was worse than total abstinence. In the midst of this solid water you might die of thirst. At some distance below the col, on the Rhône side, the musical trickle of the liquid made itself audible, and to the rocks from which it fell we repaired and refreshed ourselves. The day was far spent, the region was wild and lonely, when, beset by that feeling which has often caused me to wander singly in the Alps, I broke away from my companions, and went rapidly down the snow-field. Our guide had previously informed me that before reaching the cascade of the Rhône glacier the ice was to be forsaken, and the Grimsel, our destination, reached by skirting the base of the peak called Nägelis Grätli. After descending the ice for some time I struck the bounding rocks, and climbing the mountain obliquely, found myself among the crags which lie between the Grimsel pass and the Rhône glacier. It was an exceedingly desolate place, and I soon had reason to doubt the wisdom of being there alone. Still, difficulty rouses powers of which we should otherwise remain unconscious. The heat of the

[1] Seven years previously Mr. Huxley and myself had attempted to reach this col from the other side.

day had rendered me weary, but among these rocks the
weariness vanished, and I became clear in mind and
fresh in body through the necessity of escape before
nightfall from this wilderness.

I reached the watershed of the region. Here a
tiny stream offered me its company, which I accepted.
It received in its course various lateral tributaries, and
at one place expanded into a blue lake bounded by
banks of snow. The stream quitted this lake augmented in volume, and I kept along its side until,
arching over a brow of granite, it discharged itself
down the glaciated rocks, which rise above the Grimsel.
In fact, this stream was the feeder of the Grimsel lake.
I halted on the brow for some time. The hospice was
fairly in sight, but the precipices between me and
it seemed desperately ugly. Nothing is more trying
to the climber than cliffs which have been polished
by the ancient glaciers. Even at moderate inclinations, as may be learned from an experiment on the
Höllenplatte, or some other of the polished rocks in
Haslithal, they are not easy. I need hardly say that
the inclination of the rocks flanking the Grimsel is the
reverse of moderate. It is dangerously steep.

How to get down these smooth and precipitous
tablets was now a problem of the utmost interest to me ;
for the day was too far gone, and I was too ignorant of
the locality, to permit of time being spent in the search
of an easier place of descent. Right or left of me I saw
none. The continuity of the cliffs below me was occasionally broken by cracks and narrow ledges, with scanty
grass-tufts sprouting from them here and there. The
problem was how to get down from crack to crack and
from ledge to ledge. A salutary anger warms the mind
when thus challenged, and, aided by this warmth, close
scrutiny will dissolve difficulties which might otherwise

seem insuperable. Bit by bit I found myself getting lower, closely examining at every pause the rocks below. The grass-tufts helped me for a time, but at length a rock was reached on which no friendly grass could grow. This slab was succeeded by others equally forbidding. A slip was not admissible here. I looked upwards, thinking of retreat, but the failing day urged me on. From the middle of the smooth surface jutted a ledge about fifteen inches long and about four inches deep. Once upon this ledge I saw that I could work obliquely to the left-hand limit of the face of the rock, and reach the grass-tufts once more. Grasping the top of the rock, I let myself down as far as my stretched arms would permit, and then let go my hold. The boot-nails had next to no power as a brake, the hands had still less, and I came upon the ledge with an energy that shocked me. A streak of grass beside the rock was next attained ; it terminated in a small steep couloir, the portion of which within view was crossed by three transverse ledges. There was no hold on either side of it, but I thought that by friction the motion down the groove could be so regulated as to enable me to come to rest at each successive ledge. Once started, however, my motion was exceedingly rapid. I shot over the first ledge, an uncomfortable jolt marking my passage. Here I tried to clamp myself against the rock, but the second ledge was crossed like the first. The outlook now became alarming, and I made a desperate effort to stop the motion. Braces gave way, clothes were torn, wrists and hands were skinned and bruised, while hips and knees suffered variously. I did, however, stop myself, and here all serious difficulty ended. I was greatly heated, but a little lower down reached a singular cave in the mountain-side, with water dripping from its roof into a clear well. The ice-cold liquid soon restored me

to a normal temperature. I felt quite fresh on entering the Grimsel inn, but a curious physiological effect
manifested itself when I had occasion to speak. The
power of the brain over the lips was so lowered that I
could hardly make myself understood.

My guide Bennen reached the Grimsel the following
morning. Uncertain of my own movements, I had permitted him this year to make a new engagement, which
he was now on his way to fulfil. There was a hint of
reproach in his tone as he asked me whether his Herr
Professor had forsaken him. There was little fear of
this. A guide of proved competence, whose ways you
know, and who knows you and trusts you, is invaluable
in the Alps. Bennen was all this, and more, to me.
As a mountaineer, he had no superior, and he added to
his strength, courage, and skill, the qualities of a
natural gentleman. He was now ready to bear us company over the Oberaarjoch to the Æggischhorn. On
the morning of the 22nd we bade the cheerless Grimsel
inn good-bye, reached the Unteraar glacier, crossed its
load of uncomfortable *débris*, and clambered up the
slopes at the other side. Nestled aloft in a higher
valley was the Oberaar glacier, along the unruffled surface of which our route lay.

The morning threatened. Fitful gleams of sunlight
wandered with the moving clouds over the adjacent ice.
The Joch was swathed in mist, which now and then gave
way, and permitted a wild radiance to shoot over the
col. On the windy summit we took a mouthful of food
and roped ourselves together. Here, as in a hundred
other places, I sought in the fog for the vesicles of De
Saussure, but failed to find them. Bennen, as long as
we were on the Berne side of the col, permitted Jaun
to take the lead; but now we looked into Wallis, or

rather into the fog which filled it, and the Wallis guide came to the front.   I knew the Viesch glacier well ; it is badly crevassed, and how Bennen meant to unravel its difficulties without landmarks I knew not.   I asked him whether, if the fog continued, he could make his way down the glacier.   There was a pleasant *timbre* in Bennen's voice, a light and depth in his smile due to the blending together of conscious power and warm affection.   With this smile he turned round and said, ' Herr ! Ich bin hier zu Hause. Der Viescher Gletscher ist meine Heimath.'

Downwards we went, striking the rocks of the Rothhorn so as to avoid the riven ice.   Suddenly we passed from dense fog into clear air ; we had crossed ' the cloud-plane, and found a transparent atmosphere between it and the glacier.   The dense covering above us was sometimes torn asunder by the wind, which whirled the detached cloud-tufts round the peaks. Contending air-currents were thus revealed, and thunder, which is the common associate, if not the product of such contention, began to rattle among the crags. At first the snow upon the glacier was sufficiently heavy to bridge the crevasses, thus permitting of rapid motion ; but by degrees the fissures opened, and at length drove us to the rocks.   These in their turn became impracticable.   Dropping down a waterfall well known to the climbers of this region, we came again upon the ice, which was here cut by complex chasms.   These we unravelled as long as necessary, and finally escaped from them to the mountain-side.   The first big drops of the thunder-shower were already falling when we reached an overhanging crag which gave us shelter.   We quitted it too soon, beguiled by a treacherous gleam of blue, and were thoroughly drenched before we reached the Æggischhorn.

This was my last excursion with Bennen. In the month of February of the following year he was killed by an avalanche on the Haut de Cry, a mountain near Sion.[1]

Having work to execute, I remained at the Æggisch-horn for nearly a month in 1863. My favourite place for rest and writing was a point on the mountain-side about an hour westwards from the hotel, where the mighty group of the Mischabel, the Matterhorn, and the Weisshorn were in full view. One day I remained in this position longer than usual, held there by the fascination of the sunset. The mountains had stood out nobly clear during the entire day, but towards evening, upon the Dom, a cloud settled, which was finally drawn into a long streamer by the wind. No-thing can be finer than the effect of the red light of sunset on those streamers of cloud. Incessantly dissi-pated, but ever renewed, they glow with the intensity of flames. By-and-bye the banner broke, as a liquid cylinder is known to do when unduly stretched, forming a series of cloud-balls united together by slender fila-ments. I watched the deepening rose, and waited for the deadly pallor which succeeded it, before I thought of returning to the hotel.

On arriving there I found the waitress in tears. She conversed eagerly with the guests regarding the absence of two ladies and a gentleman, who had quitted the hotel in the morning without a guide, and who were now benighted on the mountain. Herr Wellig, the landlord, was also much concerned. 'I recom-

---

[1] A sum of money was collected in England for Bennen's mother and sisters. Mr. Hawkins, Mr. Tuckett, and myself had a small monument erected to his memory in Ernan churchyard. The supervision of the work was entrusted to a clerical friend of Ben-nen's, who, however well-intentioned, made a poor use of his trust. The monument is mean, and its inscription untrue.

mended them,' he said, ' to take a guide, but they
would not heed me, and now they are lost.' ' But they
must be found,' I rejoined ; ' at all events they must
be sought. What force have you at hand ?' Three
active young fellows came immediately forward. Two
of them I sent across the mountain by the usual route
to the Märgelin See, and the third I took with myself
along the watercourse of the Æggischhorn. After
some walking we dipped into a little dell, where the
glucking of cowbells announced the existence of châlets.
The party had been seen passing there in the morning,
but not returning. The embankment of the watercourse fell at some places vertically for twenty or thirty
feet. Here I thought an awkward slip might have
occurred, and, to meet the possibility of having to carry
a wounded man, I took an additional lithe young fellow
from the châlet. We shouted as we went along, but
the echoes were our only response. Our pace was
rapid, and in the dubious light false steps were frequent.
We all at intervals mistook the grey water for the grey
and narrow track beside it, and stepped into the stream.
We proposed ascending to the châlets of Märgelin, but
previous to quitting the watercourse we halted, and
directing our voices down hill, shouted a last shout.
And faintly up the mountain came a sound which could
not be an echo. We all heard it, though it could
hardly be detached from the murmur of the adjacent
stream. We went rapidly down the alp, and after a
little time shouted again. More audible than before,
but still very faint, came the answer from below. We
continued at a headlong pace, and soon assured ourselves that the sound was not only that of a human
voice, but of an English voice.[1] Thus stimulated, we

---

[1] We were, however, nearly thrown off the scent by a lady of
the party cooing in Australian fashion.

swerved to the left, and, regardless of a wetting, dashed
through the torrent which tumbles from the Märgelin
See. Close to the Viesch glacier we found the objects
of our search ; the two ladies, tired out, seated upon
the threshold of a forsaken châlet, and the gentleman
seated on a rock beside them.

He had started with a sprained ankle, and every
visitor knows how bewildering the spurs of the Æggischhorn are, even to those with sound tendons. He
had lost his way, and in his efforts to extricate himself, had experienced one or two serious tumbles.
Finally, giving up the attempt, he had resigned himself to spending the night where we found him. What
the consequences of exposure in such a place would
have been I know not. To reach the Æggischhorn
that night was out of the question ; the ladies were too
exhausted. I tried the châlet door and found it locked,
but my ice-axe soon hewed the bolt away, and forced
an entrance. There was some pine-wood within, and
some old hay which, under the circumstances, formed
a delicious couch for the ladies. In a few minutes a
fire was blazing and crackling in the chimney corner.
Having thus secured them, I returned to the châlets,
sent them bread, butter, cheese, and milk, and had the
exceeding gratification of seeing them return safe and
sound to the hotel next morning.

Soon after this occurrence, I had the pleasure of
climbing the Jungfrau with Dr. Hornby and Mr. Phil-
potts. Christian Almer and Christian Lauener were
our guides. The rose of sunrise had scarcely faded
from the summit when we reached it. I have sketched
the ascent elsewhere, and therefore will not refer to it
further.

On my return from the Æggischhorn in 1863, I

found my friend Huxley low in health and spirits. I
therefore carried him off to the hills of Cumberland.
Swiss scenery was so recent a memory that it was
virtually present, and I had therefore an opportunity
of determining whether it interfered with the enjoyment of English scenery. I did not find this to be the
case. I hardly ever enjoyed a walk more than that
along the ridge of Fairfield, from Ambleside to Grise-
dale Tarn. We climbed Helvellyn, and, thanks to the
hospitality of a party on the top, were enabled to survey the mountain without the intrusion of hunger. We
thought it noble. Striding Edge, Swirling Edge, the
Red Tarn, and Catchedecam, combined with the summit to form a group of great grandeur. The storm
was strong on Striding Edge, which, on account of its
associations,[1] I chose for my descent, while the better
beaten track of Swirling Edge was chosen by my more
conservative companion. At Ulleswater we had the
pleasure of meeting an eminent Church dignitary and
his two charming daughters. They desired to cross
the mountains to Lodore, and we, though ignorant of
the way, volunteered our guidance. The offer was
accepted. We made a new pass on the occasion, which
we called ' the Dean's Pass,' the scenery and incidents
of which were afterwards illustrated by Huxley.
Emerson, who is full of wise saws, speaks of the broad
neutral ground which may be occupied to their common
profit by men of diverse habits of thought; and
on the day to which I now refer there seemed no
limit to the intellectual region over which the dean
and his guides could roam without severance or collision. In the presence of these peaks and meres,

On Striding Edge was killed the traveller whose fate suggested
the fine elegy of Scott, commencing,

' I climbed the dark brow of the mighty Helvellyn.'

as well as over the oatcake of our luncheon, we were sharers of a common joy.

The gorges of the Alps interested me in 1864, as the question of their origin was then under discussion. Having heard much of the Via Mala as an example of a crack produced by an earthquake, I went there, and afterwards examined the gorge of Pfeffers, that of Bergun, the Finsteraarschlucht, and several others of minor note. In all cases I arrived at the same conclusion—namely, that earthquakes had nothing to do with the production of these wonderful chasms, but that they had been one and all sawn through the rocks by running water. From Tusis I crossed the beautiful Schien Pass to Tiefenkasten, and went thence by diligence over the Julier to Pontresina.

The scenery of the Engadin stands both in character and position between that of Switzerland and the Tyrol, combining in a high degree the grandeur of the one and the beauty of the other. Pontresina occupies a fine situation on the Bernina road, at about 6,000 feet above the sea. From the windows of the ' Krone ' you look up the Rosegg valley. The pines are large and luxuriant below, but they dwindle in size as they struggle up the heights, until they are cut off finally either by the inclemency of the air or the scantiness of their proper atmospheric food. From the earth itself these trees derive but an infinitesimal portion of their supplies, as may be seen by the barrenness of the rocks on which they flourish, and which they use almost exclusively as supports to lift their branches into the nutritive atmosphere. The valley ends in the Rosegg glacier, which is fed by the snows of a noble group of mountains.

The baths of St. Moritz are about an hour distant from Pontresina. Here every summer hundreds of

Swiss and Germans, and an increasing number of English, congregate. The water contains carbonic acid (the gas of soda water) and a trace of sulphate of iron (copperas); this the visitors drink, and in elongated tubs containing it they submerge themselves. A curious effect is produced by the collection and escape of innumerable bubbles of carbonic acid from the skin. Every bubble on detaching itself produces a little twitch, and hence a sort of prickly sensation experienced in the water. The patients at St. Moritz put me in mind of that Eastern prince whose physician induced him to kick a football under the impression that it contained a charm. The sagacious doctor knew that faith has a dynamic power unpossessed by knowledge. Through the agency of this power he stirred the prince to action, caused him to take wholesome exercise, and thus cured him of his ailments. At St. Moritz the water is probably the football—the air and exercise on these windy heights being in most cases the real curative agents. The dining-room of the Kurhaus, when Professor Hirst and I were there, was filled with guests: every window was barred, while down the chilled panes streamed the condensed vapour of respiration. The place and company illustrated the power of habit to modify the human constitution; for it was through habit that these Swiss and German people extracted a pleasurable existence out of an atmosphere which threatened with asphyxia the better-ventilated Englishman.

There was a general understanding between Hirst and myself that we should this year meet at Pontresina, and without concert as to the day both of us reached the village within the same quarter of an hour. Some theoretic points of glacier motion requiring elucidation, we took the necessary instruments with us to the Engadin; we also carried with us a quantity of other work, but our first care was to dissipate the wrecked tissues of our bodies, and to supply their place by new material.

Twenty-four years ago Mayer, of Heilbronn, with that power of genius which breathes large meanings into scanty facts, pointed out that the blood was 'the oil of life,' and that muscular effort was, in the main, supported by the combustion of this oil. The recent researches of eminent men prove the soundness of Mayer's induction. The muscles are the machinery by which the power of the food is brought into action. Nevertheless, the whole body, though more slowly than the blood, wastes also. How is the sense of personal identity maintained across this flight of molecules? As far as my experience goes, *matter* is necessary to consciousness, but the matter of any period may be all changed, while consciousness exhibits no solution of continuity. The oxygen that departs seems to whisper its secret to the oxygen that arrives, and thus, while the Nonego shifts and changes, the Ego remains intact. Constancy of *form* in the grouping of the molecules, and not constancy of the molecules themselves, is the correlative of this constancy of perception. Life is a *wave* which in no two consecutive moments of its existence is composed of the same particles.

The ancient lake-beds of the Alps bear directly upon those theories of erosion and convulsion which, in 1864, were subjects of geologic discussion. They are to be found in almost every Alpine valley, each consisting of a level plain formed by sediment, with a barrier below it, which once constituted the dam of the lake. These barriers are now cut through, a river in each case flowing through the gap. *How* cut through? was one of the problems afloat five or six years ago. Some supposed that the chasms were cracks produced by earthquakes ; and if only one or two of them existed, this hypothesis might perhaps postpone that closer examination which infallibly explodes it. But such chasms exist by hundreds in the Alps, and we could not without absurdity invoke in each case the aid of an earthquake to split the dam and drain the waters. Near Pontresina there is a good example of a rocky barrier with a lake-bed behind it, while, within the hearing of the village, a river rushes through a chasm which intersects the barrier. I have often stood upon the bridge which spans this gorge, and have clearly seen the marks of aqueous erosion from its bottom to its top. The rock is not of a character to preserve the finer traces of water action, but the larger scoopings and hollowings are quite manifest. Like all others that I have seen, it is a chasm of erosion.

The same idea may be extended to the Alps themselves. This land was once beneath the sea, and from the moment of its first emergence from the waters until now, it has felt incessantly the tooth of erosion. No doubt the strains and pressures brought into play when the crust was uplifted produced fissures and contortions, which gave direction to ice and water, the real moulders of the Alps. When the eye has been educated on commanding eminences to take in large tracts of the mountains, and when the mind has become capable of resisting the tendency to generalise from exceptional cases, conjecture grows by degrees into conviction that no other known agents than ice and water could have given the Alps their present forms. The plains at their feet, moreover, are covered by the chips resulting from their sculpture. Were they correctly modelled so as to bring their heights and inclinations in just proportions immediately under the eye, this undoubtedly is the conviction that would first force itself upon the mind. An inspection of some of the models in the Jermyn Street Museum will in part illustrate my meaning.

In connection with this question of mountain sculpture, the sand-cones of the glaciers are often instructive. The Aletsch, Unteraar, and Görner glaciers present numerous cases of the kind. On July 20 1864, I came upon a fine group of such cones upon the Morteratsch glacier. They were perfect models of the Alps. I could find among them a reduced copy of almost every mountain with which I am acquainted. One of them showed the peaks of the Mischabel to perfection. How are these miniature mountains produced? Thus: sand is strewn by a stream upon the glacier, and begins immediately to protect the ice underneath it from the action of the sun. The surrounding ice melts away, and the sand is relatively elevated. But the elevation is not mathematically uniform, for the sand is not of the same depth throughout. Some portions rise higher than others. Down the slopes little rills trickle, partially removing the sand and allowing the sun to act to some extent upon the ice. Thus the highest point is kept in possession of the thickest covering, and it rises continually in reference to the circumjacent ice. All round it, however, as it rises, the little rills are at work cutting the ice away and aiding the action of the sun, until finally the elevated hump is wrought into hills and valleys which seem a mimicry of the Alps themselves.

There is a grandeur in the secular integration of small effects here adverted to almost superior to that involved in the idea of a cataclysm. Think of the ages which must have been consumed in the execution of this colossal Alpine sculpture! The question may, of course, be pushed to further limits: Think of the ages, it may be asked, which the molten earth required for its consolidation! But these vaster epochs lack sublimity through our inability to grasp them. They bewilder us, but they fail to make a solemn impression. The genesis of the mountains comes more within the scope of the intellect, and the majesty of the operation is enhanced by our partial ability to conceive it. In the falling of a rock from a mountain-head, in the shoot of an avalanche, in the plunge of a cataract, we often see a more impressive illustration of the power of gravity than in the motions of the stars. When the intellect has to intervene, and calculation is necessary to the building up of the conception, the expansion of the feelings ceases to be proportional to the magnitude of the phenomena.

The Piz Languard is called a ladies' mountain, though it is 11,000 feet high. I climbed it on July 25, and a very grand outlook it affords. The heavens overhead were clear, but in some directions the scowl of the infernal regions seemed to fall upon the hills. The group of the Bernina was in sunshine, and its glory and beauty are not to be described. The depth of impressions upon consciousness is measured by the quantity of *change* which they involve. It is the intermittent current, not the continuous one, that tetanises the nerve, and half the interest of the Alps depends upon the caprices of the air.

The Morteratsch glacier is a very noble one to those who explore it in its higher parts. Its middle portion is troubled and crevassed, but the calm beauty of its upper portions is rendered doubly impressive by the turbulence encountered midway. Into this region, without expecting it, Hirst and myself entered one Sunday in July, and explored it up to the riven and chaotic snows which descend from the Piz Bernina and its companions. The mountains themselves were without a cloud, and, set in the blue heaven, touches of tenderness were mingled with their strength. We spent some hours of perfect enjoyment upon this fine ice-plain, listening to the roar of its moulins and the rush of its streams.

Along the centre of the Morteratsch glacier runs a medial moraine, a narrow strip of *débris* in the upper portions, but overspreading the entire glacier towards its end. How is this widening of the moraine to be accounted for ? Hirst and I set out three different rows of stakes across the glacier ; one of them high up, a second lower down, and a third still nearer to the end of the glacier. In 100 hours the central points of these three lines had moved through the following distances :

> No. 1, highest line, 56 inches.
> „ 2, middle „ 47 „
> „ 3, lowest „ 30 „

Had we taken a line still lower than No. 3, we should have found the velocity still less.

Now these measurements prove that the end, or as it is sometimes called the *snout*, of the glacier moves far less quickly than its upper portions. A block of stone, or a patch of *débris*, for example, on the portion of the glacier crossed by line No. 1, approaches another block or patch at No. 3 with a velocity of 26 inches per 100 hours. Hence such blocks and patches must be more and more crowded together as the end of the glacier is approached, and hence the greater accumulation of stones and *débris* near the end.[1]

---

[1] Above the Märgelin See the centre of the Aletsch glacier moves at the rate of 19 inches a day. A mile or so above the Bel Alp the velocity is 16 inches a day. Opposite the Bel Alp Hotel it

And here we meet point-blank an objection raised by that very distinguished man, Professor Studer, of Berne, to the notion that the glacier exerts an erosive action on its bed. He urges that at *the ends* of the glaciers of Chamouni, of Arolla, Ferpecle, and the Aar, we do not see any tendency exhibited by the glacier to bury itself in the soil. The reason is, that at the point chosen by Professor Studer the glacier is almost stationary. To observe the ploughing or erosive action of the ice we must observe it where the share is in motion, and not where it is comparatively at rest. Indeed, the snout of the glacier often rests upon the rubbish which its higher portions have dug away.

While I was staying at Pontresina, Mr. Hutchinson of Rugby, Mr. Lee Warner, and myself joined in a memorable expedition up the Piz Morteratsch. This is a very noble mountain, and nobody had previously thought of associating the idea of danger with its ascent. The resolute Jenni, by far the boldest man in Pontresina, was my guide; while Walter, the official *guide chef*, was taken by my companions. With a dubious sky overhead, we started on the morning of July 30, a little after 4 A.M. There is rarely much talk at the beginning of a mountain excursion : you are either sleepy or solemn so early in the day. Silently we passed through the pine-woods of the beautiful Rosegg valley ; watching anxiously at intervals the play of the clouds around the adjacent heights. At one place a spring gushed from the valley bottom as clear and almost as copious as that which pours out the full-formed river Albula. The traces of ancient glaciers were present everywhere, the

is about 8 inches a day ; while the measured velocity near its end is only 2 inches a day. As in the case of the Morteratsch, the moraine quite covers the lower portion of the glacier.

valley being thickly covered with the *débris* which the ice had left behind. An old moraine, so large that in England it might take rank as a mountain, forms a barrier across the upper valley. Once probably it was the dam of a lake, but it is now cut through by the river which rushes from the Rosegg glacier. These works of the ancient ice are to the mind what a distant horizon is to the eye. They give to the imagination both pleasure and repose.

The morning, as I have said, looked threatening, but the wind was good ; by degrees the cloud scowl relaxed, and broader patches of blue became visible. We called at the Rosegg châlets, and had some milk, afterwards winding round a shoulder of the hill, at times upon the moraine of the glacier, at times upon the adjacent grass slope ; then over shingly inclines, covered with the shot rubbish of the heights. Two ways were now open to us, the one easy but circuitous, the other stiff but short. Walter was for the former, and Jenni for the latter, their respective choices being characteristic of the two men. To my satisfaction Jenni prevailed, and we scaled the steep and slippery rocks. At the top of them we found ourselves upon the rim of an extended snow-field. Our rope was here exhibited, and we were bound by it to a common destiny. In those higher regions the snow-fields show a beauty and a purity of which those who linger below have no notion. We crossed crevasses and bergschrunds, mounted vast snow-bosses, and doubled round walls of ice with long stalactites pendent from their cornices. One by one the eminences were surmounted, the crowning rock being attained at half-past twelve. On it we uncorked a bottle of champagne. Mixed with the pure snow of the mountain, it formed a beverage, and was enjoyed with a gusto, which the sybarite of the city could neither imitate nor share.

We spent about an hour upon the warm gneiss-blocks on the top. Veils of cloud screened us at intervals from the sun, and then we felt the keenness of the air; but in general we were cheered and comforted by the solar light and warmth. The shiftings of the atmosphere were wonderful. The white peaks were draped with opalescent clouds which never lingered for two consecutive minutes in the same position. Clouds differ widely from each other in point of beauty, but I had hardly ever seen them more beautiful than they appeared to-day, while the succession of surprises experienced through their changes were such as rarely fall to the lot even of a practised mountaineer.

These clouds are for the most part produced by the chilling of the air through its own expansion. When thus chilled, the aqueous vapour diffused through it, which is previously unseen, is precipitated in visible particles. Every particle of the cloud has consumed in its formation a little polyhedron of vapour, and a moment's reflection will make it clear that the size of the cloud-particles must depend, not only on the size of the vapour polyhedron, but on the relation of the density of the vapour to that of its liquid. If the vapour were light and the liquid heavy, other things being equal, the cloud-particle would be *smaller* than if the vapour were heavy and the liquid light. There would evidently be more *shrinkage* in the one case than in the other. Now there are various liquids whose weight is not greater than that of water, while the weight of their vapour, bulk for bulk, is five or six times that of aqueous vapour. When those heavy vapours are precipitated as clouds, which is easily done artificially, their particles are found to be far coarser

than those of an aqueous cloud. Indeed, water is
without a parallel in this particular. Its vapour is the
lightest of all vapours, and to this fact the soft and
tender beauty of the clouds of our atmosphere is
mainly due.[1]

After an hour's halt, our rope, of which we had
temporarily rid ourselves, was reproduced, and the descent began. Jenni is the most daring man and
powerful character among the guides of Pontresina.
The manner in which he bears down all the others in
conversation, and imposes his own will upon them,
shows that he is the dictator of the place. He is a
large and rather an ugly man, and his progress uphill, though resistless, is slow. He had repeatedly
expressed a wish to make an excursion with me, and I
think he desired to show us what he could do upon
the mountains. To-day he accomplished two daring
things—the one successfully, while the other was within
a hair's-breadth of a very shocking end.

In descending we went straight down upon a berg-
schrund, which compelled us to make a circuit in
coming up. This particular kind of fissure is formed
by the lower portion of a snow-slope falling away from
the higher, a crevasse being thus formed between the
two, which often surrounds the mountain as a fosse of
great depth. Walter was here the first of our party, and
Jenni was the last. It was quite evident that Walter
hesitated to cross the chasm ; but Jenni came forward,
and half by expostulation, half by command, caused
him to sit down on the snow at some height above the
fissure. I think, moreover, he helped him with a

---

[1] Since this was written Mr. Sinclair has greatly augmented our
knowledge of cloud-formation. By a series of striking experiments
he has shown the part played by solid nuclei in the act of precipitation.

shove. At all events the slope was so steep that the guide shot down it with an impetus sufficient to carry him clear over the schrund. We all afterwards shot the chasm in this pleasant way. Jenni was behind. Deviating from our track, he deliberately chose the widest part of the chasm, and shot over it, lumbering like behemoth down the snow-slope at the other side. It was an illustration of that practical knowledge which long residence among the mountains can alone impart, and in the possession of which our best English climbers fall far behind their guides.

The remaining steep slopes were also descended by glissade, and we afterwards marched cheerily over the gentler inclines. We had ascended by the Rosegg glacier, and now we wished to descend upon the Morteratsch glacier and make it our highway home. It was while attempting this descent that we were committed to that ride upon the back of an avalanche, a description of which is given in the ' Times ' newspaper for October 1, 1864.[1]

In July 1865 my friend Hirst and myself visited Glarus, intending, if circumstances favoured us, to climb the Tödi. Checked by the extravagant demands of the guides, we gave the expedition up. Crossing the Klausen pass to Altdorf, we ascended the Gotthardt Strasse to Wasen, and went thence over the Susten pass to Gadmen, which we reached late at night. We halted for a moment at Stein, but the blossom of 1863 was no longer there and we did not tarry. Before quitting Gadmen next morning I was accosted by a guide, who asked me whether I knew Professor Tyndall. ' He is killed, sir,' said the man ; ' killed upon the Matterhorn.' I then listened to a somewhat detailed

[1] See also *Alpine Journal*, vol. i. p. 437.

account of my own destruction, and soon gathered that
though the details were erroneous something serious
had occurred. At Imhof the rumour became more
consistent, and immediately afterwards the Matterhorn catastrophe was in every mouth and in all the
newspapers. My friend and myself wandered on to
Mürren, whence, after an ineffectual attempt to cross the
Petersgrat, we went by Kandersteg and the Gemmi to
Zermatt.

Of the four sufferers killed on the Matterhorn, one
remained behind. But expressed in terms either of
mental toiture or physical pain, the suffering in my
opinion was *nil*. Excitement during the first moments
left no room for terror, and immediate unconsciousness
prevented pain. No death has probably less of agony
in it than that caused by a fall upon a mountain.
Expected it would be terrible, but unexpected, not. I
had heard, however, of other griefs and sufferings consequent on the accident, and this prompted a desire
on my part to find the remaining one and bring him
down. I had seen the road-makers at work between
St. Nicholas and Zermatt, and was struck by the
rapidity with which they pierced the rocks for blasting.
One of these fellows could drive a hole a foot deep
into hard granite in less than an hour. I was therefore determined to secure in aid of my project the
services of a road-maker. None of the Zermatt guides
would second me, but I found one of the Lochmatters
of St. Nicholas willing to do so. Him I sent to Geneva
to buy 3,000 feet of rope, which duly came on heavily-
laden mules to Zermatt. Hammers and steel punches
were prepared ; a tent was put in order, and the apparatus was carried up to the chapel by the Schwartz-See.
But the weather would by no means smile upon the
undertaking. I waited in Zermatt for twenty days,

making, it is true, pleasant excursions with pleasant
friends, but these merely spanned the brief intervals
which separated one rain-gush or thunder-storm from
another. Bound by an engagement to my friend Professor De la Rive, of Geneva, where the Swiss *savants*
had their annual assembly in 1865, I was forced to
leave Zermatt. My notion had been to climb to the
point where the men slipped, and to fix there suitable
irons in the rocks. By means of ropes attached to these
I proposed to scour the mountain along the line of the
glissade. There were peculiarities of detail which need
not now be dwelt upon, inasmuch as the weather rendered them all futile.

In the summer of 1866 I first went to Engsteln,
one of the most charming spots in the Alps. It had
at that time a double charm, for the handsome young
widow who kept the inn supplemented by her kindness and attention within doors the pleasures of the
outer world. A man named Maurer, of Meyringen,
was my guide for a time. We climbed the Titlis,
going straight up it from the Joch pass, in the track
of a scampering chamois which showed us the way.
The Titlis is a very noble mass—one of the few which,
while moderate in height, bear a lordly weight of snow.
The view from the summit is exceedingly fine, and on
it I repeated with a hand spectroscope the observations
of M. Janssen on the absorption-bands of aqueous
vapour. On the day after this ascent I quitted Engsteln, being drawn towards the Wellhorn and Wetterhorn, both of which as seen from Engsteln came out
nobly. The upper dome of heaven was of the deepest
blue, while only the faintest lightening of the colour
towards the horizon indicated the augmented turbidity
of the atmosphere in that direction. The sun was

very hot, but there was a clear rivulet at hand, deepening here and there into pebbled pools, into which I plunged at intervals, causing my guide surprise, if not anxiety. For he shared the common superstition that plunging, when hot, into cold water is dangerous. The danger, and a very serious one it is, is to plunge into cold water when *cold*. The strongest alone can then bear immersion without damage.

This year I subjected the famous Finsteraarschlucht to a close examination. The earthquake theory already adverted to was prevalent regarding it, and I wished to see whether any evidences existed of aqueous erosion. It will be remembered that the Schlucht or gorge is cut through a great barrier of limestone rock called the Kirchet, which throws itself across the valley of Hasli, about three-quarters of an hour's walk above Meyringen. The plain beyond the barrier, on which stands the hamlet of Imhof, is formed by the sediment of an ancient lake of which the Kirchet constituted the dam. This dam is now cut through for the passage of the Aar, forming one of the noblest gorges in Switzerland. Near the summit of the Kirchet is a house with a signboard inviting the traveller to visit the *Aarenschlucht*, a narrow lateral gorge which runs down to the very bottom of the principal one. The aspect of this smaller chasm from its bottom to its top proves to demonstration that water had in former ages worked there as a navigator. But it was regarding the sides of the great chasm that I needed instruction, and from its edge I could see nothing to satisfy me. I therefore stripped and waded until a point was reached in the centre of the river which commanded an excellent view of both sides of the gorge. Below me, on the left-hand side, was a jutting cliff, which caused the Aar to swerve from its direct course, and had to bear the thrust of the river.

From top to bottom this cliff was polished, rounded, and scooped. There was no room for doubt. The river which now runs so deeply down had once been above. It has been the delver of its own channel through the barrier of the Kirchet.

I went on to Rosenlaui, proposing to climb the neighbouring mountains in succession. In fact, I went to Switzerland in 1866 with a particular hunger for the heights. But the weather thickened before Rosenlaui was reached, and on the night following the morning of my departure from Engsteln I lay upon my plaid under an impervious pine, and watched as wild a thunderstorm and as heavy a downpour of rain as I had ever seen. Most extraordinary was the flicker on cliffs and trees, and most tremendous was the detonation succeeding each discharge. The fine weather came thus to an end, and next day I gave up the Wetterhorn for the ignoble Faulhorn. Here the wind changed, the air became piercingly cold, and on the following morning heavy snow-drifts buttressed the doors, windows, and walls of the inn. We broke away, sinking at some places to the hips in snow. A thousand feet made all the difference; a descent of this amount carrying us from the bleakest winter into genial summer. My companion held on to the beaten track, while I sought a rougher and more direct one to the Scheinigeplatte. We were solitary visitors there, and I filled the evening with the 'Story of Elizabeth,' which some benevolent traveller had left at the hotel.

Thence we dropped down to Lauterbrunnen, went up the valley to the little inn at Trechslawinen, and crossed the Petersgrat the following day. The recent precipitation had cleared the heavens and reloaded the heights. It was perhaps the splendour of the weather and purity of the snows, aided by the subjective effect due to contrast with a series of most dismal days, that made me think the Petersgrat so noble a standpoint for a view of the mountains. The horizontal extent was vast, and the grouping magnificent. The undoubted monarch of this unparagoned scene was the Weisshorn. At Platten we found shelter in the house of the curé. Next day we crossed the Lötschsattel, and swept round by the Aletsch glacier to the Æggischhorn.

Here I had the pleasure of meeting a very ardent climber, who entertains peculiar notions regarding guides. He deems them, with good reason, very expensive, and he also feels pleasure in trying his own powers. I would admonish him that he may go too far in this direction, and probably his own experience has by this time forestalled the admonition. Still, there is much in his feeling which challenges sympathy; for if skill, courage, and strength are things to be cultivated in the Alps, they are, within certain limits, best exercised and developed in the absence of guides. And if the real climbers are ever to be differentiated from the crowd, it is only to be done by dispensing with professional assistance. But no man without natural aptitude and due training would be justified in undertaking anything of this kind, and it is an error to suppose that the necessary knowledge can be obtained in one or two summers in the Alps. Climbing is an art, and those who wish to cultivate it on their own account ought to give themselves sufficient previous practice in the company of first-rate guides. This would not shut out expeditions of minor danger now and then without guides. But whatever be the amount of preparation, real climbers must still remain select men. Here, as in every other sphere of human action whether intellectual or physical, as indeed among the guides themselves, real eminence falls only to the lot of few.

From the Bel Alp, in company with Mr. Girdlestone, I made an attack upon the Aletschhorn. We
failed. The weather as we started was undecided, but
we hoped the turn might be in our favour. We first
kept along the Alp, with the Jaggi glacier to our right,
then crossed its moraine, and made the trunk glacier
our highway until we reached the point of confluence
of its branches. Here we turned to the right, the
Aletschhorn from base to summit coming into view.
We reached the true base of the mountain, and without
halting breasted its snow. But as we climbed the
atmosphere thickened more and more. About the Nesthorn the horizon deepened to pitchy darkness, and on
the Aletschhorn itself hung a cloud which we at first
hoped would melt before the strengthening sun, but
which instead of melting became denser. Now and
then an echoing rumble of the wind warned us that we
might expect rough handling above. We persisted,
however, and reached a considerable height, unwilling
to admit that the weather was against us; until a more
savage roar and a ruder shake than ordinary caused us
to halt and look more earnestly and anxiously into the
darkening atmosphere. Snow began to fall, and we felt
that we must yield. The wind did not increase, but the
snow thickened and fell in heavy flakes. Holding on in
the dimness to the medial moraine, we managed to get
down the glacier, and cleared it at a practicable point ;
whence, guided by the cliffs which flanked our right,
and which became visible only when we came almost
into contact with them, we hit the proper track to the
hotel.

Though my visits to the Alps already numbered
thirteen, I had never gone as far southward as the
Italian lakes. The perfectly unmanageable weather of

July 1866 caused me to cross with Mr. Girdlestone into Italy, in the hope that a respite of ten or twelve days might improve the temper of the mountains. We walked across the Simplon to the village of the same name, and took thence the diligence to Domo d'Ossola and Baveno. The atmospheric change was wonderful; and still the clear air which we enjoyed below was the self-same air that heaped clouds and snow upon the mountains. It came across the heated plains of Lombardy charged with moisture, but the moisture was reduced by the heat to the transparent condition of true vapour, and hence invisible. Tilted by the mountains the air rose, and as it expanded it became chilled, and as it became chilled it discharged its vapour as visible cloud, the globules of which were swelled by coalescence into rain-drops on the mountain flanks, or were frozen to ice-particles on their summits, the particles collecting afterwards to form flakes of snow.

At Baveno we halted on the margin of the Lago Maggiore. I could hear the lisping of the waters on the shingle far into the night. My window looked eastward, and through it could be seen the first warming of the sky at the approach of dawn. I rose and watched the growth of colour all along the east. The mountains, from mere masses of darkness projected against the heavens, became deeply empurpled. It was not as a mere wash of colour overspreading their surfaces. They blent with the atmosphere as if their substance was a condensation of the general purple of the air. Nobody was stirring at the time-and the very lap of the lake upon its shore only increased the sense of silence.

> The holy time was quiet as a nun
> Breathless with adoration.

In my subsequent experience of the Italian lakes I met

with nothing which affected me so deeply as this morning scene on the Lago Maggiore.

From Baveno we crossed the lake to Luino and went thence to Lugano. At Belaggio, which stands at the junction of the two branches of the Lake of Como, we halted a couple of days. Como itself we reached in a small sailing-boat—the sail being supplemented by oars. There we saw the statue of Volta— a prophet justly honoured in his own country. From Como we went to Milan. The object of greatest interest there is, of course, the cathedral. A climber could not forego the pleasure of getting up among the statues which crowd its roof, and of looking thence towards Monte Rosa. The distribution of the statues magnified the apparent vastness of the pile ; still, the impression made on me by this great edifice was one of disappointment. Its front seemed to illustrate an attempt to cover meagreness of conception by profusion of adornment. The interior, however, notwithstanding the cheat of the ceiling, is exceedingly grand.

From Milan we went to Orta, where we had a plunge into the lake. We crossed it subsequently and walked on to Varallo : thence by Fobello over a country of noble beauty to Ponte Grande in the Val Ansasca. Thence again by Macugnaga over the deep snow of the Monte Moro, reaching Mattmark in drenching rain. The temper of the northern slopes did not appear to have improved during our absence. We returned to the Bel Alp, fitful triumphs of the sun causing us to hope that we might still have fairplay upon the Aletschborn. But the day after our arrival snow fell so heavily as to cover the pastures for 2,000 feet below the hotel, introducing a partial famine among the herds. They had eventually to be driven below the snow-line. Avalanches were not unfrequent on slopes which a day or two previously had been covered with grass and flowers. In this condition of things Mr. Milman, Mr. Girdlestone, and myself climbed the Sparrenhorn, and found its heavy-laden Kamm almost as hard as that of Monte Rosa. Occupation out of doors was, however, insufficient to fill the mind, so I wound my plaid around my loins and in my cold bedroom studied ' Mozley upon Miracles.'

PART II.

THE pause in the middle of this article, which was written without reference to its division, has caused me to supplement these memories by looking into the notes of my first Swiss journey. In September 1849 my friend Hirst, so often mentioned in these brief chronicles, had joined me at Marburg, in Hesse Cassel, where I was then a student, and we had joyful anticipations of a journey in Switzerland together. But the death of a near relative compelled him to return to England, and the thought of the Alps was therefore given up. As a substitute, I proposed to myself a short foot-journey through the valley of the Lahn, and a visit to Heidelberg. On the 19th of September I walked from Marburg to Giessen, and thence to Wetzler, the scene of ' Werther's Leiden.' From Wetzler, I passed on to Limburg, through Diez, where the beauties of the valley began, to Nassau, reaching it after a sunset and through a scene which might have been condensed intellectually into Goethe's incomparable lines :—

> Ueber allen Gipfeln
> Ist Ruh,'
> In allen Wipfeln
> Spürest du
> Kaum einen Hauch ;
> Die Vögelein schweigen im Walde.
> Warte nur, balde
> Ruhest du auch.

The ' balde ruhest du auch ' had but a sentimental value for me at the time. The field of hope and action, which in all likelihood lay between me and it, deprived the idea of the definition which it sometimes possesses now.

From Nassau, I passed through Ems to Niederlahn-stein, where the little Lahn which trickles from the earth in the neighbourhood of Siegen (visited in 1850 by Hirst and myself) falls into the broader Rhine. Thence along the river, and between the rocks of the Lurlei, to Mayence; afterwards to Frankfort and Heidelberg. I reached my proposed terminus on the night of the 22nd, and early next morning was among the castle ruins. The azure overhead was perfect, and among the twinkling shadows of the surrounding woods, the thought of Switzerland revived. ' How must the mountains appear under such a sky ? ' That night I slept at Basel. In those days it was a pleasure to me to saunter along the roads, enjoying such snatches of scenery as were thus attainable. I knew not then the distant mountains; the attraction which they afterwards exercised upon me had not yet begun to act. I moreover did not like the diligence, and therefore walked all the way from Basel to Zürich. I passed along the lake to Horgen, thence over the hills to Zug, and afterwards along the beautiful fringe of the Zugersee to Arth. Here, on September 26, I bought my first alpenstock, and faced with it the renowned Rigi. The sunset on the summit was fine, but I retain no particular impression of the Rigi's grandeur ; and now, rightly or wrongly, I think of it as a cloudy eminence, famous principally for its guzzling and its noise.

I descended the mountain through a dreamy opalescent atmosphere, but the dreaminess vanished at Weggis as soon as the steamer from Lucerne arrived. I took the boat to Fluellen. My journal expresses

wonder at the geological contortions along the flanks of
the adjacent mountain, and truly famous examples they
happen to be. I followed the Gotthardt's-strasse over
the Devil's Bridge, the echoes of which astonished me,
to Andermatt and Hospenthal, where the road was
quitted to cross the Furka. Taking by mistake the
wrong side of the river Reuss, I was earnestly admonished by a pretty, dirty, little châlet girl that I
had gone astray. At this time there was no shelter on
the Furka, and being warned at Realp of the danger
of crossing the pass late in the evening, I halted at
that hamlet for the night. Here pastoral Switzerland
first revealed itself to me, in the songs of the *Senner*,
and the mellow music of the cow-bells at milking-time.

On the 29th I first saw the glacier of the Rhône.
Snow had fallen during the night ; the weathered ice-
peaks of the fall were of dazzling whiteness, while a
pure cerulean light issued from the clefts and hollows
of the ice. A week previously a young traveller had
been killed by falling into one of these chasms. I did
not venture upon the ice, but went down to the source
of the historic river. From this point the Mayenwand
ought to have been assailed, but the track over it was
marked so faintly on my small map that it escaped my
attention, and I therefore went down the Rhône valley.
The error was discovered before Oberwald was reached.
Not wishing to retrace my steps over so rough a track, I
inquired at Oberwald whether it would not be possible to
reach the Grimsel without returning to the Rhône glacier. A peasant pointed to a high hill-top, and informed
me that if I could reach it an erect pole would be found
there, and after it other poles, which marked the way
over the otherwise trackless heights to the Hospice. I
tucked up my knapsack, and faced the mountain. My
remarks on this scramble would make a climber smile

possibly with contempt for the man who could refer to
such a thing as difficult. The language of my journal
regarding it, however, is, ' By the Lord, I should not
like to repeat this ascent!' I found the signal poles
and reached the Grimsel. Old Zybach and his fine
daughters were still there. He had not yet, by setting
fire to the house, which belonged to the commune, condemned himself to the life of a felon.

That night I slept at Gutannen, and next day halted
on the Great Scheideck. Heavy rain fell as I ascended,
but the thick pines provided shelter. Vapours leaped from
the clefts of the mountains, and thunder rattled upon the
heights. At every crash I looked instinctively upwards,
expecting to see the rocks sent down in splinters. On
the following day I crossed the Wengern Alp, saw the
avalanches of the Jungfrau, and heard the warble of
her echoes. Then swiftly down to Lauterbrunnen, and
through the valley of Interlaken, with hardly a hope
of being able to reach Neuhaus in time to catch the
steamer. I had been told over and over again that it
was hopeless, but I thought it a duty to *try*. The
paddles were turning, and a considerable distance
already separated the steamer and the quay when I
arrived. This distance was cleared at a bound, under
a protest on the part of the captain and the bystanders,
and that night I bivouacked at Thun.

On the following day I drove to Berne, and walked
thence through Solothurm to Basel. The distant aspect
of the Alps appeared to be far more glorious than the
nearer view. From a distance the *Vormauer*, or spurs,
and the highest crests were projected against a common background, the apparent height of the mountains
being thereby enormously augmented. The aqueous
air had also something to do with their wonderful
illumination. The railway station being then at Effrin-

gen, a distance of some miles from Basel, I set out to walk there, but on crossing the frontier was intercepted by two soldiers. I had a passport, but it had not been *viséd*, and back to Berne it was stated I must go. The fight at Rastatt had occurred a short time previously, and the Prussians, then the general insurgent-crushers of Germany, held possession of the Grand Duchy of Baden. I was detained for some hours, being taken from one official to another, neither logic nor entreaty appearing to be of any avail. The Inspector at Leopoldshöbe was at first polite but inexorable, then irate; happily, to justify his strictness, he desired me to listen while he read his instructions. They were certainly very emphatic, but they were directed against ' Deutsche Flüchtlinge.' I immediately drew his attention to the words, and flatly denied his right to detain me. I appealed to my books, my accent, and my shirt collars, none of which at the time had become German. A new light seemed to dawn upon the inspector ; he admitted my plea, and let me go. Thus ended my first Swiss expedition, and until 1856 I did not make a second. The reminiscences of humanity which these old records revive interest me more than those of physical grandeur. The little boys and girls and the bright-eyed maidens whom I chanced to meet, and who at times ministered to my wants, have stamped themselves more vividly and pleasantly on my memory than the Alps themselves.

Grindelwald was my first halting-place in the summer of 1867 ; I reached it, in company with a friend, on Sunday evening, July 7. The air of the glaciers and the fare of the Adler Hotel rendered me rapidly fit for mountain work. The first day we made an excursion along the lower glacier to the Kastenstein, crossing, in returning, the Strahleck branch of the glacier above the ice-fall, and coming down by the Zasenberg. The second day was spent upon the upper glacier. The sunset covered the crest of the Eiger with indescribable glory that evening, causing the dinner-table to be forsaken while it lasted. It gave definition to a vague desire which I had previously entertained; and I arranged forthwith with Christian Michel, a famous old roadster, to attempt the Eiger, engaging Peter Bauman, a strong and gallant climber, to act as second guide.

At half-past one o'clock on the morning of the 11th we started from the Wengern Alp. No trace of cloud was visible in the heavens, which were sown broadcast with stars. Those low down twinkled with extraordinary vivacity, many of them flashing in quick succession lights of different colours. When an opera-glass was pointed to one of these flashing stars, and shaken, the line of light described by the image of the star resolved itself into a string of richly-coloured beads: rubies and emeralds were hung thus together on the same curve. The dark intervals between the beads corresponded to the moments of extinction of the star through the 'interference' of its own rays in our atmosphere. Over the summit of the Wetterhorn the Pleiades hung like a diadem, while at intervals a solitary meteor shot across the sky.

We passed along the Alp, and then over the balled snow and broken ice, shot down from the end of a glacier which fronted us. Here the ascent began; we passed by turns from snow to rock and from rock to snow. The steepness for a time was moderate, the only thing requiring caution being the thin crusts of ice upon the rocks over which water had trickled the previous day. The east gradually brightened, the stars became paler and disappeared, and at length the crown of the adjacent Jungfrau rose out of the twilight into the purple of the rising sun. The bloom crept gradually downwards over the snows, until the whole mountain world partook of the colour. It is not in the night nor in the day—it is not in any statical condition of the atmosphere—that the mountains look most sublime. It is during the few minutes of transition from twilight to full day through the splendours of the dawn.

Seven hours' climbing brought us to the higher slopes, which were for the most part ice, and required deep step-cutting. The whole duty of the climber on such slopes is to cut his steps deeply, and to stand in them securely. At one period of my mountain life I looked lightly on the possibility of a slip, having full faith in the resources of him who accompanied me, and very little doubt of my own. Experience has qualified this faith in the power even of the best of climbers upon a steep ice-slope. A slip under such circumstances must not occur. The Jungfrau began her cannonade of avalanches very early : five of them thundered down her precipices before eight o'clock in the morning. Bauman, being the youngest man, undertook the labour of step-cutting, which the hardness of the ice rendered severe. He was glad from time to time to escape to the snow-cornice which, unsupported save by its own tenacity, overhung the Grindelwald side of the mountain, checking himself at intervals by looking over the edge of the cornice, to assure himself of its sufficient thickness to bear our weight. A wilder precipice is hardly to be seen than this wall of the Eiger. Viewed from the cornice at its top it seems to drop sheer for eight thousand feet down to Grindelwald. When the cornice became unsafe, Bauman retreated, and step-cutting recommenced. We reached the summit before nine o'clock, and had from it an outlook over as glorious a scene as this world perhaps affords.

On the following day, accompanied by Michel, I
went down to Lauterbrunnen, and afterwards crossed
the Petersgrat a second time to Platten, where the door
of the curé being closed against travellers, we were
forced into dirty quarters in an adjacent house. From
Platten, instead of going as before over the Lötschsattel,
we struck obliquely across the ridge above the Nesthorn,
and down upon the Jaggi glacier, making thus an
exceedingly fine excursion from Platten to the Bel Alp.
Thence, after a brief halt, I pushed on to Zermatt.

I have already mentioned Carrel, *dit le bersaglier*,
who accompanied Bennen and myself in our attempt
upon the Matterhorn in 1862, and who in 1865 reached
the summit of the mountain. With him I had been
in correspondence for some time, and from his letters
an enthusiastic desire to be my guide up the Matterhorn might be inferred. From the Riffelberg I crossed
the Theodule to Breuil, where I saw Carrel. He had
naturally and deservedly grown in his own estimation.
In the language of philosophy, his environment had
changed and he had assumed new conditions of
equilibrium, but they were decidedly unfavourable to
the climbing of the Matterhorn. His first condition
was that I should take three guides at 150 francs
apiece, and these were to be aided by porters as far as
the cabin upon the Matterhorn. He also objected to
the excellent company of Christian Michel. In fact,
circumstances had produced their effect upon my friend
Carrel, and he was no longer a reasonable man. To do
him justice, I believe he afterwards repented, and sent
his friends Bich and Meynet to speak to me while he
kept aloof. A considerable abatement was soon made
in their demands; and without arranging anything definitely, I quitted Breuil on the understanding that I

should return if the weather, which was then unfit for
the Matterhorn, improved.

I waited at the Riffel for twelve days, making small
excursions here and there. But though the weather
was not so abominable as it had been last year, the
frequent snow-discharges on the Matterhorn kept it unassailable. In company with Mr. Craufurd Grove, who
had engaged Carrel as his guide, Michel being mine,
I made the pass of the Trift from Zermatt to Zinal.
Carrel led and, on the rocks, acquitted himself admirably. He is a first-rate rockman. I could understand
and share the enthusiasm experienced by Mr. Hinchliff
in crossing this truly noble pass. It is certainly one
of the finest in the whole Alps. For that one day
moreover the weather was magnificent. Next day we
crossed to Evolena, going far astray, and thus converting a light day into a heavy one. From Evolena we
purposed crossing the Col d'Erin back to Zermatt, but
the weather would not let us. This excursion had
been made with the view of allowing the Matterhorn
a little time to arrange its temper; but the temper
continued sulky, and at length wearied me out. We
went round by the valley of the Rhône to Zermatt,
and finding matters there worse than ever, both Mr.
Grove and myself returned to Visp, intending to quit
Switzerland together. Here he changed his mind and
returned to Zermatt; on the same day the weather
changed also, and continued fine for a fortnight. He
succeeded in getting with Carrel to the top of the
Matterhorn, being therefore the first Englishman that
gained the summit from the southern side. A ramble
in the Highlands, including a visit to the Parallel
Roads of Glenroy, concluded my vacation in 1867.

PASSAGE OF THE MATTERHORN.

Call not waste that barren cone
Above the floral zone;
Where forests starve
It is pure use.
What sheaves like those which here we glean and bind
Of a celestial Ceres and the Muse? [1]

The ' oil of life ' burnt very low with me in June
1868. Driven from London by Dr. Bence Jones, I
reached the Giessbach Hotel on the Lake of Brientz
early in July. No pleasanter position could be found for
an invalid. My friend Hirst was with me, and we made
various little excursions in the neighbourhood. The
most pleasant of these was to the Hinterburger See, a
small and lonely lake high up among the hills, fringed
on one side by pines, and overshadowed on the other by
the massive limestone buttresses of the Hinterburg. It
is an exceedingly lovely spot, but rarely visited. The
Giessbach Hotel is an admirably organised establishment. The table is served by well-brought-up Swiss
girls in Swiss costume, fresh, handsome, and modest,
who come there not as servants, but to learn the mysteries of housekeeping. And among her maidens moved
like a little queen the graceful daughter of the host;
noiseless, but effectual in her rule and governance. [2]   I
went to the Giessbach with a prejudice against its artificial illumination. The crowd of spectators may suggest
the theatre, but the lighting up of the water is fine.
The colourless light pleased me best ; it merely intensified the contrast revealed by ordinary daylight between the white foam of the cascades and the black
surrounding pines.

From the Giessbach we went to Thun, and thence

---

[1] Emerson's poems.   [2] All this is now changed.

up the Simmenthal to Lenk. Over a sulphur spring a large hotel has been recently erected, and here we found a number of Swiss and Germans, who thought the waters did them good. In one large room the liquid gushes from a tap into a basin, diffusing through the place the odour of rotten eggs. The patients like this smell ; indeed, they regard its foulness as a measure of their benefit. The director of the establishment was intelligent and obliging, sparing no pains to meet the wishes and promote the comfort of his guests. We wandered, while at Lenk, to the summit of the Rawyl pass, visited the Siebenbrünnen, where the river Simmen bursts full-grown from the rocks, and we should have clambered up the Wildstrubel had the weather been tolerable. From Lenk we went to Gsteig, a finely-situated hamlet, but not celebrated for the peace and comfort of its inn ; and from Gsteig to the Diablerets hotel. While there I clambered up the Diablerets mountain, and was amazed at the extent of the snow-field upon its tabular top. The peaks, if they ever existed, have been shorn away, and miles of flat névé, unseen from below, overspread their section.

From the Diablerets we drove down to Aigle. The Traubenkur had not commenced, and there was therefore ample space for us at the excellent hotel. We were compelled to spend a night at Martigny. I heard the trumpet of its famous musquito, but did not feel its attacks ; still, the itchy hillocks on my hands for some days afterwards reported the venom of the insect. The following night was more pleasantly spent on the cool col of the Great St. Bernard. On Tuesday, July 21, we reached Aosta, and, in accordance with previous telegraphic arrangement, met there the Chanoine Carrel. Jean Jaques Carrel, the old companion of Mr. Hawkins and myself, and others at Breuil, had been

greatly dissatisfied with the behaviour of the *bersaglier*
last year, and this feeling the Chanoine shared. He
wrote to me during the winter, stating that two new
men had scaled the Matterhorn, and that they were
ready to accompany me anywhere. He now drove, with
Hirst and myself, to Chatillon, where at the noisy and
comfortless inn we spent the night. Here Hirst quitted
me, and I turned with the Chanoine up the valley to
Breuil.

At Val Tournanche I saw a maiden niece of the
Chanoine who had gone high up the Matterhorn, and
who, had the wind not assailed her petticoats too
roughly, might, it was said, have reached the top. I
can believe it. Her wrist as I shook her hand seemed
like a weaver's beam, and her frame a mass of potential
energy. The guides recommended to me by the Chanoine were the brothers Joseph and Pierre Maquignaz
of Val Tournanche, his praise of Joseph as a man of
unshaken courage and proved capacity as a climber
being particularly strong. Previous to reaching Breuil
I saw this Joseph, who seemed to divine by instinct my
name and aim.

Carrel was there, looking very gloomy, while Bich
petitioned for a porter's post; but I left the arrangement of these matters wholly in the hands of Maquignaz. He joined me in the evening, and on the
following day we ascended one of the neighbouring
summits, discussing as we went our chances on the
Matterhorn. In 1867 the chief precipitation took
place in a low atmospheric layer, the base of the
mountain being heavily laden with snow, while the
summit and the higher rocks were bare. In 1868 the
distribution was inverted, the top being heavily laden
and the lower rocks clear. An additional element of
uncertainty was thus introduced. Maquignaz could

not say what obstacles the snow might oppose to us above, but he was resolute and hopeful. My desire had long been to complete the Matterhorn by making a pass over its summit from Breuil to Zermatt. In this attempt my guide expressed his willingness to aid me, his interest in the project being apparently equal to my own.

He however only knew the Zermatt side of the mountain through inspection from below; and he acknowledged that a dread of it had taken possession of him during the previous year. That feeling however had disappeared, and he reasoned that as Mr. Whymper and the Taugwalds had safely descended, we should be able to do the same. On the Friday we climbed to the Col de la Furka, examined from it the northern face of the pyramid, and discovered the men who were engaged in building the cabin on that side. We worked afterwards along the ridge which stretches from the Matterhorn to the Theodule, crossing its gulleys and scaling all its heights. It was a pleasant piece of discipline on ground new to both my guide and me.

On the Thursday evening a violent thunderstorm had burst over Breuil, discharging new snow upon the heights but also clearing the oppressive air. Though the heavens seemed clear in the early part of Friday, clouds showed a disposition to meet us from the south as we returned from the Theodule. I inquired of my companion whether in the event of the day being fine, he was willing to start on Sunday. His answer was a prompt negative. In Val Tournanche, he said, they always 'sanctified the Sunday.' I referred to Bennen, my pious Catholic guide, whom I permitted and encouraged to attend his mass on all possible occasions, but who nevertheless always yielded without a murmur to the demands of the weather. The reasoning had its effect. On Saturday Maquignaz saw his confessor, and arranged with him to have a mass at 2 A.M. on Sunday ; after which, unshaded by the sense of duties unperformed, he would commence the ascent.

The claims of religion being thus met, the point of next importance, that of money, was immediately arranged by my accepting, without hesitation, the tariff proposed by the Chanoine Carrel. The problem being thus reduced to one of muscular physics we pondered the question of provisions, decided on a bill-of-fare, and committed its execution to the mistress of the hotel.

A fog impenetrable to vision had filled the whole of the Val Tournanche on Saturday night, and the mountains were half concealed and half revealed by this fog when we rose on Sunday morning. The east at sunrise was lowering, and the light which streamed through the cloud-orifices was drawn in ominous red bars across the necks of the mountains. It was one of those uncomfortable Laodicean days, which engender indecision— threatening, but not sufficiently so to warrant postponement. Two guides and two porters were considered necessary for the first day's climb. A volunteer joined us, who carried a sheepskin as part of the furniture of the cabin. To lighten their labour the porters took a mule with them as far as the quadruped could climb, and afterwards divided the load among themselves. While they did so I observed the weather. The sun had risen with power and had broken the cloud-plane to pieces. The severed clouds gathered themselves into masses more or less spherical and were rolled grandly over the ridges into Switzerland. Save for a swathe of fog which now and then wrapped its flanks, the Matterhorn itself remained clear, and strong hopes were entertained that the progress of the weather was in the right direction.

We halted at the base of the Tête du Lion, a bold precipice formed by the sudden cutting down of the ridge which flanks the Val Tournanche to the right. From its base to the Matterhorn stretches the Col du Lion, crossed for the first time in 1860, by Mr. Hawkins, myself, and our two guides. We were now beside a snow-gully which was cut by a deep furrow along its centre, and otherwise scarred by the descent of stones. Here each man arranged his bundle and himself so as to cross the gully in the minimum of time. The passage was safely made, a few flying shingle only coming down upon us. But danger declared itself where it was not expected. Joseph Maquignaz led the way up the rocks. I was next, Pierre Maquignaz next, and last of all the porters. Suddenly a yell issued from the leader : ' *Cachez-vous!* ' I crouched instinctively against the rock which formed a by no means perfect shelter, when a boulder buzzed past me through the air, smote the rocks below me, and with a savage hum flew down to the lower glacier. Thus warned we swerved to an arête, and when stones fell afterwards they plunged to the right or left of us.

In 1860 the great couloir which stretches from the Col du Lion downwards was filled with a deep névé. But the atmospheric conditions, which have caused the glaciers of Switzerland to shrink so remarkably during the last ten years,[1] have swept away this névé. We had descended it, in 1860, hip-deep in snow, and I was now reminded of its steepness by the inclination of its bed. Maquignaz was incredulous when I pointed

[1] I should estimate the level of the Lower Grindelwald glacier, at the point where it is usually entered upon to reach the Eismeer, to be nearly 100 feet vertically lower in 1867 than it was in 1856. I am glad to find that the question of ' Benchmarks ' to fix such changes of level is now before the Council of the British Association. [The shrinking of the glaciers continues—1889]

out to him the line of our descent, to which we had been committed in order to avoid the falling stones of the Tête du Lion. Bennen's warnings on the occasion had been very emphatic, and I could understand their wisdom now better than I did then.

An admirable description of the difficulties of the Matterhorn, up to a certain elevation, has been given by Mr. Hawkins, in ' Vacation Tourists for 1860.'[1] At that time, however, a temporary danger, sufficient to quell for a while the enthusiasm even of our lion-hearted guide, was added to the permanent ones. Fresh snow had fallen two days before ; it had quite oversprinkled the Matterhorn, converting the brown of its crags into an iron-grey ; this snow had been melted and re-frozen, forming upon the rocks a coating of ice. Besides their physical front, moreover, in 1860 the rocks presented a psychical one, derived from the rumour of their savage inaccessibility. The crags, the ice, and the character of the mountain all conspired to stir the feelings. Much of the wild mystery has now vanished, especially at those points which in 1860 were places of virgin difficulty, but down which ropes now hang to assist the climber. The grandeur of the Matterhorn is, however, not to be effaced.

After some hours of steady climbing we halted upon a platform beside the tattered remnant of one of my tents, had a mouthful of food, and sunned ourselves for an hour. We subsequently worked upwards, scaling the crags and rounding the bases of those wild and wonderful rock-towers into which the weather of ages has hewn the southern arête of the Matterhorn. The work here requires knowledge, but with a fair amount of skill it is safe work. I can fancy nothing more fascinating to a man given by nature and habit to such things, than a

[1] Macmillan and Co.

climb *alone* among these crags and precipices. He
need not be *theological* but, if complete, he must be
religious with such an environment. To the climber
amongst them, the southern cliffs and crags of the Matterhorn are incomparably grander than those of the
north. Majesty of form and magnitude, and richness of
colouring, combine to ennoble them.

Looked at from Breuil, the Matterhorn presents two
summits: the one, the summit proper, a square rock-
tower in appearance; the other, which is really the end
of a sharp ridge abutting against the rock-tower, an
apparently conical peak. On this peak Bennen and
myself planted our flagstaff in 1862, and with it, which
had no previous name, Italian writers have done me the
honour of associating mine. At some distance below it
the mountain is crossed by an almost horizontal ledge
always loaded with snow, which from its resemblance to
a white necktie has been called the *Cravatte*. On the
ledge a cabin was put together last year. It stands
above the precipice where I quitted my rope in 1862.
Up this precipice, by the aid of a thicker—I will not
say a stronger—rope we now scrambled, and following
the exact route pursued by Bennen and myself five years
previously, we came to the end of the Cravatte. At
some places the snow upon the ledge fell steeply from
its junction with the cliff. Here steps were necessary.
Deep step-cutting was also needed where the snow had
been melted and recongealed. The passage was soon
accomplished along the Cravatte to the cabin, which
was almost filled with snow.

Our first inquiry now had reference to the supply
of water. We could of course always melt the snow,
but this would involve a wasteful expenditure of heat.
The cliff at the base of which the hut was built overhung, and from its edge the liquefied snow fell in

showers beyond the cabin. Four ice-axes were fixed on the ledge, and over them was spread the residue of a second tent which I had left at Breuil in 1862. The water falling upon the canvas flowed towards its centre. Here an orifice was formed, through which the liquid descended into vessels placed to receive it. Some modification of this plan might probably be employed with profit for the storing up of water in droughty years by the farmers of England.

I lay for some hours in the warm sunshine in presence of the Italian mountains, watching the mutations of the air. But when the sun sank the air became chill and we all retired into the cabin. We had no fire, though warmth was much needed. A lover of the mountains and of his kind had contributed an india-rubber mattress to the cabin. On this I lay down, a light blanket being thrown over me, while the guides and porters were rolled up in sheepskins. The mattress was a poor defence against the cold of the subjacent rock. I bore this for two hours, unwilling to disturb the guides, but at length it became intolerable. The little circles with a speck of intensified redness in the centre, which spotted the neck of our volunteer porter, had prevented me from availing myself of the warmth of my companions, so I lay alone and suffered the penalty of isolation. On learning my condition, however, the good fellows were soon alert, and folding a sheepskin round me restored me gradually to a pleasant temperature. I fell asleep, and found the guides preparing breakfast and the morning well advanced when I opened my eyes.

It was past six o'clock when the two Maquignazs and myself quitted the cabin. The porters deemed their work accomplished, but they halted for a time to ascertain whether we were likely to be driven back or

to push forward. We skirted the Cravatte, and reached
the ridge at its western extremity. This we ascended
along the old route of Bennen and myself to the conical
peak already referred to which, as seen from Breuil,
constitutes a kind of second summit of the Matterhorn.
From this point to the base of the final crag of the
mountain stretches an arête, terribly hacked by the
weather, but on the whole horizontal.[1] When I first
made the acquaintance of this savage ridge it was
almost clear of snow. It was now loaded, the snow
being bevelled to a sharp edge. The slope to the left
falling towards Zmutt was exceedingly steep, while the
precipices on the right were abysmal. No part of the
Matterhorn do I remember with greater interest than
this. It was terrible, but its difficulties were fairly
within the grasp of human skill, and this association is
more elevating than where the circumstances are such
as to make you conscious of your own helplessness. On
one of the sharpest teeth of the Spalla Joseph Maquignaz halted, and turning to me with a smile, remarked, ' There is no room for giddiness here, sir.' In
fact, such possibilities, in such places, must be altogether excluded from the chapter of accidents of the
climber.

It was at the end of this ridge, where it abuts
against the last precipice of the Matterhorn, that my
second flagstaff was left in 1862. I think there must
have been something in the light falling upon this precipice that gave it an aspect of greater verticality when
I first saw it than it seemed to possess on the present
occasion. Or, as remarked in my brief account of our
attempt in the ' Saturday Review,' we may have been
dazed by our previous exertion. I cannot otherwise
account for our stopping short without making some

[1] By Italian writers this ridge is called the ' Spalla ' (shoulder).

attempt upon the precipice. It looks very bad, but no climber with his blood warm would pronounce it without trial insuperable. Fears of this rock-wall, however, had been excited long before we reached it. At three several places upon the arête I had to signalise points in advance, and to ask my companions in French (which Bennen alone did not understand) whether they thought these points could be reached without peril. Thus bit by bit we moved along the ridge to its end, where farther advance was declared to be impossible. It was probably the addition of the psychical element to the physical, the reluctance to encounter new dangers on a mountain which had hitherto inspired a superstitious fear, that quelled further exertion.

To assure myself of the correctness of what is here stated I have turned to my notes of 1862. The reperusal of them has interested me, and a portion of them may possibly interest some of my readers. Here then they are, rapidly thrown together. They embrace our passage from the crags adjacent to the Col du Lion to the point where we were compelled to retreat.

'We had gathered up our things and bent to the work before us, when suddenly an explosion occurred overhead. Looking aloft, in mid-air was seen a solid shot from the Matterhorn describing its proper parabola through the air. It split to pieces as it hit one of the rock-towers, and its fragments came down in a kind of spray, which fell wide of us, but was still near enough to compel a sharp look-out. Two or three such explosions occurred afterwards, but we crept along the back fin of the mountain from which the falling boulders were speedily deflected right and left. Before the set of sun we reached our place of bivouac. A tent was already there. Its owner had finished a prolonged

attack upon the Matterhorn and had kindly permitted
the tent to remain, thus saving me the labour of carrying
up one of my own. I had with me a second and
smaller tent, made for me under the friendly supervision of Mr. Whymper, which the exceedingly nimble-
handed Carrel soon placed in position upon a platform
of stones. Both tents stood in the shadow of a great
rock which effectually sheltered us from all projectiles
from the heights.

'As the evening advanced fog, the enemy of the
climber, came creeping up the valley, and heavy
flounces of cloud draped the bases of the hills. The
fog thickened through a series of changes which only a
mountain land can show. Sudden uprushings of air
would at one place carry the clouds aloft in vertical
currents, while at other places horizontal gusts wildly
tossed them to and fro. Impinging upon each other
at oblique angles they sometimes formed whirling
cyclones of cloud. The air was tortured in its search
for repose. Explosive peals above us, succeeded by the
sound of tumbling rocks, were heard from time to
time. We were swathed in the densest fog when
we retired to rest, and had scarcely a hope that the
morrow's sun would be able to dispel the gloom.
Throughout the night I heard the intermittent roar of
the stones as they rushed down an adjacent couloir.
Looking at midnight through a small hole in the canvas of my tent I saw a star. I rose and found the
heavens without a cloud; while above me the black
battlements of the Matterhorn were projected against
the fretted sky.

'It was 4 A.M. before we started. We adhered to
the hacked and weather-worn spine, until its disintegration became too vast. The alternations of sun and frost
have made wondrous havoc on the southern face of the

Matterhorn ; cutting much away but leaving brown-red masses of the most imposing magnitude behind—pillars and towers and splintered obelisks, cut out of the mountain—grand in their hoariness, and softened by the colouring of age. At length we were compelled to quit the ridge for the base of a precipice which seemed to girdle the mountain like a wall. It was a clean section of rock, with cracks and narrow ledges here and there. We sought to turn this wall in vain. Bennen swerved to the right and to the left to make his inspection complete. There was no alternative—over the precipice we must go or else retreat. For a time it was manifest our onset must be desperate. We grappled with the cliff. Walters, an exceedingly powerful climber, went first. Close to him was Bennen, with arm and knee and counsel ready in time of need. As usual, I followed Bennen, while the two porters brought up the rear. The behaviour of all of them was admirable. A process of reciprocal lifting continued for half an hour, when a last strong effort threw Walters across the brow of the precipice and rendered our progress thus far secure.

' After scaling the precipice we found ourselves once more upon the ridge, with safe footing on the ledges of gneiss. We approached the conical peak seen from Breuil, while before us and, as we thought, assuredly within our grasp was the proper summit of the renowned Matterhorn. To test Bennen's feelings I remarked, " We shall at all events reach the lower peak." There was a kind of scorn in his laugh as he replied, stretching his arm towards the summit, " In an hour, sir, the people of Zermatt will see our flagstaff planted yonder." We went upward, this spirit of triumph forestalled making the ascent a jubilee.

' We reached the first summit, and on it fixed our

flag. But already doubt had begun to settle about the final precipice. Walters once remarked, " We may still find difficulty there." It was perhaps the pressure of the same thought upon my own mind that caused its utterance to irritate me. So I grimly admonished Walters and we went on. The nearer, however, we came to the summit, the more formidable did the precipice appear. From the point where we had planted our flag-staff a hacked and extremely acute ridge (the Spalla), with ghastly abysses right and left of it, ran straight towards the final cliff. We sat down upon the ridge and inspected the precipice. Three out of the four men shook their heads and muttered "Impossible." Bennen was the only man amongst them who refused from first to last to utter the word.

' Resolved not to push them beyond the limits of their own clear judgment, I was equally determined to advance until that judgment should pronounce the risk too great. I therefore pointed to a tooth at some distance from the place where we sat, and asked whether it could be reached without much danger. " We think so," was the reply. " Then let us go there." We did so, and sat down again. The three men murmured, while Bennen growled like a foiled lion. " We must give it up," was here repeated. " Not yet," was my answer. " You see yonder point quite at the base of the precipice; do you not think we might reach it?" The reply was " Yes." We moved cautiously along the arête and reached the point aimed at. So savage a spot I had never previously visited, and we sat down there with broken hopes. The thought of retreat was bitter. We may have been dazed by our previous efforts and thus rendered less competent than fresh men would have been to front the danger before us. As on other occasions, Bennen sought to fix on me the onus of returning, but with the usual result. My reply was, "Where you go I follow, whether it be up or down." It took him half an hour to make up his mind. Had the other men not yielded so utterly, he would have tried longer. As it was our occupation was gone, and backing a length of six feet from our ladder, we planted it on the spot where we halted.' So much is due to the memory of a brave man.

Six hundred feet, if the barometric measurement can be trusted, of very difficult rock-work now lay above us. In 1862 this height had been under-estimated by both Bennen and myself. Of the 14,800 feet of the Matterhorn, we then thought we had accomplished 14,600. If the barometer speaks truly, we had only cleared about 14,200. Descending the end of the arête we crossed a narrow cleft and grappled with the rocks at the other side of it. Our ascent was oblique, bearing to the right. The obliquity at one place fell to horizontality, and we had to work on the level round a difficult protuberance of rock. We cleared the difficulty without haste, and then rose straight against the precipice. Joseph Maquignaz drew my attention to a rope hanging down the cliff, left there by himself on the occasion of his first ascent. We reached the end of this rope, and some time was lost by the guide in assuring himself that it was not too much frayed by friction. Care in testing it was doubly necessary, for the rocks, bad in themselves, were here crusted with ice. The rope was in some places a mere hempen core surrounded by a casing of ice. Over this the hands slid helplessly. With the rope in this condition it required a considerable effort to get to the top of the precipice, and we willingly halted there to take a minute's breath. The ascent was now virtually accomplished, and a few minutes' more of rapid climb-

ing placed us upon the crest of the mountain. Thus
ended an eight years' war between myself and the
Matterhorn.

The day thus far had swung through alternations of
fog and sunshine. While we were on the ridge below
the air at times was blank and chill with mist; then
with rapid solution the cloud would vanish, and open up
the abysses right and left of us. On our attaining the
summit a fog from Italy rolled over us, and for some
minutes we were clasped by a cold and clammy atmosphere. But this passed rapidly away, leaving above us
a blue heaven and far below us the sunny meadows
of Zermatt. The mountains were almost wholly unclouded, and such clouds as lingered amongst them only
added to their magnificence. The Dent d'Erin, the
Dent Blanche, the Gabelhorn, the Mischabel, the range
of heights between it and Monte Rosa, the Lyskamm,
and the Breithorn were all at hand, and clear; while
the Weisshorn, noblest and most beautiful of all, shook
out towards the north, a banner formed by the humid
southern air as it grazed the crest of the mountain.

The world of peaks and glaciers surrounding this
immediate circle of giants was also open to us to the
horizon. Our glance over it was brief, and our enjoyment of it intense. It was eleven o'clock, and the
work before us soon claimed all our attention. I found
the *débris* of my former expedition everywhere—below,
the fragments of my tents, and on the top a piece of my
ladder fixed in the snow as a flagstaff. The summit of
the Matterhorn is a sharp horizontal arête, and along
this we now moved eastward. On our left was the roof-
like slope of snow seen from the Riffel and Zermatt, on
our right were the savage precipices which fall into
Italy. Looking to the farther end of the ridge the snow
there seemed to have been trodden down, and I drew my

companions' attention to the apparent footmarks. As
we approached the place it became evident that human
feet had been there two or three days previously. I think
it was Mr. Elliot [1] who had made this ascent—the first
accomplished from Zermatt since the memorable one of
1865. On the eastern end of the ridge we halted to
take a little food ; not that I seemed to need it. It was
the remonstrance of reason rather than the consciousness of physical want that caused me to do so.

Facts of this kind illustrate the amount of force
locked up in the muscles which may be drawn upon
without renewal. I had quitted London ill, and when
the Matterhorn was attacked I was by no means well. In
fact, this climb was one of the means adopted to drive
the London virus from my blood. The day previous I
had taken scarcely any food, and on starting from the
cabin half a cup of bad tea, without any solid whatever,
constituted my breakfast. Still, during the five hours'
climb from the cabin to the top of the Matterhorn,
though much below par physically and mentally, I felt
neither faint nor hungry. This is an old experience of
mine upon the mountains. The Weisshorn, for example, was climbed on six meat lozenges, though it
was a day of nineteen hours. Possibly this power of
long-continued physical effort, without eating, may be
a result of bad digestion which deals out stingily, and
therefore economically, to the muscles the energy of the
food previously consumed.

We took our ounce of nutriment and gulp of wine
and stood for a moment silently and earnestly looking
down towards Zermatt. There was a certain official for-
mality in the manner in which the guides turned to me
and asked, ' *Etes-vous content d'essayer ?* ' A quick responsive ' *Oui !* ' set us immediately in motion. It was

[1] Lost the following year upon the Schrackhorn.

nearly half-past eleven when we quitted the summit.
The descent of the roof-like slope already referred to
offered no difficulty; but the gradient very soon became
more formidable. One of the two faces of the Matterhorn pyramid seen from Zermatt falls towards the
Zmutt glacier, and has a well-known snow-plateau at
its base. The other face falls towards the Furgge
glacier. We were on the former. For some time, however, we kept close to the arête formed by the intersec-
tion of the two faces of the pyramid, because nodules of
rock jutted from it which offered a kind of footing.
These rock protuberances helped us in another way:
round them an extra rope which we carried was frequently
doubled, and we let ourselves down by the rope as far as
it could reach, liberating it afterwards (sometimes with
difficulty) by a succession of jerks. In the choice and
use of these protuberances the guides showed both judgment and skill. The rocks became gradually larger
and more precipitous; a good deal of time being consumed in dropping down and doubling round them.
Still we preferred them to the snow-slope at our left as
long as they continued practicable.

This they at length ceased to be, and we had to
commit ourselves to the slope. It was in the worst possible condition. When snow first falls at these great
heights it is usually dry, and has no coherence. It
resembles to some extent flour, or sand, or sawdust.
Shone upon by a strong sun it shrinks and becomes
more consolidated, and when it is subsequently frozen it
may be safely trusted. Even though the melting of the
snow and its subsequent freezing may be only partial,
the cementing of the granules adds immensely to the
safety of the footing; but then the snow must be
employed before the sun has had time to unlock the
rigidity imparted to it by the night's frost. We were

on the steepest Matterhorn slope during the two hottest
hours of the day, and the sun had done his work effectually. The snow seemed to offer no foothold whatever;
with cautious manipulation it regelated, but to so small
an extent that the resistance due to regelation was insensible to the foot. The layer of snow was about
fifteen inches thick. In treading it we came immediately upon the rock, which in most cases was too
smooth to furnish either prop or purchase. It was on
this slope that the Matterhorn catastrophe occurred : it
is on this slope that other catastrophes will occur, if
this mountain should ever become fashionable.

Joseph Maquignaz was the leader of our little party,
and a cool and competent leader he proved himself to
be. He was earnest and silent, save when he answered
his brother's anxious and oft-repeated question, ' *Es-tu
bien placé, Joseph?* ' Along with being perfectly cool
and brave, he seemed to be perfectly truthful. He did
not pretend to be ' *bien placé* ' when he was not, nor
avow a power of holding which he knew he did not
possess. Pierre Maquignaz is, I believe, under ordinary
circumstances an excellent guide, and he enjoys the
reputation of being never tired. But in such circumstances as we encountered on the Matterhorn he is not
the equal of his brother. Joseph, if I may use the term,
is a man of high boiling-point ; his constitutional *sang-
froid* resisting panic ebullition. Pierre, on the contrary, shows a strong tendency to boil over in perilous
places.

Our progress was exceedingly slow but it was steady
and continuous. At every step our leader trod the snow
cautiously, seeking some rugosity on the rock beneath
it. This however was rarely found, and in most cases
he had to establish practicable attachments between the
snow and the slope which bore it. No semblance of a

slip occurred in the case of any one of us; had a slip
occurred I do not think the worst consequences could
have been avoided. I wish to stamp this slope of the
Matterhorn with the character that really belonged to
it when we descended it, and I do not hesitate to express
the belief that the giving way of any one of our party
would have carried the whole of us to ruin. Why, then,
it may be asked, employ the rope? The rope, I reply,
all its possible drawbacks under such circumstances
notwithstanding, is the safeguard of the climber. Not
to speak of the moral effect of its presence, an amount
of help upon a dangerous slope that might be measured
by the gravity of a few pounds is often of incalcul-
able importance; and thus, though the rope may be not
only useless but disastrous if the footing be clearly lost,
and the glissade fairly begun, it lessens immensely the
chance of this occurrence.

With steady perseverance, difficulties upon a mountain, as elsewhere, come to an end. We were finally
able to pass from the face of the pyramid to its rugged
edge, feeling with comfort that honest strength and fair
skill, which might have gone for little on the slope,
were here masters of the situation.

Standing on the arête at the foot of a remarkable
cliff-gable seen from Zermatt, and permitting the vision
to range over the Matterhorn, its appearance from above
was exceedingly wild and impressive. Hardly two things
can be more different than the respective aspects of the
mountain from above and from below. Seen from the
Riffel or from Zermatt it presents itself as a compact
pyramid, smooth and steep, and defiant of the weather-
ing air. From above it seems torn to pieces by the frosts
of ages, while its vast facettes are so foreshortened as to
stretch out into the distance like plains. But this under-
estimate of the steepness of the mountain is checked by

the deportment of its stones. Their discharge along the
side of the pyramid was incessant, and at any moment
by detaching a single boulder we could let loose a
cataract of them, which flew with wild rapidity and with
a clatter as loud as thunder down the mountain. We
once wandered too far from the arête, and were warned
back to it by a train of these missiles sweeping past us.

As long as the temperature of our planet differs from
that of space so long will the forms upon her surface
undergo mutation, and as soon as equilibrium has been
established we shall have, not peace, but death. Life is
the product and accompaniment of change, and the self-
same power that tears the flanks of the hills to pieces is
the mainspring of the animal and vegetable worlds.
Still, there is something chilling, if not humiliating, in
the contemplation of the irresistible and remorseless
character of those infinitesimal forces whose summation
through the ages pulls down even the Matterhorn.
Hacked and hurt by time, the aspect of the mountain
from its higher crags saddened me. Hitherto the impression it had made was that of savage strength, but
here we had inexorable decay.

This notion of decay implied a reference to a period
of prime when the Matterhorn was in the full strength
of mountainhood. Thought naturally ran back to its
possible growth and origin. Nor did it halt there, but
wandered on through molten worlds to that nebulous
haze which philosophers have regarded, and with good
reason, as the proximate source of all material things.
Could the blue sky above be the residue of that haze?
Would the azure which deepens on the heights sink
into utter darkness beyond the atmosphere? I tried
to look at this universal cloud, containing within itself
the prediction of all that has since occurred; I tried
to imagine it as the seat of those forces whose action

was to issue in solar and stellar systems, and all that they involve. Did that formless fog contain potentially the *sadness* with which I regarded the Matterhorn ? Did the *thought* which thus ran back through the ages simply return to its primeval home ? If so, had we not better recast our definitions of matter and force ? for if life and thought be the very flower of both, any definition which omits life and thought must be inadequate, if not untrue. Are questions like these warranted ? Are they healthy ? Ought they not to be quenched by a life of action ? Healthy or unhealthy, *can* we quench them ? And if the final goal of man has not been yet attained, if his development has not been yet arrested, who can say that such yearnings and questionings are not necessary to the opening of a finer vision, to the budding and the growth of diviner powers ? When I look at the heavens and the earth, at my own body, at my strength and imbecility of mind, even at these ponderings, and ask myself is there no being or thing in the universe that knows more about these matters than I do ; what is my answer ? Does antagonism to theology stand with none of us in the place of a religion ? Supposing our theologic schemes of creation, condemnation, and redemption to be dissipated ; and the warmth of denial, which as a motive force can match the warmth of affirmation, dissipated at the same time ; would the undeflected mind return to the meridian of absolute neutrality as regards these ultra-physical questions ? Is such a position one of stable equilibrium ? The channels of thought being already formed, such are the questions without replies which could run through the mind during a ten minutes' halt upon the weathered spine of the Matterhorn.

We shook the rope away from us, and went rapidly down the rocks. The day was well advanced when we reached the cabin, and between it and the base of the pyramid we lost our way. It was late when we regained it, and by the time we reached the ridge of the Hörnli we were unable to distinguish rock from ice. We should have fared better than we did if we had kept along that ridge and felt our way to the Schwarz-See, whence there would have been no difficulty in reaching Zermatt. But we left the Hörnli to our right, and found ourselves incessantly checked in the darkness by ledges and precipices, possible and actual. We were afterwards entangled in the woods of Zmutt, but finally struck the path and followed it to Zermatt, which we reached between one and two o'clock in the morning.

[In the woods of Zmutt I was beset by overpowering sleepiness, which disappeared in the open. Madame Seiler divined the meaning of my knocking for admittance to the Monte Rosa Hotel. 'It is the Professor,' she said, 'who has come over the Matterhorn.' While food was preparing, Mr. Seiler asked me whether, in view of future ascents, it would not be wise to place ropes or chains at the dangerous points. 'By doing so,' I replied, 'you will save life, but you will spoil the mountain.' I made the acquaintance of Seiler thirty-three years ago. To the sorrow of his friends, his well-known figure will be seen at Zermatt no more.—*October* 1891.]

A MORNING ON ALP LUSGEN.

THE sun has cleared the peaks and quenched the flush
Of orient crimson with excess of light.
The tall grass quivers in the rhythmic air
Without a sound ; yet each particular blade
Trembles in song, had we but ears to hear.
The hot rays smite us, but a quickening breeze
Keeps languor far away.   Unslumbering,
The soul enlarged takes in the mighty scene.

The plummet from this height must sink afar
To reach yon rounded mounds which seem so small.
They shrink in the embrace of vaster forms,
Though, placed amid the pomp of Cumbrian Fells,
These hillock crests would overtop them all.
Steep fall the meadows to the vale in slopes
Of freshest green, scarred by the humming streams,
And flecked by spaces of primeval pine.
Unplanted groves ! whose pristine seeds, they say,
Were sown amid the flames of nascent stars—
How came ye thence and hither ?   Whence the craft
Which shook these gentian atoms into form,
And dyed the flower with azure deeper far
Than that of heaven itself on days serene ?
What built these marigolds ?   What clothed these
          knolls
With fiery whortle leaves ?   What gave the heath

Its purple bloom—-the Alpine rose its glow ?
Shew us the power which fills each tuft of grass
With   sentient   swarms ? — the   art   transcending
        thought,
Which paints against the canvas of the eye
These crests sublime and pure, and then transmutes
The picture into worship ?   Science dumb—-
Oh babbling Gnostic ! cease to beat the air.
We yearn, and grope, and guess, but cannot know.

Low down, the yellow shingle of the Rhone
Hems in the scampering stream, which loops the
        sands
In islands manifold.   Beyond, a town,
Whose burnished domes flash back the solar blaze—
Proud domes for town so small !   But here erewhile
Unfurled itself the Jesuit oriflamme,
And souls were nurtured in the tonic creed
Of Loyola.   Grand creed ! if only true.
Oh ! sorrowing shade of him,[1] who preached through
        life
Obedience to the Highest ! could men find
That Highest much were clear ! Yon tonsured monk
Will face the flames obedient to a power
Which   he   deems   highest,   but   which   you   deem
        damned.

Cut by a gorge, the vale beyond the town
Breaks into squares of yellow and of green—·
Of rye and meadow.   Through them winds the road
Which opened to the hosts of conquering France
Lombardian plains—sky-touching Simplon Pass—
Flanked by the Lion Mountain to the left,

[1] Carlyle.

While to the right the mighty Fletschorn lifts
A beetling brow, and spreads abroad its snows.
Dom, Cervin—Weisshorn of the dazzling crown—
Ye splendours of the Alps! Can earth elsewhere
Bring forth a rival? Not the Indian chain,
Though shouldered higher o'er the standard sea,
Can front the eye with more majestic forms.

From one vast brain yon noble highway came;
' Let it be made,' he said, and it was done.
In one vast brain was born the motive power
Which swept whole armies over heights unscaled,
And poured them, living cataracts, on the South.
Or was it force of faith—faith warranted
By antecedent deeds, that nerved these hosts
And made Napoleon's name a thunderbolt?
What is its value now? This man was called
' A mortal God!' Oh, shade before invoked,
You spoke of Might and Right; and many a shaft
Barbed with the sneer, ' He preaches force—brute
      force,'
Has rattled on your shield. But well you knew
Might, to be Might, must base itself on Right,
Or vanish evanescent as the deeds
Of France's Emperor. Reflect on this,
Ye temporary darlings of the crowd.
To-day ye may have peans in your ears;
To-morrow ye lie rotten, if your work
Lack that true core which gives to Right and Might
One meaning in the end.

Spottiswoode & Co. Printers, New street Square, London